3rd International Conference on Functional Materials Science 2016 (3rd ICFMS 2016)

IOP Conference Series: Materials Science and Engineering
Volume 196

Sanur-Bali, Indonesia
19 - 20 October 2016

Editors:

Risdiana
A.A. Nugroho
Kuwat Triyana

ISBN: 978-1-5108-4073-7
ISSN: 1757-8981

Printed from e-media with permission by:

Curran Associates, Inc.
57 Morehouse Lane
Red Hook, NY 12571

Some format issues inherent in the e-media version may also appear in this print version.

Printed by Curran Associates, Inc. (2017)

For permission requests, please contact the Institute of Physics
at the address below.

Institute of Physics
Dirac House, Temple Back
Bristol BS1 6BE UK

Phone: 44 1 17 929 7481
Fax: 44 1 17 920 0979

techtracking@iop.org

Additional copies of this publication are available from:

Curran Associates, Inc.
57 Morehouse Lane
Red Hook, NY 12571 USA
Phone: 845-758-0400
Fax: 845-758-2633
Email: curran@proceedings.com
Web: www.proceedings.com

Table of contents

Volume 196

3rd International Conference on Functional Materials Science 2016

19–20 October 2016, Sanur-Bali, Indonesia

Accepted papers received: 27 April 2017
Published online: 20 May 2017

Preface

Invited lectures

Oral Presentations

Preface

This volume of *IOP Conference series: Materials Science and Engineering* contains selected and peer reviewed papers presented at the 3rdInternational Conference on Functional Materials Science 2016 (3rd ICFMS 2016). The 3rd ICFMS 2016 is our third activities for gathering and discussing some research in material sciences held in Bali, Indonesia, October 19-20, 2016. The main theme of ICFMS 2016 is "Trend in Functional Materials: From Fundamental to Applications". This conference is organized by Department of Physics, Faculty of Mathematics and Natural Sciences, Padjadjaran University together with Bandung Institute of Technology (ITB), University of Indonesia (UI), Sepuluh Nopember Institute of Technology (ITS) and Gadjah Mada University (UGM).

The conference aims to improve and continue communication among the researcher in material sciences and related field. About 80 papers are presented in the range of material sciences including Advanced and Functional Materials, Energy Conversion Materials and Devices, New Materials for Energy Biomaterials, Theoretical/Modeling/Computer Simulations of Functional Materials, Spectroscopy for Advanced Materials, Hybrid and Composite Materials and Magnetic Materials.

Editors and committees would like to thanks our collaborators, reviewers, invited speakers and contributed speakers from University of Tokyo, Sophia University, Argon National Laboratory USA, Osaka University, National University of Singapore, Tohoku University, Saitama University, Padjadjaran University (UNPAD), Bandung Institute of Technology (ITB), University of Indonesia (UI), Sepuluh Nopember Institute of Technology (ITS), Gadjah Mada University (UGM) and Tokyo Denki University, Japan.

Risdiana
A.A. Nugroho
Kuwat Triyana
Guest Editors

IOP Conf. Series: Materials Science and Engineering **196** (2017) 011001 doi:10.1088/1757-899X/196/1/011001

IOP Conf. Series: Materials Science and Engineering **196** (2017) 011001 doi:10.1088/1757-899X/196/1/011001

IOP Conf. Series: Materials Science and Engineering **196** (2017) 011002 doi:10.1088/1757-899X/196/1/011002

Peer review statement

All papers published in this volume of *IOP Conference Series: Materials Science and Engineering* have been peer reviewed through processes administered by the proceedings Editors. Reviews were conducted by expert referees to the professional and scientific standards expected of a proceedings journal published by IOP Publishing.

Calorimetric and Neutron Scattering Studies on Glass Transitions and Ionic Diffusions in Imidazolium-based Ionic Liquids

O Yamamuro[1] and M Kofu[1,2]

[1] Institute for Solid State Physics, University of Tokyo, 5-1-5 Kashiwanoha, Kashiwa, Chiba 277-8581, Japan
[2] Present address: Neutron Science Section, J-PARC Center, Tokai, Ibaraki 319-1195, Japan

E-mail : yamamuro@issp.u-tokyo.ac.jp

Abstract. Glass transition is one of the central research issuesof ionic liquids (ILs). In particular, the most typical ILs, imidazolium-basedones (ImILs) are readily supercooled and exhibit glass transitions below room temperature. We have measured the heat capacities of several ImILs,encoded as C*n*mimX (*n*: alkyl carbon number, *n* = 2–8, X: anion, X = Cl, I, $FeCl_4$, TFSI)using an adiabatic calorimeter. We found that most of ImILs exhibit glass transitions with large C_p jumps in a temperature range between 170 K and 230 K. The large C_p jumps reflect that these ILs are fragile liquids that exhibit large structural change depending on temperature near the glass transition temperature T_g. It is also revealed that T_g does not depend much on *n* but on the anion radius. We have investigated the dynamics of C*n*mimX (*n* = 2–8, X = Cl, NO_3, PF_6, TF, FSI, TFSI) by means of a quasielastic neutron scattering (QENS) technique. It was clarified that the ionic diffusion is directly associated with the viscosity and glass transition. The activation energy ΔE_a of the ionic diffusion increases with decreasing anion size but remains almost unchanged with *n*as found for T_g. These systematic change of T_g and ΔE_a can be explained well by taking account the nano-domain structure which is the most characteristic feature of ImILs.

1. Introduction

"Ionic liquid" (IL) is a generic chemical name for novel ionic compounds which are in liquid state around room temperature. It is not usual that ionic materials with strong and long-range interaction have fusion temperature T_{fus}as low as room temperature; cf. T_{fus} = 1074 K for NaCl. The cations of ILs are alkylimidazolium, alkylprydinium,alkylammonium, etc., which consist of core partswith positive electric charges and alkyl-chains with hydrophobicinteractions. On the other hand, halogens and various ionic groups, e.g., NO_3^-, BF_4^-, PF_6^-can be the anions of ILs.ILs exhibit many unique physical properties, e.g., negligible vapor pressure, non-combustibility, amphiphilicity, high ionic conductivity, and so they are remarked as novel functional materials such as non-volatile solvents, actuators, electrochemical materials. Another merit of ILs is that both cations and anions can be designed to produce various physical and chemical properties one wants.

We have studied the thermodynamic properties of alkylimidazolium-based ILs (ImILs) by means of adiabatic calorimetry [1-3], and also their structures and dynamics by using X-ray diffraction [4],inelastic neutron scattering[5],and quasielasticneutron scattering (QENS) [6-10] techniques. ImILs are the most typical ILs, and so to investigate their basic physical properties should be meaningful not only for basic scientific interests but also for various applications of ILs. In this article, we will show some of new heat capacity data and review our previous calorimetric and quasielasticworks especially focusing on the glass transitions and related ionic diffusions.

2. Experimental

The heat capacitiesof several ImILs encoded as C*n*mimX (*n*: alkyl carbon number, $n = 2$–8, X: anion, X = Cl, I, $FeCl_4$, TFSI) were measured with an in-house built adiabatic calorimeter [11] in the temperature range from 5 K to 340 K. Here, TFSI is a well-known large multi-atomic anionbis(trifluoromethanesulfonyl)imide, $(CF_3SO_2)_2N^-$. The samples of ca. 2 g were loaded into the copper cell with helium gas which facilitates thermal equilibration inside the cell. The measurement was carried out using a standard intermittent heating method, i.e., repetition of equilibration and energizing intervals. The accuracy of the heat capacity measurement was better than 1% at $T < 10$ K, 0.5% at 10 K $< T <$ 30 K, and 0.1% at $T >$ 30 K.

The ionic diffusion of C*n*mimX ($n = 2$–8, X = Cl, NO_3, PF_6, TF, FSI, TFSI) was investigated by QENS experiments using a backscattering spectrometer HFBS [12]at the NISTCenter for Neutron Research in USA.Here, TF and FSI denote trifluoromethanesulfonate, $(CF_3SO_3)^-$ and bis(fluorosulfonyl)imide, $(FSO_2)_2N^-$, respectively. The energy and momentum transfer *Q*windows and energy resolution of HFBS are -17 μeV $<E<$17μeV, 0.25Å$^{-1}<Q<$ 1.75 Å^{-1}, and ΔE = 0.8 μeV, respectively. HFBS can cover a relaxation time range from 50ps to 5 ns.

3. Results and discussion

Figure 1 (left) shows the heat capacities C_p of several ImILs; the heat capacities of C2mimTFSI [3], C4mimCl [1], C8mimCl [7], and C4mim$FeCl_4$ [2] were reported previously, while those of C*n*mimI are shown first in this paper. All of the plotted samples exhibit glass transitions with large C_p jumps in a temperature range between 170 K and 230 K. Around the glass transition, an exothermic effect followed by an endothermic one appeared as usually observed in the glass transitions of molecular liquids [13,14].The large C_p jumps reflects that these ILs are fragile liquids [15-18] that exhibit large structural change depending on temperature near T_g. It is rather unexpected that "ionic materials", which have strong inter-ionic interaction, are of fragile liquids. This may be due to the conformational disorder of alkyl-chains and the orientational disorder of anions. Actually, C8mimCl, C2mimTFSI, and C4mim$FeCl_4$ have larger C_p jumps than other ImILs.

Figure 1 (right)shows the glass transition temperature T_g and fusion temperature T_{fus}, which were determined from the heat capacity measurements, plotted against the carbon numbers of alkyl-chains. The plots are made for three ImILs with typical anions (X = Cl, I, TFSI) which have been studied well. The data for C1mimI [19], C2mimI [20], C1mimTFSI [21], C4mimTFSI [22], C6mimCl [4], C6mimTFSI [23], and C8mimTFSI [24] are from literature. Some ImILs have no T_g data because of

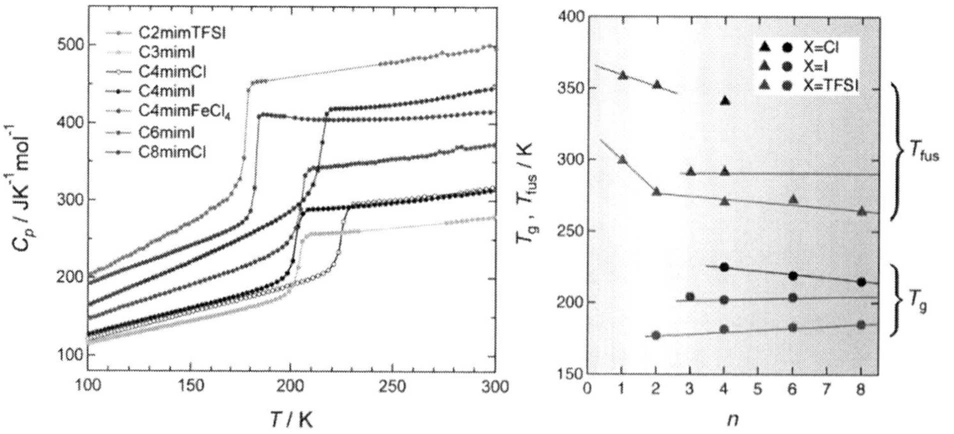

Figure 1. (left) Heat capacities of several ImILs around glass transition temperatures. (right) Glass transition and fusion temperatures plotted against the alkyl-carbon number.

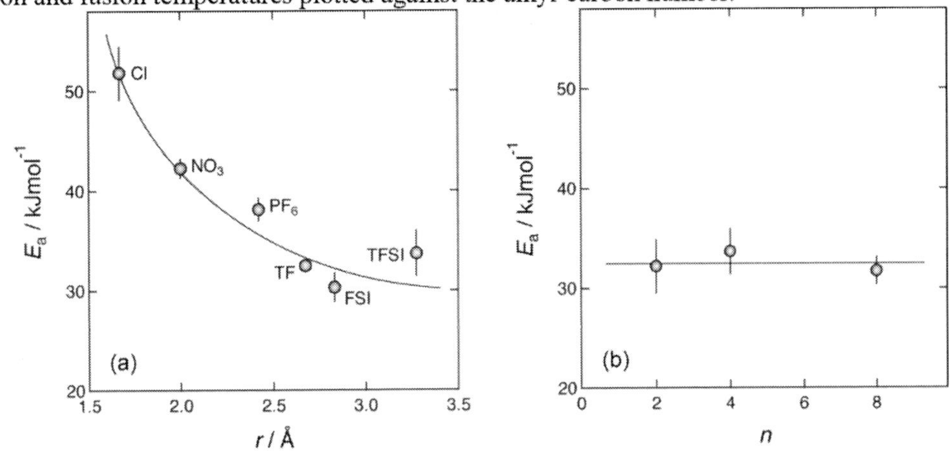

Figure 2. Activation energies for the ionic diffusions of several ImILs plotted (a) against the radius of anions for C4mimX and (b) against the alkyl-carbon number for C*n*mimTFSI. The lines in the figures are guides for eyes.

inevitable crystallization on cooling, and some ImILs have no T_{fus} data because of difficulty of crystallization. We found that there is a boundary at around$n$ = 2; C*n*mimX are normal ionic crystals on the left side of the boundary while they are ionic liquids on the right side. The relation between T_g and T_{fus} almost follows an empirical rule $T_g = 2/3T_{fus}$ [25]. It is noteworthy that T_g does not depend much on n but clearly on the anion species. The size of anion is in an order of TFSI > I > Cl. The larger the anion size is, the lower T_g becomes.

From the QENS measurements, we have obtained the relaxation time $\tau$of the ionic diffusion for several ImILs as a function of temperature T.The Arrhenius plot (logτ vs $1/T$) covering both neutron and viscosity time ranges revealed that the ionic diffusion is directly associated with the viscosity and glass transition. The activation energy of the ionic diffusion was obtained from the slopes of the Arrhenius plot. Figure 2 shows the activation energy as functions of (a) the radius r of anionX for C4mimX, and (b) the alkyl-carbon number$n$ for C*n*mimTFSI. Obviously, E_a does not depend much on n but clearly tends to decrease with an increase in$r$. Thisresult corresponds well to the $T_g$vs $r$and T_g vs $n$relations, being consistent with that both T_g and E_a are associated with the ionic diffusion that is the α-relaxation in the language of glass physics.

The existence of nanometer-size domain structure (nanostructure) is now believed by most ofIL researchers. Our X-ray [4] and neutron [7,9] diffraction experiments suggest that ImILs have the nanostructuresas shown in Figure 3. This structure is similar to the smectic A (SmA) phase of liquid crystals. There are polar domains consisting of the imidazolium cores (positive charge) and anions (negative charge) and nonpolar domains consisting of the alkyl-chains. This structure may bestabilizedby the local ordering due to theelectrostatic (between positive-charge imidazolium cores and negative anions) and van der Waals (between alkyl-chains) interactions. The electrostatic interaction, which is much stronger than van der Waals one, is of course proportional to the inverse square of interionic distance. When the anion size becomes larger, the interionic distance becomes longer and so the electrostatic interaction becomes weaker. As a consequence,T_g becomes lower.

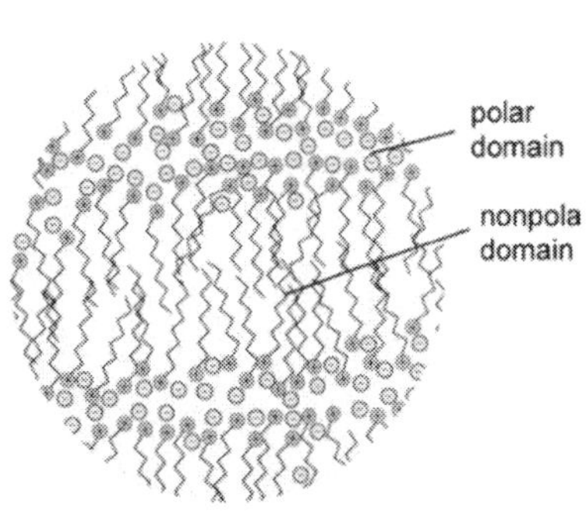

Figure 3. Schematic view of the nanostructure of ImILs.

4. Conclusion

Most of ImILs exhibit glass transitions with large C_p jumps in a temperature range between 170 K and 230 K. The large C_p jumps are due to the conformational disorder of alkyl-chains and the orientational disorder of anions, being related to fragile nature of ImILs. The relaxation times for the ionic diffusions, which are determined by QENS, are directly associated with the viscosities and glass transitions of ImILs. The$T_g$and the activation energy for the ionic diffusion do not depend much on the alkyl-chain length but on the anion radius. These properties on the glass transition and ionic diffusion are closely associated with the nanostructure which is the most characteristic feature of ImILs.

References

[1] Yamamuro O, Minamimoto Y, Inamura Y, Hayashi S and Hamaguchi H 2006 *Chem. Phys. Lett.***423** 371
[2] Yamamuro O, Inamura Y, Hayashi S and Hamaguchi H 2006 *AIP Conference Proceedings* **832** 73
[3] Yamamuro O, Someya T, Kofu M, Ueki T, Ueno K and Watanabe M 2012 *J. Phys. Chem. B* **116** 10935
[4] Nemoto F, Kofu M and Yamamuro O 2015 *J. Phys. Chem. B* **119** 5028
[5] Kofu M, Inamura Y, Moriya Y, Podlesnyak A, Ehlers G and Yamamuro O 2015 *J. Mol. Liq.***210** 164
[6] Inamura Y, Yamamuro O, Hayashi S and Hamaguchi H 2006 *Physica B* **385-386** 732
[7] Yamamuro O, Yamada T, Kofu M, Nakakoshi M and Nagao M 2011 *J. Chem. Phys.***135** 054508
[8] Kofu M, Someya T, Tatsumi S, Ueno K, Ueki T, Watanabe M, Matsunaga T, Shibayama, Garcia Sakai V, Tyagi M and Yamamuro O 2012 *Soft Matter* **8** 7888
[9] Kofu M, Nagao M, Ueki T, Kitazawa Y, Nakamura Y, Sawamura S, Watanabe M and Yamamuro O 2013 *J. Phys. Chem. B* **117** 2773
[10] Kofu M, Tyagi M, Inamura Y, Miyazaki K and Yamamuro O 2015*J. Chem. Phys.* **143** 234502
[11] Yamamuro O, Oguni M, Matsuo T and Suga H 1987 *Bull. Chem. Soc. Jpn.* **60** 1269
[12] Meyer A, Dimeo R M, Gehring P M and Neumann D A 2003 *Rev. Sci. Instrum.***74** 2759
[13] Suga H and Seki S 1980 *Faraday Discuss. R. Soc. Chem.***69** 221
[14] Suga H 1985 *J. Chim. Phys. Phys. Chim. Biol.* **82** 275
[15] Angell C A 1991 *J. Non-Cryst. Solids* **131-133** 13
[16] Ito K, Moynihan C T and Angell C A 1999 *Nature* **398** 492
[17] Martinez L -M, Angell C A 2001 *Nature* **410** 663
[18] Wang L –M, Angell C A, Richert R 2006 *J. Chem. Phys.***125** 074505
[19] Brünig T, Krekic K, Bruhn C and Pietschnig R 2016 *Chem. Eur. J.* **22** 1
[20] Every H A, Bishop A G, MacFarlane D R, Orädd G and Forsyth M 2001 *J. Mater. Chem.* **11** 3031
[21] Tokuda H, Tsuzuki S, Susan M A B H, Hayamizu K and Watanabe M 2006 *J. Phys. Chem. B* **110** 19593
[22] Blokhin A V, Paulechka Y U, Strechan A A and Kabo G J 2008 *J. Phys. Chem. B***112** 4357
[23] Shimizu Y, Ohte Y, Yamamura Y, Saito K and Atake T 2006 *J. Phys. Chem. B***110** 13970
[24] Paulechka Y U, Blokhin A V, Kabo G J and StrechanAA 2007 *J. Chem. Thermodyn.***39** 866
[25] Turnbull D, Cohen M H 1960 *Modern Aspect of the Vitreous State I*ed J D Mackenzie (London: Butterworth) p 38

Acknowledgments

This work is financially supported by the Grant-in-Aid for Scientific Research on Priority Area No. 17073004, MEXT, Japan. The experiments at NCNR was financially supported by the U.S.-Japan Cooperative Program on Neutron Scattering.

IOP Conf. Series: Materials Science and Engineering **196** (2017) 012002 doi:10.1088/1757-899X/196/1/012002

Electronic Properties of Alkali Metals Loaded into Channel-Type Zeolite L

P T Thi[1,2], T Nakano[2], Y Sakamoto[2,3] and Y Nozue[2]

[1]Center for Innovative Materials & Architectures (INOMAR), Vietnam National University, Ho Chi Minh City (VNU-HCM), Quart 6, Linh Trung Ward, Thu Duc District, Ho Chi Minh City, Vietnam
[2]Department of Physics, Graduate School of Science, Osaka University, Toyonaka, Osaka 560-0043, Japan
[3]Precursory Research for Embryonic Science and Technology (PRESTO), Japan Science and Technology Agency (JST), Kawaguchi, Saitama 332-0012, Japan

E-mail : nozue@phys.sci.osaka-u.ac.jp

Abstract. Rb metal was loaded into dehydrated Rb-form zeolite L. The loading density of Rb atoms per unit cell, n, was changed up to 7.1. A resonant optical absorption-reflection band is observed at ≈1.0 eV in all samples. A new reflection band appears at ≈1.7 eV for $n > 5$. The 1.0 and 1.7 eV bands are assigned to the respective excitations from 1s to 1p and from 1p to 1d quantum states of s-electrons one-dimensionally confined in main-channels with an inside diameter of ≈1 nm. A temperature-dependent semiconducting resistivity is observed at $n > 2$, and is assigned to the inter-channel polaron hopping between metallic 1s and 1p states. Besides, a nearly metallic resistivity with the low thermal activation energy is observed at $n > 5$. The metallic conductivity is assigned to three-dimensional s-electrons in the 1d state. The metallic property in Rb-loaded zeolite L is more remarkable than those in K-loaded zeolite L. The difference is explained by the smaller ionization energy of Rb atom, which leads to a shallower one-dimensional confinement potential of main-channel and a smaller polaron effect.

1. Introduction

Alkali metals loaded into zeolite crystals display novel properties of s-electrons, such as a ferromagnetism and an insulator-to-metal transition, depending on the kind of alkali metals, the loading density and the structure type of zeolites [1,2]. The quantum confinement of electrons in the bundle of cylindrical potentials has attracted deep interest in their novel electronic states [3,4]. An electron in a quantum cylinder has the two terms in the kinetic energy [3,4]. The first term is given by the radial confinement, and forms 1s, 1p and 1d quantum states in the increasing order of energy. The second term is given by the free motion along the cylinder. When a cylinders has a finite potential depth, the wave function of an electron spills out [5]. The hopping-type incoherent motion of polarons increases with the increase in temperature [6,7]. If the mutual overlap of wave function increases furthermore, the band-type coherent motion occurs as the three-dimensional (3D) metallic phase [3,4].

Aluminosilicate zeolite L has the LTL framework structure (IUPAC nomenclature), as shown in Figure 1. The one-dimensional (1D) main-channels with an inside diameter of ≈1 nm are bundled along the *c*-axis. The framework is negatively charged, and cations are distributed at the space of framework. K metal can be loaded into K-form zeolite L [4,8,9]. The loading density of K atoms per unit cell, *n*, was changed up to 9.6, and a resonant optical absorption-reflection band has been observed at ≈1.1 eV in all K-loaded samples [4,9]. Another reflection band appears at ≈1.8 eV for $n > 5$. These 1.1 and 1.8 eV bands are assigned to the respective excitations from 1s to 1p and from 1p to 1d quantum states of s-electrons confined in main-channels of zeolite L. An increase in the mid-infrared absorption coefficient at 0.3 eV has been observed at $n > 3$. This absorption is assigned to a tail part of the Drude term of the 1D metallic state of s-electrons confined in main-channels. A temperature-dependent semiconductive electrical conductivity has been observed at $n > 3$. A nearly metallic conductivity is observed at $n > 9$. The total conductivity is given by the sum of two terms with the high and the low thermal activation energies [4]. The semiconductive conductivity is assigned to the polaron hopping between main-channels in 1s and 1p quantum states. The metallic conductivity is assigned to the 3D state of s-electrons in 1d quantum states. A temperature independent spin-susceptibility extracted from the integrated intensity of ESR signal increases at $n > 5$. This term is assigned to the Pauli paramagnetism of metallic states. These properties are interpreted systematically in terms of an insulating states of 1s and 1p states at $n < 5$, 1D metallic states of 1s and 1p states at $n > 5$, and a 3D metallic state of 1d state at $n > 9$.

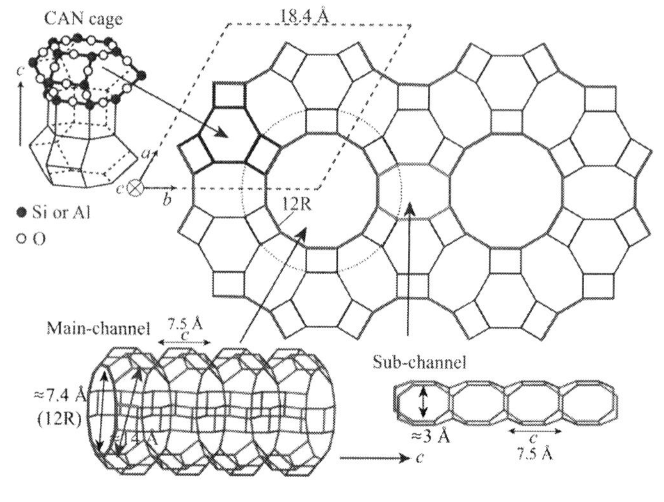

Figure 1. Schematic illustration of the LTL framework structure of zeolite L. Main-channels have the inside diameters of ≈14Å at the widest part and ≈7.4Å at 12-membered ring (12R). Sub-channels are located between main-channels, and share elliptic 8-membered rings with main-channels. Each CAN cages are surrounded by sub-channels and main-channels. Cations are not shown here.

The metallic property in Rb-system is expected to be more remarkable than those in K, because of the smaller ionization energy of Rb atom, which leads to a shallower 1D confinement potential and a smaller polaron effect. In the present study, we prepared Rb-loaded zeolite L at various Rb-loading densities, and measured their optical and electrical properties.

2. Experimental Procedures

K cations of synthetic zeolite L powder (Tosoh Corporation) were exchanged to Rb cations, and Rb-rich zeolite L with the chemical formula $K_2Rb_7Al_9Si_{27}O_{72}$ per unit cell was prepared. Rb metal was loaded into dehydrated Rb-rich zeolite at the loading density of *n*Rb atoms per unit cell. Hereafter, the samples are denoted by Rb_n/K_2Rb_7-L. We performed optical and electrical properties. Sample preparations and characterizations are detailed in Ref. 4.

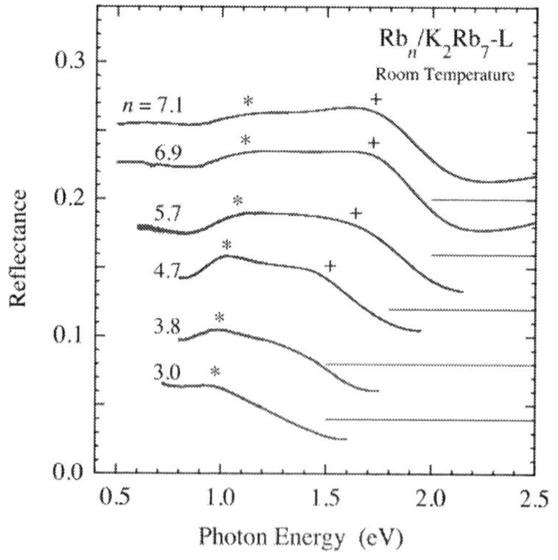

Figure 2. Reflection spectra of Rb_n/K_2Rb_7-L at room temperature.

3. Experimental Results and Discussions

In dilutely Rb-loaded Rb_n/K_2Rb_7-L, an absorption band clearly appears at ≈1.0 eV [9]. Figure 2 shows the reflection spectra of Rb_n/K_2Rb_7-L at room temperature. A reflection band at ≈1.0 eV with asterisk is seen all samples. New band at ≈1.7 eV with plus sign appears around $n > 5$ and grows with n. Similar reflection bands have been observed in K-loaded K-form L (K_n/K_9-L) [4]. These reflection bands can be assigned to the excitations from 1s to 1p and from 1p to 1d quantum states of s-electrons in the main-channel of zeolite L. Differently from the localized states in clusters (zero-dimensional), the excitation from 1s to 1p states appears together with that from 1p to 1d states, reflecting the continuous 1D energy band [4]. An increase in the mid-infrared absorption coefficient at 0.3 eV is observed at $n > 2$ in Rb_n/K_2Rb_7-L [9]. This absorption is assigned to a tail part of the Drude term in 1D metallic s-electrons in main-channels. At $n < 2$, s-electrons may become small bipolarons by the interaction with the cation displacement, and insulating properties with a finite optical gap are observed.

The temperature dependence of electrical resistivity ρ in Rb_n/K_2Rb_7-L is shown in Fig. 3. The resistivity increases with the decrease in temperature at lower loading densities. The resistivity remarkably decreases with the loading density, and the increase at low temperatures becomes weak. These temperature dependence is explained by the following conductivity σ with two terms [4],

$$\sigma = \sigma_H \exp\left(-\frac{E_H}{k_B T}\right) + \sigma_L \exp\left(-\frac{E_L}{k_B T}\right), \quad (1)$$

where E_H and E_L are high and low thermal activation energies, respectively, and σ_H and σ_L the respective conductivities at the infinite temperature. k_B and T are the Boltzmann constant and the temperature, respectively. ρ is given by $1/\sigma$. The fitting results are indicated by solid curves in Figure 3, where the data at $n \leq 3.8$ and $n = 7.1$ are fitted by one term with the high and the low activation energies in Eq. 1, respectively. Estimated values of E_H, E_L, σ_H and σ_L are plotted in Figure 4 by closed squares, together with those in K_n/K_9-L [4] by open circles. Data of Rb_n/K_2Rb_7-L is clearly shifted to lower values of n compared with those of K_n/K_9-L.

The electrical conductivity is usually given by

$$\sigma = e\mu N, \quad (2)$$

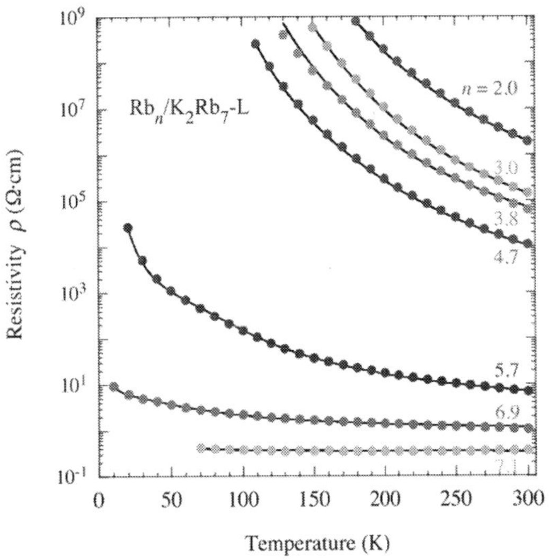

Figure 3. Temperature dependence of electrical resistivity in Rb_n/K_2Rb_7-L (closed circles). Solid curves are fitting results.

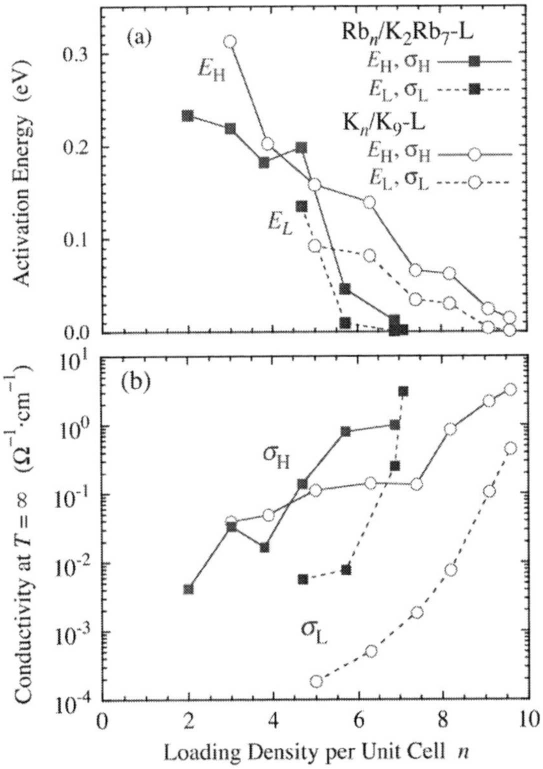

Figure 4. Parameters of conductivity in Eq. 1, E_H, E_L, σ_H and σ_L, estimated in Rb_n/K_2Rb_7-L (closed squares) and K_n/K_9-L (open circles).

where e, μ and N are the elementary electric charge, the carrier mobility and the number density of carriers, respectively. The temperature dependence of conductivities are given by μ and/or N. In the bundle of channels, s-electrons in 1D quantum states can move to other channels by the hopping process with the increase in temperature. This mechanism contributes to the thermal excitation of mobility with the constant carrier density. The high thermal excitation term in Figure 4 can be assigned to the hopping of s-electrons in the 1s and 1p states. With the increase in n, low thermal activation term dominates the conductivity in Figure 4. This n-dependence is assigned to the thermal excitation of carriers with the band-type conduction in the 1d state, because a larger overlap of 1d wave function can be expected to form the 3D energy band.

In order to overview, the electronic state in Rb_n/K_2Rb_7-L as the function of the Rb loading density, the density of states (DOS) is illustrated schematically in Figure 5. At $n < 2$, most of s-electrons in main-channels are localized as small bipolarons, and an insulating properties are observed. At $n > 2$, electrons in main-channels become 1D-metallic, and the hopping motion of them provides the conductivity with the high thermal activation energy. At $n > 6$, some electrons are distributed in the 1d state, and the band-type 3D-conduction with low thermal activation energy is observed below the mobility edge of the Anderson localization. At $n > 7$, the 3D metallic state is realized above the mobility edge. In K_n/K_9-L, the 1D-metallic state is assigned at $n > 5$, and the 3D-conduction is assigned at $n > 9$. The mechanism in Rb_n/K_2Rb_7-L is similar but the corresponding values of n is effectively smaller than those in K_n/K_9-L, as clearly indicated in Figure 4. This difference can be explained by the smaller ionization energy of Rb atom.

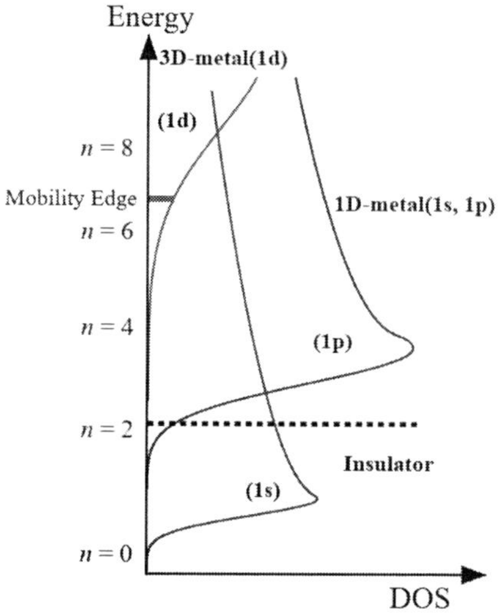

Figure 5. The density of states (DOS) of s-electrons in the bundle of main-channels in Rb_n/K_2Rb_7-L.

4. Summary

Rb metal was loaded into Rb-rich zeolite L at various loading densities. Optical and electrical properties vary systematically with n. Observed results are explained by the quantum confinement of s-electrons in 1s, 1p and 1d states of main-channels bundled in zeolite L. Results are compared with those in K_n/K_9-L.

References

[1] Nakano T and Nozue Y 2017 *Adv. Phys. X* **2** 254
[2] Nakano T, Hanh D T, Nozue Y, Nam N H, Duan T C and Araki S 2013 *J. Korean Phys. Soc.* **63** 699
[3] Kelly M J 1995 *J. Phys.: Condens. Matter* **7** 5507
[4] Thi P T, Nakano T, Sakamoto Y, Nozue Y 2016 *J. Phys. Soc. Jpn.* **85** 024703-1
[5] Van den Broek M and Peeters F M 2001 *Physica E* **11** 345
[6] Holstein T, 1959 *Ann. Phys. (Leipzig)* **8** 343 [Reprint **281** (2000) 725-773]
[7] Sumi H 1972 *J. Phys. Soc. Jpn.* **33** 327
[8] Anderson P A, Armstrong A R, Porch A, Edwards P P and Woodall L J 1997 *J. Phys. Chem. B* **101** 9892
[9] Thi P T 2016 *Doctor Thesis* Osaka University

The Role of graphene in Fe_3O_4/graphene Composites on the Adsorption of Methylene Blue and Their Kinetic Study

A Taufik[1,2], **R Saleh**[1,2]

[1]Integrated Laboratory of Energy and Environment, Fakultas MIPA-Universitas Indonesia, Kampus UI Depok, Depok 16424, Indonesia
[2]Departemen Fisika, Fakultas MIPA-Universitas Indonesia, Kampus UI Depok, Depok 16424, Indonesia

E-mail : rosari.saleh@ui.ac.id

Abstract. In this study Fe_3O_4 with addition of graphene materials were investigated as an adsorbent for removing organic dye from the aqueous solution. Fe_3O_4-graphene were synthesized using hydrothermal method. The contents of graphene were varied from 5, 10, and 15 weight percent (wt%). Their physical properties were investigated using X-ray diffraction (XRD), thermogravimetric analysis (TGA), and surface area analysis. The cubic spinel of the samples was detected which attributed to the existence of Fe_3O_4 materials. The removal of organic dye was investigated using adsorption process. Methylene blue was selected as a model of organic pollutant. Their kinetic reaction indicates that the samples follow the pseudo second order kinetic reaction with maximum adsorption capacity was 48 mg/g. The samples performed good stability for removing methylene blue using adsorption process with 95 % efficiencies.

1. Introduction

During the last few decades, the organic dye removal from waste water is extensively investigated using various method. However, mostof these dyes exhibit high resistance to photodegradation and biodegradation which may transfer the primary pollutant to secondary onethat may be more toxic. Adsorption was found as a simple and low cost processfor removing various hazardous pollutantsfrom wastewater [1].

There are many materials that have been used as catalyst in the adsorption process such as zeolite [2], clay [3], silica [4], and activated carbon [5]. However, the difficulty of separation between catalyst and the organic dye solution is another challenge for improving the adsorption efficiencies [6]. Many strategies have been developed to improve the separation efficiency of the samples such as centrifugation [7]. However thus process is expensive and complex system. Using magnetic catalyst as adsorbent is very promising for enhance the separation efficiency due to its ability to attract by external magnetic that provide magnetic separation process [8]. Among others magnetic materials, magnetite (Fe_3O_4) has been considered as a promising candidate as catalyst due to its unique magnetic properties and also high adsorption capacity.

In practical experiment, Fe_3O_4 adsorption process is still has limitation due to the small specific surface area. Many method has been tried to improve the specific surface area of Fe_3O_4. Recently the incorporation of graphene materials with theoritical surface area about $2600 m^2/g$ on semiconductors have been proved could improve the specific surface area of the samples [9].

IOP Conf. Series: Materials Science and Engineering **196** (2017) 012003 doi:10.1088/1757-899X/196/1/012003

Based on that reason in this reasearch we try to evaluate the influence of Fe_3O_4 modified by graphene materials for enhance the adsorption ability of the samples. The kinetic calculation wasaso observed to understand the adsorption behaviour of the samples and the reusability was also observed for investigated the stability of the samples.

2. Experimental Detail

2.1. Materials

$Fe(SO)_4.7H_2O$ (Iron Sulfate Heptahydrate), NaOH (Sodium Hydroxide), CH3COOH (Acetic Acid) were purchased from merk and used without further purifications, while graphene were purchased from Angstron Materials.

2.2. Synthesis and Characterization

The Fe_3O_4 nanoparticles were prepared based on our previous study [10]. The Fe_3O_4-graphene were synthesized using hydrothermal method. Typically, graphene powder was dispersed in mixture water and ethanol solution. After that, the graphene dispersion was sonicated using ultrasonic bath operated at 40 kHz frequency for 2h. The Fe_3O_4 nanoparticles were added into the graphene dispersion and stirred for 30 min. The mixture then heated at 120°C for 30 min to deposit Fe_3O_4 on graphene sheet. After that the samples were collected using centrifugation method and dried under vacuum condition The all prepared samples were characterized using X-ray diffraction (XRD),Thermogravimetric analysis (TGA), *Brunauer–Emmett–Teller* (BET) surface area analysis.

2.3. Adsorption Experiment

The adsorption process was carried out for removal methylene blue from the aqueous solution. Typically, 60 mg/L methylene blue solution was prepared followed by the incorporation of 0.03 g adsorbent. The methylene blue concentration was monitored by UV-Vis spectroscopy. The decolorization of methylene blue and the adsorption capacity of the samples was calculated using the following equation:

$$Degradation: \frac{C_t}{C_0}, \tag{1}$$

$$q_e = \frac{(C_o - C_e)V}{m} \tag{2}$$

where C_0 is the initial methylene blue concentration (mg/L), C_t is the methylene blue concentration after time (t) (mg/L), qe represents the equilibrium adsorption capacity of the Fe_3O_4 nanocomposite (mg/g); and C_e are the initial and equilibrium concentrations of methylene blue (mg/L) in aqueous solutions, respectively; V is the volume of the aqueous MB (mL); and m is the amount of Fe_3O_4/ graphene used (g).

To understand the underlying adsorption mechanisms in this study, two models, including pseudo-first order and pseudo-second order models expressed in Eq. (2) and (3),

$$\ln(q_e - q_t) = \ln q_e - K_1 t \tag{3}$$

$$\frac{t}{q_t} = \frac{1}{K_2 q_e^2} + \frac{t}{q_e} \tag{4}$$

were applied to extract the useful information from the experimental kinetic data. In these equations, qe and qt are the adsorption capacities at equilibrium and time t, respectively. K_1 is the pseudo-first order kinetic constant (min^{-1}), and K_2 is pseudo-second order constant (g/(mg min)).

3. Result and Discussion

Figure 1a shows the XRD spectra of Fe_3O_4 modified by graphene materials. As comparison, the XRD spectra of Fe_3O_4 nanoparticles, as well as graphene materials, were also plotted. The XRD result for all samples were detected peak at 2θ = 30.14°, 35.49°, 43.28°, 53.76°, 57.20°, and 62.83°. These peak were

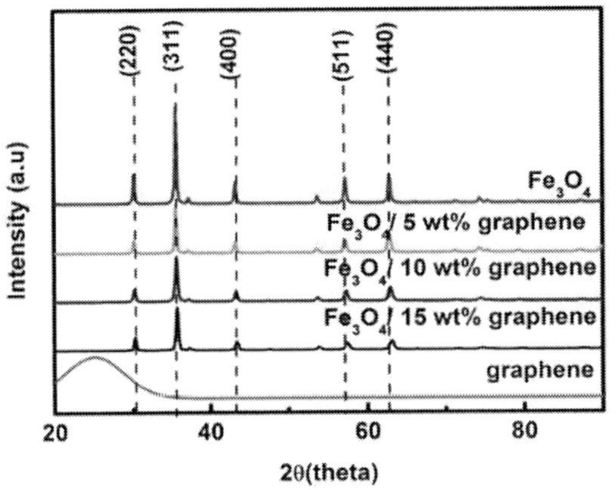

(a)

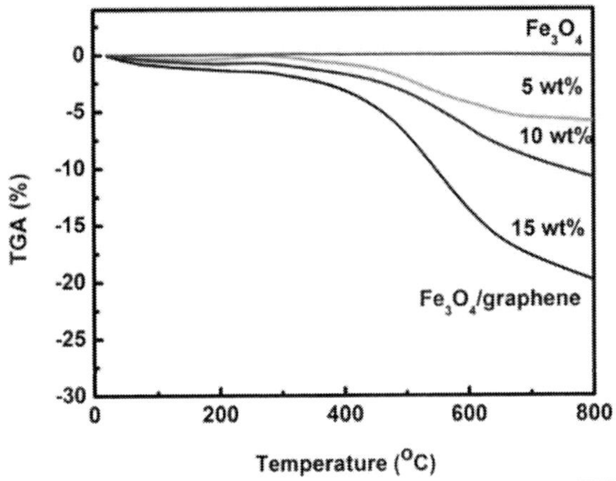

(b)

Figure 1. (a) XRD spectra and (b) TGA measurement of Fe_3O_4 nanoparticles and Fe_3O_4/graphene nanocomposites with different weight percent.

Table 1. Lattice parameter, grain size and specific surface area of Fe_3O_4 nanoparticles and Fe_3O_4/graphene composites with different weight percent.

Sample	Lattice parameter (Å) a=b=c	Grain size (nm)	Surface area (m^2/g)
Fe_3O_4	8.363	45	2.3
Fe_3O_4/5 wt% graphene	8.357	42	65.0
Fe_3O_4/10 wt% graphene	8.354	39	95.0
Fe_3O_4/15 wt% graphene	8.332	39	106.9

attributed to (220), (311), (400), (442), (511), and (440) plane of cubic spinel structures from Fe3O4 nanoparticles. No undesired peaks were detected that indicated the successfully formation of Fe_3O_4/graphene composites. The lattice parameter and grain size of the samples were tabulated in Table 1. It can be seen that the grain size value of Fe_3O_4/graphene was slightly decrease with the increasing of graphene weight percent. It can be due to the existence of graphene inhibits the crystallization of Fe_3O_4.No diffraction peak was detected from graphene in Fe_3O_4/graphene composites that probably due to the small amount and broad peak of graphene.

The specific surface area of the samples was tabulated in Table 1. The specific surface area of the samples was gradually increase with the incorporation of graphene from 2.3 m^2/g to106.9m^2/g. The higher specific surface area is beneficial for the enhancement of adsorption process due to its ability to increase the space contact between adsorbent and adsorbate.

The thermal stability of the samples was investigated in the range of room temperature until 800 °C. Results were plotted in Figure 1b. as can be seen from the figure that Fe_3O_4 nanoparticles exhibit good stability until 800 °C. however the incorporation of graphene reduces its stability starting from 400 °C due to the combustion of carbon. The higher graphene weight percent resulting the lower stability of the samples due to the combustion of carbon [11].

The adsorption ability of the samples was recorded on the degradation of methylene blue from the aqueous solution. Figure 2a shows the relative adsorption concentration of methylene blue after and before adsorption process. The removal of methylene blue using Fe3O4 nanoparticles was only 14 %. The incorporation of graphene from 5, 10, and 15 wt% the removal ability of the samples was increase from 53 %, 64 %, and 73 %. It probably due to the increasing of specific surface area of the samples.

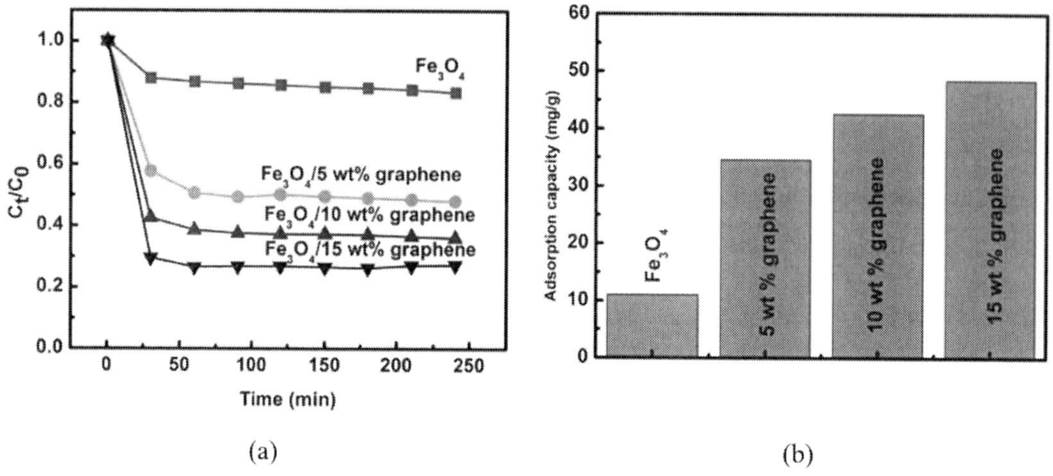

Figure 2. (a) Decolorization ability and (b) adsorption capacity of Fe_3O_4 nanoparticles and Fe_3O_4/graphene composites with different graphene weight percent.

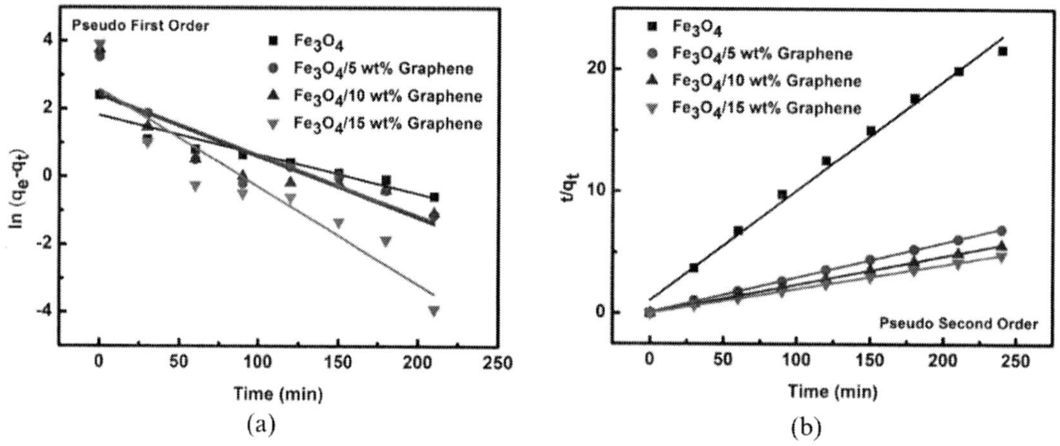

Figure 3. (a) Pseudo first order and (b) second order kinetic reactions

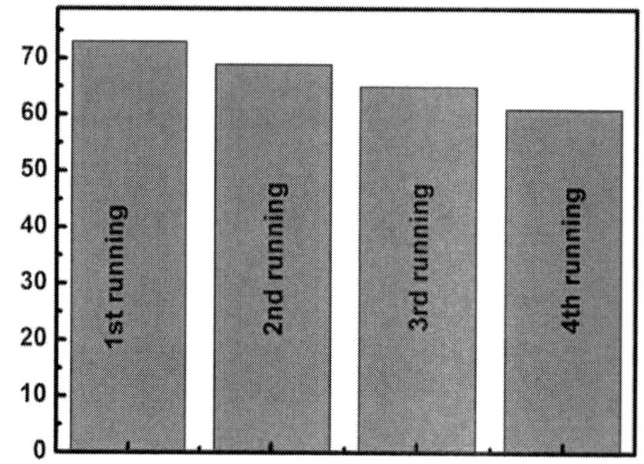

Figure 4. Reusability of adsorbent.

Figure 2b shows the adsorption capacity for all samples. As can be seen from the figures that the samples with the higher graphene content exhibits the higher adsorption ability. With maximum adsorption capacity about 48 mg/g. It is probably due to the higher surface area of adsorbent that promoted the better contact between adsorbent and adsorbate.

The kinetic model of adsorption was calculated using pseudo first and second order kinetic reaction. The fitted data were plotted in Figure 3. As can be seen in the figure that the linear plot of the experimental data well fitted with the pseudo second order kinetic reaction with R^2 are about 98 however using pseudo first order kinetic reaction the R^2 are only about 88. It indicates that pseudo first order kinetic model failed to described the adsorption kinetic of the samples while pseudo second order kinetic reaction successfully described the adsorption kinetic of the samples.

Stability of adsorbent is an important parameter for practical applications. Figure 4 shows the reusability of the Fe_3O_4/graphene with 15 wt% graphene. As can be seen from the figure that the sample performed good stability. After 4 time cycling process the removal ability of the samples was decrease about 95 % from the initial experiment. It indicates that Fe_3O_4/graphene is potential as adsorbent for removing waste water.

4. Conclusion

The Fe_3O_4/graphene composites were successfully synthesized using hydrothermal method for removing methylene blue using adsorption process. The incorporation of graphene in Fe_3O_4 significantly enhance the adsorption ability, with the highest graphene content resulting the highest removal efficiencies. The kinetic calculation indicates that the samples followed pseudo second order kinetic reactions.

References

[1] Salem A N M, Ahmed M A, Shahat M F E 2016 *J. Mol. Liq.* **219** 780
[2] Alver E, Metin AÜ 2012 *Chem. Engin. J.* **200–202** 59
[3] Pan Y, Cai P, Farahani M F, Li Y, Hou X, Xiao H 2016 *Appl. Surf. Sci.* **385** 333
[4] Perullini M, Jobbágy M, Japas M L, Bilmes S A 2014 *J. Colloid Interf.Sci.* **425** 91
[5] Demirbas A 2009 *J. Hazard. Mater.* **167** 1–9.
[6] Zhao J, Liu J, Li N, Wang W, Nan J, Zhao Z, Cui F 2016 *Chem. Eng. J.* **304** 737
[7] Jia X, Dai R, Lian D, Han S, Wu X, Song H 2017 *Appl. Surf. Sci.* **392** 268
[8] Kalantari K, Ahmad M B, Masoumi H R F, Shameli K, Basri M, Khandanlou R 2015 *Journal of the Taiwan Institute of Chemical Engineers* **49** 192
[9] Gadipelli S, Guo Z X 2015 *Progress in Materials Science* **69** 1
[10] Taufik A, Saleh R 2016 *Mater. Sci. Forum* **864** 106
[11] Lin C C, Ho J M 2014 *Ceram. Int.* **40** 10275

IOP Conf. Series: Materials Science and Engineering **196** (2017) 012004 doi:10.1088/1757-899X/196/1/012004

Crystal Structure Characterization of Thin Layer Zinc Oxide

Aris Doyan[1], Susilawati[1], Siti Azizatul Fitri[1], Sukainil Ahzan[2]

[1]Magister Science, Post Graduate of Mataram University, Mataram, Indonesia
[2]Physcish Education Department, FPMIPA IKIP Mataram, Mataram, Indonesia

E-mail : arisdoyan@yahoo.co.id

Abstract. In this research the characterization of the crystal structure of a thin layer of ZnO (zinc oxide) were synthesized by sol - gel method and spin coating deposited on a glass substrate. The samples were divided into three sol concentrations of 0.1, 0.3, 0.5 Molar and two deposition temperature is 350 °C, and 550 °C. UV-Vis. spectrophotometer results showed that in the spectrum of visible light (wavelength range 300-800 nm) has a transmittance value of which increases with increasing concentration and temperature deposition of zinc oxide, otherwise the value of the absorption and the band gap energy decreases with the addition of concentration and deposition temperature. The transmittances value of the highest and lowest absorption was 93.5% and 0.03 is at a concentration of 0.1 M and zinc oxide deposition temperature of 550 °C, with a value of band gap energy of 2.98 eV. The XRD results showed that the zinc oxide crystal orientation in the field of 013 with a crystal grain size 14.4472 nm. SEM results showed the surface morphology of zinc oxide such as rod-like.

1. Introduction

Nano-meter sized material has now attracted the attention of researchers science because electrical, physical, chemical, and magnetic properties of potential in the development of technology [1]. One of the material in question is a metallic material. Metal material which widely used in application today is in the form of a thin layer, for example, tungsten oxide (WO_3), indium (In_2O_3), tin (SnO_2), titania (TiO_2), zinc oxide (ZnO) and many other materials. Among the metal materials, metal ZnO has the greatest abundance in nature [2]. Moreover, ZnO metal inexpensive and versatile [3]. A thin layer of zinc oxide (ZnO) pure and doping has been regarded as an excellent semiconductor material for transparent conductive oxide (TCO), because it has a high conductivity, good optical transmittance (> 90%), and low cost of fabrication. A thin layer of ZnO shows the thermal stability, and are not easily change shape if the process of dislocation of material structure [4]. Some of the parameters that can be observed in the manufacture of thin layers by the method of sol-gel spin coating, among other sol solution concentration, the concentration of doping, deposition temperature, the speed of the spin coating, and aging time. Ongoing research to produce the layer characteristics as expected. The characteristics of the optical properties of ZnO thin layers in the range of the visible spectrum (370-800 nm). ZnO thin layer has an optical transmittance values decreasing from 90 to 25 % when the concentration of sol added 0.25 M to 1 M, and a smaller band gap energy values of 3.30 eV to 3.22 eV at a concentration increased from 0.25 M to 0.75

M. However the band gap energy increased from 0.22 eV to 3.27 eV at a concentration of 0.75 M to 1 M [3]. The optical properties of ZnO seen transmitansinya 91% to 95 % in the range 400 to 800 nm. Crystal size in the range of 18 to 24 nm [5]. Furthermore in this research conducted characterization of thin layer of zinc oxide to variations in the concentration of the sol and the temperature of deposition. The characterizations are the optical properties (transmittance, absorption, energy band gap) of thin layer. Then, determine the crystal structure and surface morphology of the sample of the best of a layer of zinc oxide.

2. Materials and Methods

Substrate materials that used in this research is a glass slide that cut to a size of (10 x 5) mm^2, and a thickness of 3 mm. Before use of the glass slide is washed in stages. Laundering gradually aims to remove oil and debris on the glass. Leaching the first stage, the glass substrate is inserted into a measuring cup containing a mixture of water and detergent. Furthermore the glass substrate is rinsed with water until clean. Washing at the second stage, the substrate is inserted into a glass beaker containing alcohol. The glass substrate is dried in a furnace at 100 °C for one hour. The glass substrate that is clean, then stored in plastic clip. The process of making sol-gel using basic materials zinc acetate dehydrate (ZnAc), ethanol and MEA (Mono Ethanolamine), each of which serves as the solute, solvent and stabilizer. The sol-gel materials that synthesized divided into three concentrations, namely 0.1 M, 0.3 M, and 0.5 M. The solution ZnAc and ethanol stirrer on a hot plate at a temperature of 70 °C for 10 minutes, or until the solution homogeneously mixed. Layer growth process begins with a drop of sol-gel materials on a glass substrate is approximately 3 drops. Then spin for two minutes with a rotation speed of 2000 rpm. Once the surface evenly coated substrate, then the substrate is dried in a furnace at 100 °C for 10 minutes. The heating process is done in three stages. First heating at a temperature of 150 °C (calcinations) for one hour, aiming to eliminate water and residual solvent content in the coating gradually. A second heating at a temperature of 350 °C for one hour. The increase in the regulated temperature slowly from room temperature to 350 °C. This stage is regarded as pre-heating phase that serves to remove ethanol, water, and MEA, and facilitating change ZnOH into ZnO. The next stage is the post-heating or heating end at 550 °C, for one hour. The same with the process of pre-heating, regulated temperature rise slowly from room temperature to 550 °C. Post-heating function to form ZnO particles with a uniform crystal orientation, and eliminating pores.

The characterization using UV-Vis spectrophotometer is intended to identify how much the value of transmittance and absorption of ZnO thin film for some variations in the concentration (molar sol gel solution). Characterization using X-ray diffraction (XRD) is intended to identify the different forms of the crystal structure in a thin layer of ZnO quantitatively. The identification results obtained by comparing the diffractogram of the ZnO thin layer of the unknown with standard data that contains a reference pattern. The characterization using Scanning Electron Microscopy (SEM) is intended to identify the micro-structure surface of a thin layer of ZnO crystals. The identification results obtained in the form of a thin layer of ZnO surface profiles such as how and where the direction of the slope ramps that can be built using an image processing program in the computer. Determining the value of the energy gap of a semiconductor material needed to know to provide an analysis of the optical absorption coefficient associated with the frequency of photons. Energy band gap of ZnO thin layer (zinc oxide) is calculated by following equation [6].

$$(\alpha h v)^{\frac{1}{n}} = A(E - E_g) \tag{1}$$

where α is the absorption coefficient, hv is the photon energy, A is a constant, and E_g is the band gap energy. Exponent n depends on the type of transition, where the transition to direct, indirect, prohibited direct and indirect forbidden row has a value of 1/2 , 2, 3/2, and 3.

Results and Discussion

The measurement using equipment UV-Visible, a thin layer of ZnO can transmit visible light at wave number range of 300 nm-800 nm. For a concentration of 0.1 M, 0.3 M, 0.5 M at temperature deposition at 350 °C with the highest and lowest transmittance values are (83.1 % and 21.7 %), (80.4 % and 10.6 %), (78.7 % and 2.7 %), respectively. For temperature deposition 550°C with the highest and lowest transmittance values are (93.5 % and 9.6 %), (87.6 % and 8.4 %), (84.9 % and 5.9 %) respectively as shown in Figure 1 (a) and (b).

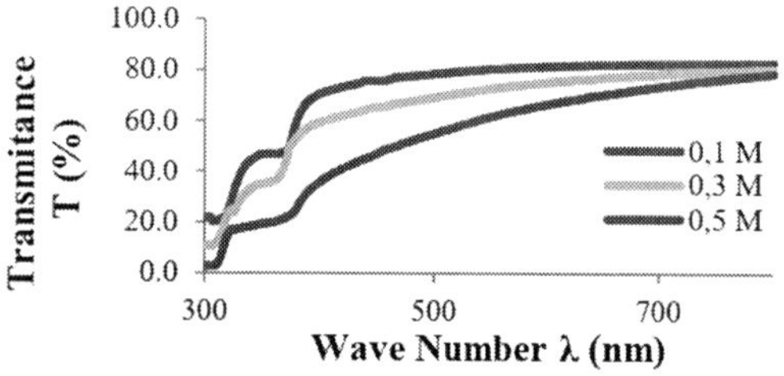

Figure 1(a). Graph the value of optical transmittance T (%) as a function of wavelength λ (nm) thin layer of ZnO temperature deposition at 350 °C

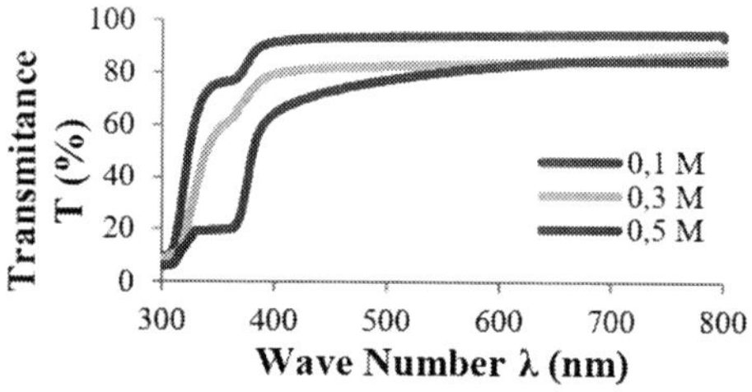

Figure 1(b). Graph the value of optical transmittance T (%) as a function of wavelength λ (nm) thin layer of ZnO temperature deposition at 550 °C

Furthermore, the value of the highest optical transmittance of ZnO thin film for the wavelength range of 300 nm-800 nm was 93.5%. That value shows the condition of ZnO concentration of 0.1 M and for a deposition temperature of 550 °C. While the lowest value of the optical transmittance of ZnO thin layer to the wavelength range of 300 nm-800 nm was 2.7%. That value shows the condition of ZnO concentration of 0.5 M and a deposition temperature of 350 °C. Thus the optical absorption values (A%) as a function of wavelength (λ) for the variation of the concentration of ZnO is 0.1 M, 0.3 M, 0.5 M respectively and temperature deposition variations was 350 °C and 550 °C as shown in Figure 2 (a) and (b).

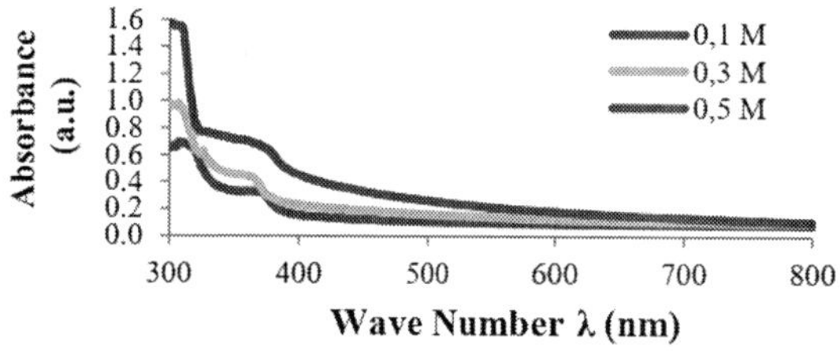

Figure 2(a). Graph the value of a optical absorption (a.u.) as a function of wavelength λ (nm) thin layer of ZnO temperature deposition at 350°C

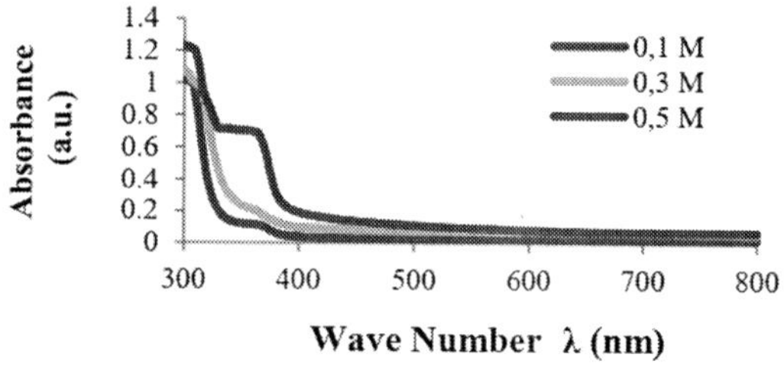

Figure 2(b). Graph the value of a optical absorption (a.u.) as a function of wavelength λ (nm) thin layer of ZnO temperature deposition at 550 °C

Figure 2. shows that for deposition temperature 350 °C and a concentration of 0.1M, 0.3 M, 0.5 M respectively, a thin layer of ZnO can absorption of visible light in range 300nm -800 nm with the lowest and highest values are (8 % and 66 %), (9 % and 97 %), (10 % and 157 %) respectively. Furthermore for deposition temperature 550 °C and a concentration of 0.1M, 0.3 M, 0.5 M respectively, a thin layer of ZnO can absorption of visible light in range 300nm -800 nm with the lowest and highest values are (3 % and 102 %), (6 % and 108 %) (7 % and 124 %) respectively.

Thus, the value of the highest optical absorption of ZnO thin film for the wavelength range of 300 nm to 800 nm is 157 %, i.e on condition ZnO concentration of 0.5 M and temperature deposition at 350 °C. While the lowest value of the optical transmittance of a thin layer of ZnO for the wavelength range of 300 to 800 nm is 3%, that on condition of ZnO concentration of 0.1 M and temperature deposition at 550 °C. The values of gap energy of ZnO thin film obtained for various concentrations (0.1, 0.3, 0.5 M) and the variation of the deposition temperature (350 and 550°C) can be seen in Figure 3 (a) and (b).

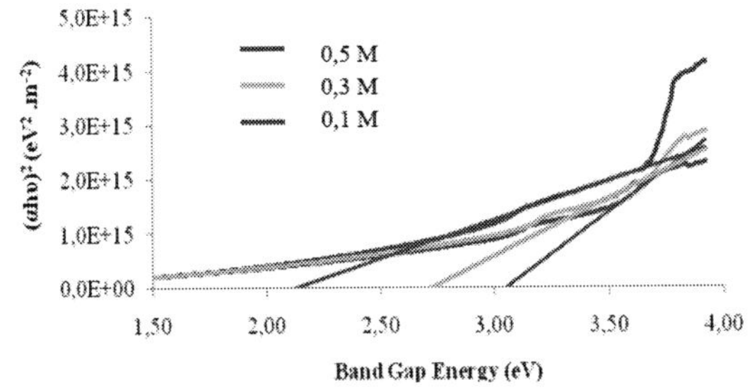

Figure 3(a). Graph The values of gap energy of ZnO thin film obtained for various concentrations (0.1 M, 0.3 M, 0.5 M) and the variation of the temperature deposition at 350°C

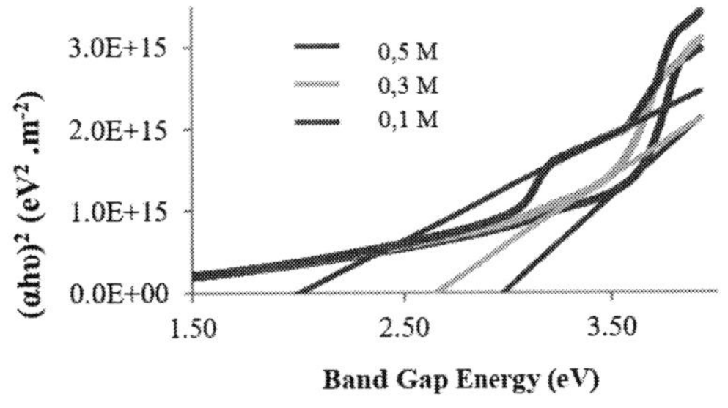

Figure 3(b). Graph The values of gap energy of ZnO thin film obtained for various concentrations (0.1 M, 0.3 M, 0.5 M) and the variation of the temperature deposition at 550°C

Energy gap of thin layer ZnO at different concentration of precursor and temperature deposition was calculated using Equation 1 and shown in Table 1.

Table 1. The value of gap energy for variations in concentration and temperature deposition of thin layer ZnO

The concentration of ZnO (M)	The Value of Gap Energy (eV)	
	Temperature (°C)	
	350°C	550°C
0,1	3,06	2,98
0,3	2,75	2,67
0,5	2,18	2,03

Figure 4 shows the X-Ray Diffraction (XRD) measurement of thin layer ZnO obtained from concentration of precursor 0.1 M and temperature deposition at 550°C. The diffraction pattern of ZnO crystal planes is (013) and hexagonal ZnO crystal, formed at positions 2 theta 64.4415 and the grain size 14.4472 nm. Based on XRD measurement, the grain size and Miller indices could be seen on Table 2

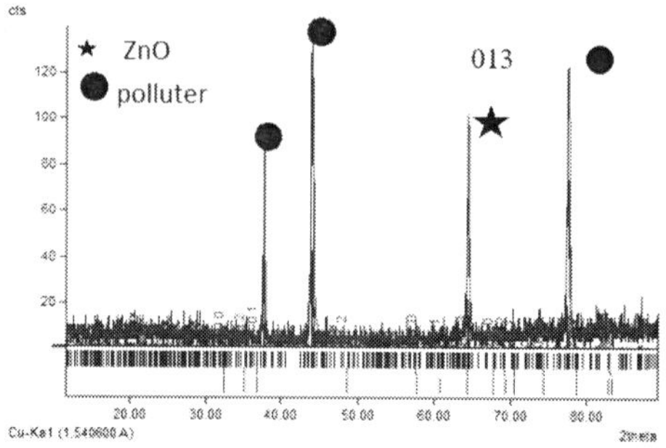

Figure 4. Thin layer ZnO, 0.1 M and temperature deposition at 550 °C of crystal diffraction intensity as a function of 2θ

Table 2. Parameters Crystal Structure of thin layer ZnO, 0.1 M at temperature deposition at 550 °C

No.	2Theta (deg)	The grain size (nm)	FWHM (deg)	Intensity (counts)	Miller indices
1	64,4415	14,4472	0,20860	57	0 1 3

The crystallites sizes (D) of thin layer ZnO is estimated using the Scherer formula [6].

$$D = \frac{k\lambda}{\beta_{2\theta} \cos\theta} \qquad (2)$$

where k is a constant taken to be 0.94, λ is the wavelength of X-Ray used (λ = 1.54 Å) and $\beta_{2\theta}$ is the full width at half maximum of (013) peak of XRD pattern, Bragg angle 2θ is around 64.44°. The average value of grain size is found to be 14.44 nm.

Based on Scanning Electron Microscope (SEM) image, surface morfologi of thin layer ZnO synthesized with concentration of precursor 0.1 M and temperature deposition at 550°C show in Figure 5.

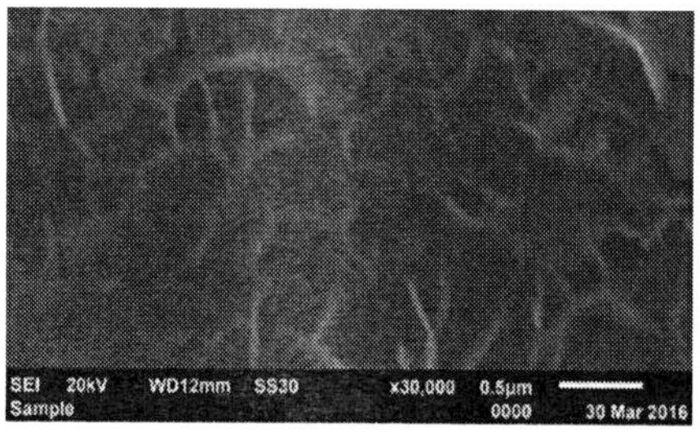

Figure 5. The surface of thin layer of ZnO 0.1 M and temperature deposition at 550 °C

The photos of thin layer ZnO crystal surface morphology seen that the growth of ZnO is not symmetrical. The surface morphology obtained in this research when compared to another result [6] are quite similar, meaning that the results of SEM in this study were nearing the properties of ZnO crystals morphology. The growth of ZnO is not symmetrical look like a stalk or stem is commonly known as nano rods. In a review of the others can be seen as the root of the tree (root like). The results of SEM testing has been done before [7] can be seen in Figure 6.

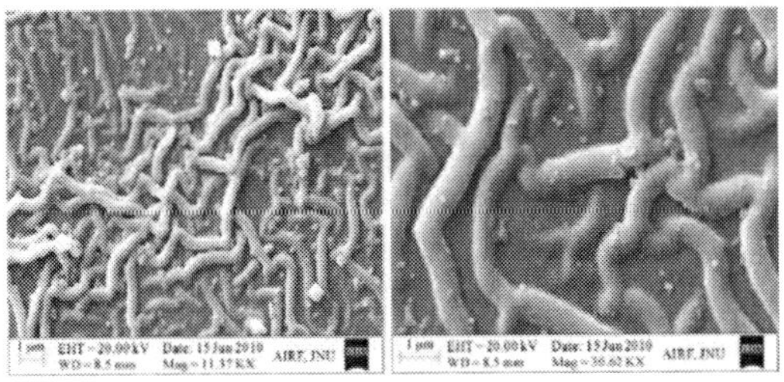

Figure 6. Surface Morphology of thin layer ZnO at concentration of 0.5 M [7]

Conclusion

The characterization of optical structure of thin layer ZnO, obtained that the increasing of concentration thin layer ZnO show that the decreasing the transmittance value, increasing the absorption, and the smaller the value of the gap energy. Furthermore the increasing temperature deposition of thin layer ZnO, show that the increasing of the transmittance value, decreasing the value of absorption, and the smaller the value of the gap energy. Based on field diffraction pattern has a diffraction peak intensity at 14.4472 nm grain size of thin layer ZnO of 0.1 M and temperature deposition at 550 °C. The surface morphology of thin layer ZnO such as rod-like.

Acknowledgement

Thank you very much to the education department of research and technology for the fundamental grant research in years 2015 and 2016. A big thank you to the students the material physics research groups have tried to help prepare samples and data retrieval. Thank you to all staff and laboratory analytical chemistry Mataram University and laboratory physics Department Dipenogoro University and all those who have helped in this research.

References

[1] Pawar B G, Pinjari P, Divak V K, Sanjay S P, Aniruddha B, and Sung H 2012 *Chemical Enginering* **954869** 1
[2] George A, Kumari P, Soin N, Roy S S, McLaughlin J A 2010 *Materials Chemistry and Physics* 123 634
[3] Nagayasamy N, Gandhimathination S, Veerasamy V 2013 *Metal Scientific Research* **3** 8
[4] Weng Min-Hang P, Cheng-Tang Y, Ru-Yuan H, Chun C 2011 *Ceramics International Journal* **43** 1
[5] De A 2015 *J. Thin. Film. Sci. Tec.* **4**, 83
[6] Khan Z, Raza K M, Shoeb Z, Mohammad K M, Shahid 2010 *Materials Science and Engineering B* **174** 1
[7] Khan Z, Raza K M, Shoeb Z, Mohammad K M, Shahid 2011 *Materials Sciences and Applications* **2** 340

IOP Conf. Series: Materials Science and Engineering **196** (2017) 012005 doi:10.1088/1757-899X/196/1/012005

Photoluminescence of Reduced Graphene Oxide Prepared from Old Coconut Shell with Carbonization Process at Varying Temperatures

Dwi Noor Jayanti[1], Ananda Yogi Nugraheni[1], Kurniasari[1], Malik Anjelh Baqiya[1] and Darminto[1]

[1] Advanced Materials Research Group, Department of Physics, Faculty of Mathematics and Natural Science, Institut Teknologi Sepuluh Nopember (ITS), Sukolilo, Surabaya 60111, Indonesia

E-mail : dwi_nj92@yahoo.com

Abstract. Reduced graphene oxide (rGO) powder has been prepared from coconut shells by carbonization process at 400°C, 600°C, 800°C and 1000°C for 5 hours at ambient air. In this study the exfoliation rGO was added into distilled water with variation of concentration solution using the sonication process for 3 hours and centrifugation at 4000 rpm for 20 minutes. The characterization were performed by using XRD and photoluminescence (PL) spectroscopy. The photoluminescence or rGO showed the peak of excitation and emission at wavelengths ranging from 340 nm to 800 nm. The PL emission spectra are at wavelength ranging from UV to visible region approaching red. Observation showed that the photoluminescence intensity was significantly increased by the increasing content of rGO in the solution. The influence of the varying temperature on the PL spectra will also be discussed in this study.

1. Introduction

Graphite is an allotrope of carbon that consists of thin sheets of bonds between carbon atoms in hexagonal structure. Graphene, a two-dimensional (2D) building block for other sp^2 carbon allotropes with other dimensionality, is a single layer sheet of carbon atoms arranged as a hexagonal honey-comb lattice structure. It has unique properties which are promising for application in various fields such as supercapasitor, solar cell, sensors and optoelectronic materials [1]. Ideal graphene is a zero bandgap semiconductor that contains sp^2 clusters which are not favorable for luminescence application [2]. Like graphene, graphene oxide (GO) is formed by the oxidation of graphite, with still sp^2 carbon atoms in its structure [3]. Meanwhile, reduced graphene oxide (rGO) is the loss of oxygen and hydrogen atoms in the structures that resemble graphene. The GO and rGO are still hot topics in the research and development of graphene.

The materials that can be used in producing the carbon compounds is coconut shell, which is promising, low-cost, simple, and environmental-friendly. Based on the previous research [4], the old coconut shell heated at 400°C and 600°C has produced rGO phase having diffraction peak of XRD at $2\theta = 24°$ and 43° in all variation of the atmosphere. Further development of the rGO is the process in order to obtain a thinner sheet size, even a single layer of rGO, by means of exfoliation process. Exfoliation process consisting of the sonication and centrifugation in water is the most common. There are various approaches used for GO to synthesize of rGO, such as chemical vapor deposition (CVD), chemical method, mechanical exfoliation and cleavage, and thermal annealing [5]. The reduction of

GO is definitely a key topic, and different reduction processes results in different properties that in turn affect the final performance of materials or devices composed of rGO [6].

Previous studies have established that rGO films exhibited decreased photoluminescence (PL) spectra and were blue-shifted as compared to GO films, resulting from the newly formed small sp^2 clusters after reduction [2]. Red-to-near-infrared PL emitted from chemically-reduced GO has been reported, consistent with increasing disorder length scale with reduction [7]. The rGO has been widely applied in energy storage devices, photovoltaics devices, corrosion protection, and lubricants [8]. In this study, we are investigating the optical properties of rGO prepared from old coconut shell using photoluminescence (PL) spectroscopy.

2. Experiment

The main materials used in this study are old coconut shell and distilled water. The equipment used in this study are furnace (as main heating treatment), ultrasonic washer (as dissolving the sample) and centrifuge (as precipitation to produce the solution). The first step was the process of burning coconut shell to produce charcoal, and then it was crushed into powder. Subsequently, it was heated at a temperature 400°C, 600°C, 800°C and 1000°C for 5 hours at ambient air. Subsequently, the resulted powder was added into distilled water with various solution concentrations of 0.001 mg/ml, 0.002 mg/ml and 0.003 mg/ml. Exfoliation was achieved by ultrasonic washer of the dispersion for 3 hours. Finally, the sample was collected by centrifugation (4000 rpm) for 20 minutes to remove any unexfoliated graphite [9]. Characterization were performed by using XRD (*Philips X'pert*) to identify the existence of rGO phase in the resulted powder after heated at varying temperatures and photoluminescence (PL) spectroscopy (*Perkin Elmer LS-55*) to determine the optical properties of the sample.

3. Results and Discussion

The following is one of the X-Ray Diffraction (XRD) pattern for coconut shell charcoal sample heated at 400°C shown in Figure 1. A further analysis by comparing with reference [10] confirmed that the sample has the similar wide peaks with rGO phase at $2\theta \sim 23°$ and 43°. The peak width indicates the small crystal size of rGO and its structure of layer arranged in short range of the layer stack. The XRD result is related to the processes of removing intercalated water molecules and the oxide groups. The bond between carbon and oxygen in the form reduced graphene oxide phase decreases with increasing temperatures.

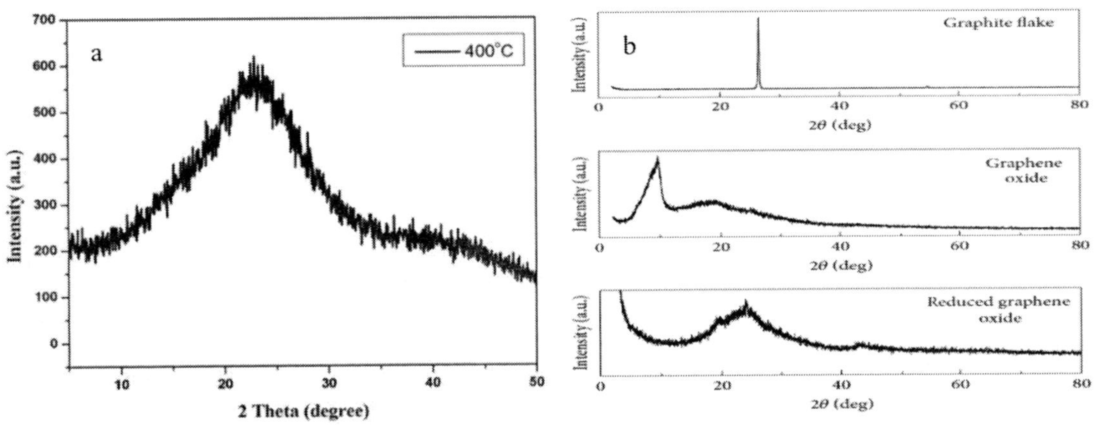

Figure 1. XRD pattern of (a) coconut shell charcoal sample in heating temperatures 400°C, (b) Graphite flake, GO, and rGO [10]

In order to investigate the optical properties of rGO, a photoluminescence study was carried out. Figure 2 (a,b,c) shows one of the results of PL emission spectra with different excitation wavelengths (λ_{ex}) at 800°C in all concentrations solution. With the increasing excitation wavelength from 296 to 352 nm, the emission peak has been red-shift with a gradual decrease in intensity. Almost at every temperature produces the same excitation wavelengths, but at temperatures of 800°C, the resulting emission spectra are more stable than others. Figure 2 (d) shows the PL spectra of rGO at 800°C in all concetrations of solution with λ_{ex} = 343 nm, it can be seen that the PL intensity tends to increase with increasing concentration of the solution.

Figure 3 represents the PL spectra at λ_{ex} = 343 nm of rGO with each temperatures prepared by exfoliation process at three different content of solutions. The higher temperature in each of concentration solution has led to an increased emission peak (red-shifted). It is supported by the results of XRD, that the higher temperatures, the reduction graphene oxide phases tends to decrease. Otherwise, the decrease of temperatures leads to a decrease in the emission peak (approaching blue-shifted). The rGO sheets exhibiting quenching in PL emission spectra are blue-shifted due to increased sp^2 clusters after reduction [3]. In the room temperature PL spectra, the defect peak is suppressed in the sample reduced at higher temperature [3].

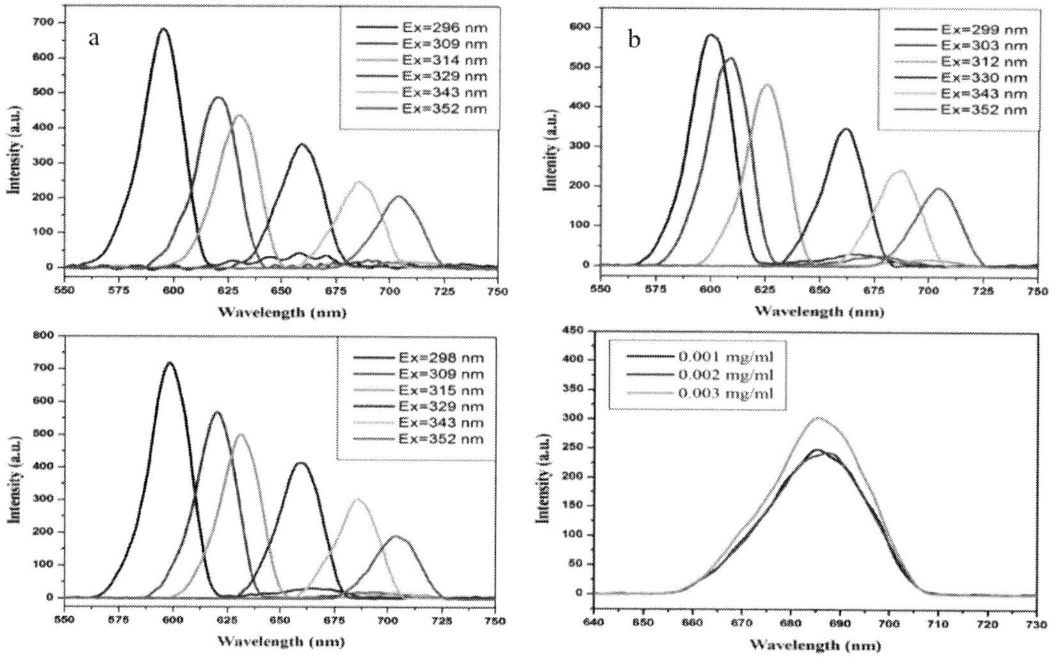

Figure 2. PL spectra of rGO at 800°C with concentration of solution: (a) 0.001 mg/ml, (b) 0.002 mg/ml, and 0.003 mg/ml. (d) PL spectra of rGO at 800°C in all concetrations of solution with λ_{ex} = 343 nm

The optical properties of sp^2/sp^3 combined materials can be influenced by the size and fraction of the chains and clusters [11]. PL in amorphous carbon originates from the radiative recombination of electrons and holes in the band-tail states created by sp^2 clusters. The photoluminescence of rGO due to the disappearance of functional oxygens which is due to the restoration of more number of sp^2 clusters, the newly formed sp^2 clusters in rGO can provide percolation pathways between sp^2 clusters already present [3]. Therefore, the electron-hole recombination of these sp^2 states improves blue flourescence.

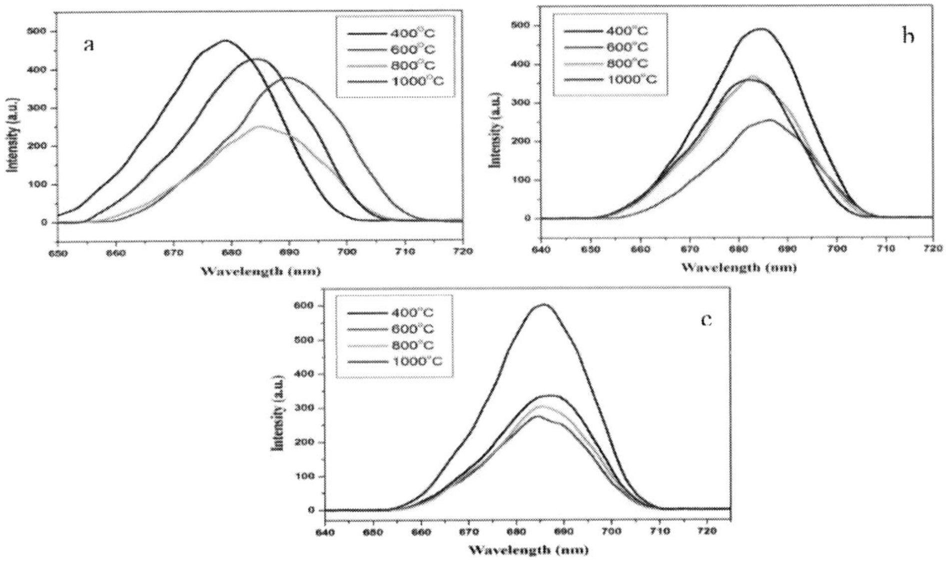

Figure 3. PL spectra of rGO at $\lambda_{ex} = 343$ nm with concentration of solution: (a) 0.001 mg/ml, (b) 0.002 mg/ml, and 0.003 mg/ml

4. Conclusion

In conclusion, we have prepared the rGO from coconut shell charcoal at three different concentration solutions with varying heating temperatures. In addition, PL spectra of rGO exhibits that the lower temperatures carbonization will produce the decrease of emission peak (approaching blue-shifted). The photoluminescence intensity was significantly increased by the increasing content of rGO in the solution.

Acknowledgement

The authors would appreciate partial support of this research by "Hibah Kompetensi" research grant, DRPM – Kemenristekdikti, 2016.

References

[1] Sakthivel T, Gunasekaran V and Kim S J 2014 *Mater. Sci. Semicond. Process* **19** 174
[2] Han F, Yang S, Jing W, Jiang Z, Liu H and Li L 2015 *Appl. Surf. Sci.* **345** 18
[3] Krishnamoorthy K, Veerapandian M, Mohan R and Kim S J 2011 *Appl. Phys. A* **106** 501
[4] Nugraheni A Y, Nasrullah M, Prasetya F A, Astuti F and Darminto 2015 *Mater. Sci. Forum* **827** 285
[5] Loryuenyong V, Totepvimarn K, Eimburanapravat P, Boonchompoo W and Buasri A 2013 *Adv. Mater. Sci. Eng.* **2013** e923403
[6] Pei S and Cheng H M 2012 *Carbon* **50** 3210
[7] Luo Z, Vora P M, Mele E J, Johnson A T C and Kikkawa J M 2009 *Appl. Phys. Lett.* **94** 111909
[8] Wong C. P. P., Lai C. W.,. Lee K. M, and Hamid S. B. A. *Materials*, vol. 8, no. 10, pp. 7118–7128, Oct. 2015.
[9] Bansal P, Panwar A S and Bahadur D 2014 *Int. J. Mater. Mech. Manuf.* 18
[10] Cho D, Choi D, Pyo Y, Pawar R C, Kim Y, Yoon E H and Lee C S 2016 *J. Nanomater.* **2016** e1050183
[11] Xin G, Meng Y, Ma Y, Ho D, Kim N, Cho S M and Chae H 2012 *Mater. Lett.* **74** 71

IOP Conf. Series: Materials Science and Engineering **196** (2017) 012006 doi:10.1088/1757-899X/196/1/012006

Electron Microscopy of the Tin-oxide Nanolayer Formed on the Surface of Sn-Ag-Cu Alloys

H Sosiati[1], N Kuwano[2] and S Hata[3]

[1]Department of Mechanical Engineering, Faculty of Engineering, Universitas Muhammadiyah Yogyakarta, Yogyakarta 55183, Indonesia.
[2]Mechanical Precision Engineering, Malaysia-Japan International Institute of Technology (MJIIT), University of Technology Malaysia (UTM) International Campus, Jalan Semarak 54100 Kualalumpur, Malaysia.
[3]Interdisciplinary Graduate School of Engineering Sciences, Kyushu University, Kasuga, Fukuoka 816-8580, Japan

E-mail : hsosiati@gmail.com

Abstract. Sn-Ag-Cu alloy used in the present study is commercial Sn-3.0Ag-0.5Cu solder ball alloys with a diameter of 400 and 300 μm which were long term atmospheric oxidized for about 6 years (specimen-1) and under high temperature/humidity at 85°C and relative humidity of 85% for 2140 h, respectively. Morphologies and nanostructure of the oxide nanolayers formed on the surface of Sn-Ag-Cu alloys were studied from the interface of the oxide film and the tin substrate by transmission electron microscopy (TEM) to verify the oxidation mechanism. Cross-sectional TEM specimens were prepared using a focused-ion-beam (FIB) micro-sampling technique. Before the FIB fabrication, the specimen surface was coated with carbon (C) and tungsten (W) films. Inhomogeneous thickness of tin-oxide nanolayer formed on specimen-1 and specimen-2 were fluctuated between 20-40 nm and 40-50 nm, respectively. The nanolayer on specimen-1, however, consists of polycrystalline SnO and SnO_2, whereas the one on the specimen-2 comprises of polycrystalline SnO_2. High resolution (HRTEM) image and fast Fourier transformation (FFT) spectra corresponding to the interface and the substrate areas have confirmed those results. The results verify that at very long atmospheric oxidation Sn was gradually oxidized to be SnO (Sn^{2+}) and then SnO_2 (Sn^{4+}), in which SnO is present at the region closed to interface between Sn-substrate and the tin-oxide layer. At high temperature oxidation, however, Sn was completely oxidized to be SnO_2.

1. Intorduction

The presence of lead (Pb) contained in conventional solder material Sn-Pb is known as a hazardous substance that potentially provides effects on the environment and health. Lead poisoning is particularly baneful to the neurological development of children such as damage to the brain, liver and kidney, and hyperactivity effect [1]. Since 2006, therefore, European directives RoHS (restriction of hazardous substances) has declared the regulation to eliminate Pb from the electronic components and devices. This led to the development and execution of Pb-free solders.

Since the last few years Sn-Ag-Cu alloy was considered as one of the most promising lead-free solder alloys for the replacement of conventional solder of Sn-Pb due to its good solderability and relatively high mechanical properties. The Sn-3.0Ag-0.5Cu alloy which is also known as SAC 305 was proposed [2] as one of candidate materials for solder balls grid arrays (BGAs) or solder joints used in electronic packaging (flip-chip packaging) applications. However, Sn-3.0Ag-0.5Cu alloy is easily oxidized after multiple reflows or high aging temperature [3, 4]. The oxygen potential of Sn-oxide formation is actually lower than that of Ag and Cu oxides [5]. This is a key point explaining that Sn is oxidized easily.

Thus, the formation of tin-oxide layer on the surface of solder alloy would inhibit interconnection/bonding between the solder surface and connectors as schematically illustrated in Fig. 1. The growth of tin-oxide would influence the interface bonding quality. That means when the formation of tin-oxide layer is very thin, good bonding between the two surfaces may be possible [3]. Increase of the thickness of tin-oxide layer will decrease the bond strength.

The oxidation of tin in the solder materials is not a new phenomenon, but studies to overcome the oxidation and wetting problems have been continued until now by adding various alloying elements into the Sn-Ag-Cu alloy such as Sn-Ag-Cu-In [4], Sn-Ag-Cu-Zn [6], Sn-Ag-Cu-Ge [7] Besides, nanoparticles (nanospherical, nanowires, nanorods) of Pb-free solder alloy of Sn-Ag-Cu have been recently investigated to suppress the melting point of solder material with respect to reduce the oxidation rate of Sn [8-10].

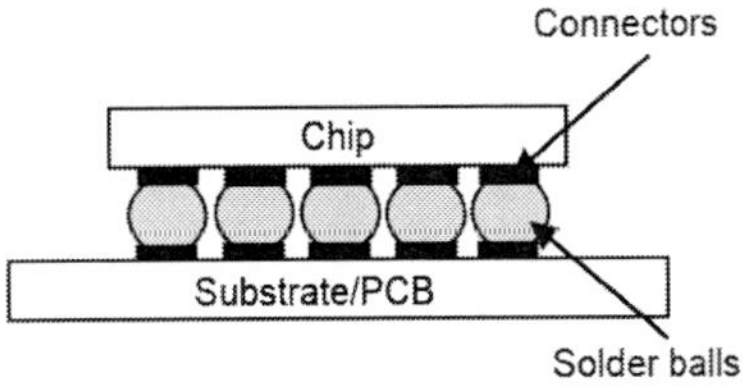

Figure 1. Side-view schematic of a typical flip chip

According to various studies on the Sn-Ag-Cu solder alloy especially for solder joint, oxidation of tin is one of significant problems in the interface bonding: i.e. oxidation at room temperature and at reflow/multi-reflow conditions. In this study, the solder balls of Sn-3.0Ag-0.5Cu alloy were oxidized at atmospheric air for about 6 years and, at 85°C and relative-humidity of 85% for 2140 h. Detailed characterization on the tin-oxide formation performed from the cross-section views has not been previously reported. Therefore, morphologies and nanostructure of the oxide nanolayers formed on the surface of Sn-Ag-Cu alloy were studied from the interface between the oxide film and the tin substrate to verify the oxidation mechanism and gain understanding the tin-oxide behavior. This is considered to be important from the viewpoints of both for the fundamental science and industrial applications.

2. Experimental methods

Material used in this study was commercial solder balls of Sn-3.0Ag-0.5Cu alloy produced by NIPPON MICROMETAL Co. The solder balls with the diameter of 400 µm or 300 µm were naturally oxidized (at atmospheric air) for about 6 years and at 85°C, relative-humidity of 85% for 2140 h, respectively. They were designated as specimen-1 and specimen-2, respectively. Cross-sectional specimens of the solder ball surfaces were prepared using a focused ion beam (FIB, FB-2000K, Hitachi) equipped with a microsampling unit. Prior to the FIB fabrication, the surface of the solder balls was coated with carbon and tungsten films to prevent the solder ball surface from of Ga^+ ion bombardment.

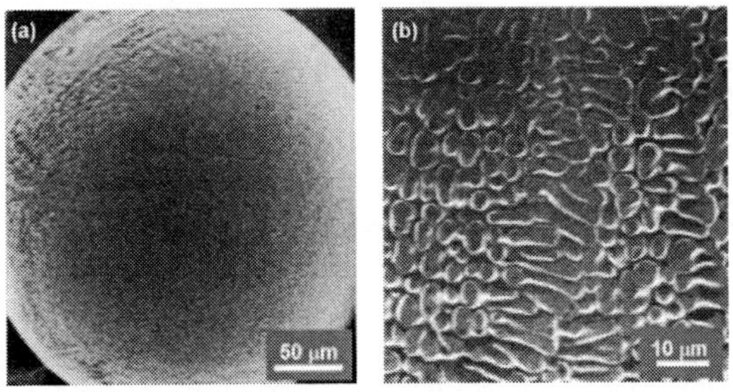

Figure 2. SIM image of a solder ball of Sn-3.0Ag-0.5Cu alloy (a) showing the dendritic structure on the surface (b).

Plan-view microstructures of the solder ball surface were observed with FIB-scanning ion microscope (FIB-SIM) as shown in Fig. 2. The dendrite structures covering a solder ball surface may be resulted from solidification during the fabrication process of the solder ball material. Cross-sectional microstructures of the tin-oxide films formed on the solder ball surface were examined by TEM (FEI TECNAI F-20) equipped with scanning TEM (STEM) and the tin-oxide phase was identified by electron diffraction analysis. The areas of tin-oxide layer, interface between the tin-oxide layer and the substrate, and the substrate were also examined by High resolution TEM (HRTEM) imaging and analyzed by fast Fourier transformation (FFT) to confirm the thickness of the tin-oxide layer and tin-oxide phases, respectively.

3. Results and discussion

Characterization of tin-oxide layer on the solder ball specimen-1. A cross-sectional bright field (BF)-TEM image of specimen-1 is shown in Fig. 3a. A magnified image of the small rectangular area in Fig. 3a demonstates a very thin film formed on the substrate with inhomogeneous thickness less than 50 nm (Fig. 3b). The thickness of carbon (C)-coating and tungsten (W)-deposition are also seen. Indexing an electron diffraction pattern (EDP) (Fig. 3A‘) obtained from a circle area A in Fig. 3a has identified that the thin film was the polycrystalline film consists of SnO and SnO_2. Diffuse intensity appears in the EDP indicating that some amorphous phases of the tin-oxide are present. An EDP (Fig. B‘ in (a)) obtained from a circle area B in Fig. 3a confirmed that the substrate was β-Sn.

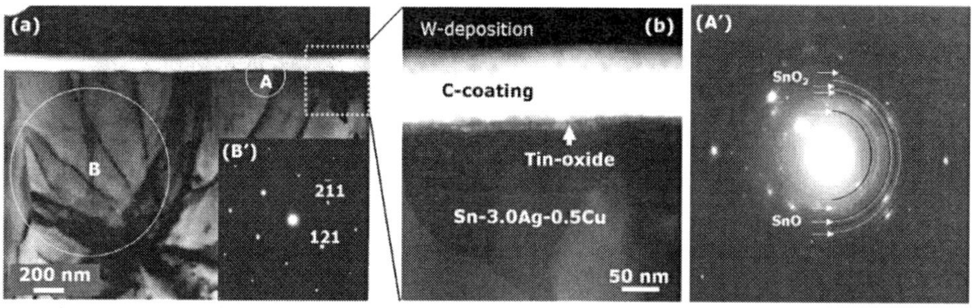

Figure 3. (a) A cross-sectional TEM image of specimen-1, (b) a magnified image of a small rectangular area in (Fig. 3a), (A′) and (B′) are EDPs obtained from circle areas A and B in (a), respectively.

In order to confirm the presence of SnO phase in the oxide nanolayer, an HRTEM image taken from the subtrate and the interface between the substrate and the oxide layer is clearly shown in Fig. 4a. The thickness of tin-oxide layer observed in Fig. 4a is around 37 nm.

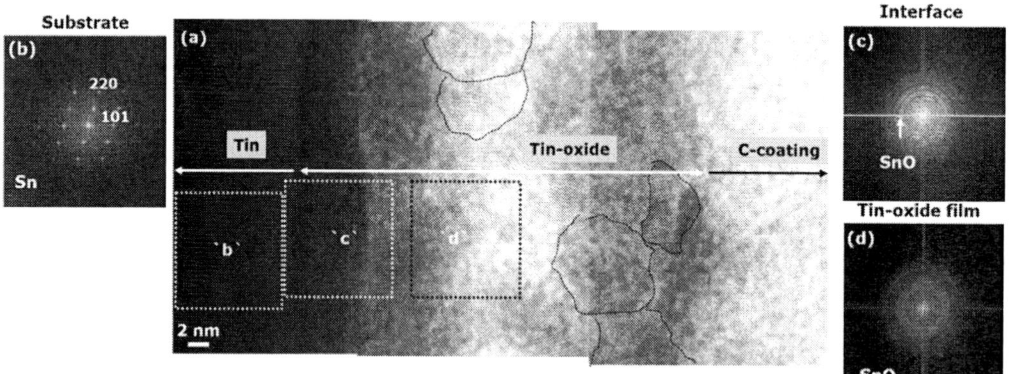

Figure 4. (a) A HRTEM image of specimen-1 taken from the area between the tin substrate and carbon coating layer, (b), (c) and (d) Fourier power spectra obtained from areas of the tin substrate (‘b‘), interface between the substrate and the tin oxide film (‘c‘) and the tin-oxide film (‘d‘), respectively.

Lattice fringes formed within the tin-oxide layer reveal slightly unclear polycrystalline particle with an average size of about 10 nm. The Fourier power spectra (Fig. 4b, 4c and 4d) related to the areas ‘b‘, ‘c‘ and ‘d‘, respectively have confirmed that SnO is identified at the interface between the substrate and the tin-oxide layer (Fig. 4c). This sugests that SnO is present in the vicinity of the tin substrate surface.

Characterization of tin-oxide layer on the solder ball specimen-2. Typical cross-sectional BF-TEM image of solder ball specimen oxidized at 85°C for 2140 h with the incident beam along the [001] zone axis of the tin-substrate (Fig. 5) clearly demonstrated the thin film about 50 nm in thickness (see a white arrow).

Indexing EDPs (Fig. 5b and 5c) obtained from circle areas A and B in Fig. 5a, respectively, indicated that the thin film is made of polycrystalline SnO_2 formed on the β–tin substrate. No SnO phase was identified.

Based on the entire microstructural observation, thickness of the tin-oxide layer formed on the surface of specimen-2 varies from 40 nm to 50 nm and cracks are developed in some areas of the outer surface of the tin-oxide layer (Fig. 5a). The formation of cracks is not affected by specimen preparation using FIB, but it may be due to stress built-up within the tin-oxide layer. An HRTEM image (Fig. 5d) obtained from the area between the tin-substrate and the carbon-coating layer identified the thickness of tin-oxide layer around 47 nm. Lattice fringes formed in the tin-oxide layer reveal polycrystalline particles in average size (~5-10 nm) as shown in Fig 4a.

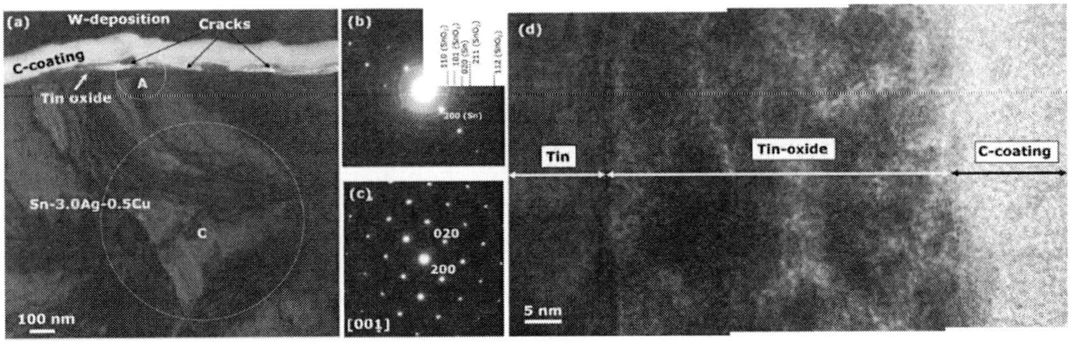

Figure 5. (a) A cross-sectional TEM image of specimen-2, (b) and (c) EDPs corresponding to circle areas A and C, respectively, and (d) a HRTEM taken from tin-substrate to carbon layer.

Pure tin and/or tin alloys are very sensitive to oxygen under all conditions even at room temperature because of low oxygen potential of tin ($\Delta G \sim -110$ Kcal/g.mol O_2 at 373 K) [5]. When tin is exposed to atmosphere, the oxygen atoms from the atmosphere would diffuse into the tin-surface to form a tin-oxide layer. The formation of non-uniform thickness of tin-oxide layers both on the surface of specimen-1 and specimen-2 is related to the different behavior along the tin-surface. Oxidation of tin nanoparticles at room temperature has indicated that the oxidation rate is particle size-dependence [11]. Combination of XPS and TEM analyzing results have reported that few nanometers (~3 nm) of amorphous SnO layer covered the tin nanoparticles (< 100 nm) surface at room temperature oxidation for about 60 days. This is inconsistent with a result based on Sn nuclear magnetic resonance spectroscopy reported by Kravchyk *et al.* [12] that showed the formation of amorphous SnO_2 layer covered the colloidal Sn nanocrystals at room temperature oxidation. On the other hand, the tin-oxide thickness of about 3.5 nm formed at the interface betwen Sn-Ag-Cu bump and the Sn-Ag film after room temperature oxidation for 60 min, but no chemical state of tin-oxide was reported [3].

It is considered that there is a relationship between the chemical state of tin-oxide and the thickness of tin-oxide layer. Polycrystalline SnO and dominantly SnO_2 are formed within the tin-oxide layer under short-term and long-term room temperature oxidation. Further exposure to air does not change the chemical state of Sn^{2+} (SnO) present in the region near the interface between the substrate and the oxide layer, but SnO leads to gradually oxidize to be SnO_2 in the vicinity toward the surface of the oxide layer exposed to air (gas). It is possible that due to the existence of SnO in the tin-oxide film, the interface bonding strength between tin-oxide film and tin-layer would become weak, and the tin-oxide film would be easily broken.

On the other hand, when oxidation of tin is done at high temperature, Sn would be completely oxidized to be SnO_2. No SnO was identified due to SnO is metastable oxide phase. In this case, high temperature accelerates the migration of anion (O^{2-}) across the oxide layer leading higher reaction rate between Sn^{4+} and O^{2-} to form SnO_2.

4. Summary

TEM studies have verified the morphology and nanostructure of tin-oxide nanolayer formed on the tin surface by room and given temperatures oxidation as follows.

(1) Long-term room temperature oxidation (~6 years) on Sn-3.0Ag-0.5Cu alloy produced inhomogeneous thickness of tin-oxide layer between 20-40 nm. The tin-oxide nanolayer is made of

polycrystalline SnO and predominantly SnO_2 where SnO present in the region near the interface between the substrate and the oxide layer, but SnO leads to gradually oxidize to be SnO_2 in the area toward the surface of the oxide layer.

(2) Oxidation at 85°C and relative humidity of 85% for 2140 h also produced non-uniform thickness of tin-oxide layer varies between 40-50 nm which is made of polycrystalline SnO_2. At high temperature oxidation Sn was completely oxidized to be SnO_2.

Acknowledgement

This work was supported in part by Nanotechnology Support Project of the Ministry of Education, Culture, Sports, Science and Technology (MEXT), Japan and Research Promotion Bureau, MEXT, Japan.

References

[1] Information on http://www.extraordinaryroadtrip.org/research-library/airpollution/understanding-air-pollution/lead/health.asp
[2] Kim K S, Huh S H and Suganuma K 2003 *J. Alloys Compd.* **352** 226
[3] Wang Y H, Howlader M R, Nishida K, T. Kimura T and Suga T 2015 *Mater. Trans.* **46** 2431
[4] Lee H and Cho S W 2014 *Appl. Sci. Convergence. Technol.* **23** 345
[5] Greenwood N N and Earnshaw A 1997 Chemistry of the Elements, second ed. Butterworth- Heinemann ISBN 0080379419.
[6] Zhang L, Han J G, He C W and Guo Y H 2012 *J Mater Sci: Mater Electron* **23** 1950
[7] Cho S W, Han K, Yi Y, Kang S J, Yoo K H, Jeong K and Whang C N 2006 *Adv. Eng. Mater.* **8** 111
[8]. Zou C D, Gao Y L, YANG B, Xia X Z, Zhai Q J, Andersson C and Liu J 2009 *J. Electron. Mater.* **38** 351
[9]. Gaoa F, Rajathuraia K, Cuia Q, Zhoub G, Acha I N, Z. Gu Z 2012 *Appl. Surf. Sci.* **258** 7507
[10]. Sun L and Zhang L 2015 *Adv. Mater. Sci. Eng.* **2015** 1
[11]. Sutter E, Barcelo F I and Sutter P 2014 *Part. Syst. Charact* 1
[12]. Kravchyk K, Protesescu L, Bodnarchuk M I, Krumeich F, Yarema M, Walter M, Guntlin C and M.V. Kovalenco M V 2013 *J. Am. Chem. Soc.* **135** 4199

Structural Analysis on Reduced Graphene Oxide Prepared from Old Coconut Shell by Synchrotron X-Ray Scattering

Ananda Yogi Nugraheni[1], Dwi Noor Jayanti[1], Kurniasari[1], Siriwat Soontaranon[2], Edy Giri Rachman Putra[3], and Darminto[1]

[1] Advanced Materials Research Group, Department of Physics, Faculty of Mathematics and Natural Science, Institut Teknologi Sepuluh Nopember (ITS), Sukolilo, Surabaya 60111, Indonesia
[2] Synchrotron Light Research Institute, 111 University Venue, Muang District, Nakhon Ratcashima 30000, Thailand
[3] Center for Science and Technology of Advanced Materials, National Nuclear Energy Agency of Indonesia, Kawasan Puspitek Serpong, Tangerang 15134, Indonesia

E-mail : darminto@physics.its.ac.id

ABSTRACT. This work is aimed to identify the distribution of size and structure of the reduced graphene oxide (rGO) particles before and after the mechanical exfoliation process from coconut shell. The burnt sample was heated at 400^{0}C in ambient air, followed by sonication and centrifugation method for various concentration. The characterization were performed by using wide angle X-ray scattering (WAXS) and synchrotron small angle X-ray scattering (SAXS). Based on SAXS result, the structure of rGO showed that in low range of momentum transfer (Q), the particle size was too big to be measured using SAXS. In the high Q, the graph shown the apparent absence of change in curvature which indicates that the small particles are typically seen on high Q range, having aggregate with a new shape and big particles. The average fitting slope is -3.3, implying that the particles structure tends to be fractal (rough) surface, where the particles form aggregates to become bigger size and irregular structure. Further, the WAXS characterization shows that the exfoliation step has not affected the rGO phase formed.

1. Introduction

Graphene is a burgeoning material with a honeycomb-like two dimensional crystal structure formed by carbon atoms in sp^2 hybrid junction, exhibits incomparable electronic, thermal and mechanical properties [1]. The most important issue to realize the potensial of graphene is to achieve its mass and controllable production [2]. There have different names of grapheme, such as chemically modified graphene, functionalized graphene, chemically converted graphene or simply graphene. Reduced grapheme oxide (rGO) is a promising material for many applications such as in the development of energy-storage capacitors, field effect transistors sensor etc [4].

Separation of thin layers of graphene (a 2D material) from a 3D layered bulk graphite results in an increase of its electrical conductivity. Graphite is an available in large quantities from natural sources and consists of a stack of flat graphene shets standing together by weak Van der Waals forces. The carbon atom plane in graphite oxide is heavily decorated by oxygen-containing groups, which expand the interlayer distance and make the atomic thick layers hydrophilic. These oxidized layers can be exfoliated in water under moderated ultrasonication [4]. This sonication produces a thinner layer of rGO.

The polydispersity of nanoparticles is of importance in controlling properties related to their small size and high specific surface area. Current techniques for the determination of particle size distribution on nanoscale can be categorized into some measurements, such as small angle X-ray scattering (SAXS). The determination of primary particle size distributions for nanoparticles that are aggregated into mass-fractal structures is challenging due to the overlap of structural features [5]. The SAXS is a well established technique, long used for colloidal system and later polymers, which has seen a resurgence because of the availability of high intensity synchrotron sources and the current trend towards the investigation and application of nanoscale structures [6].

In the prevous research, it has been conducted characterization of an old coconut shell charcoal treated with heating at a certain temperature and atmosphere and found that there was a phase of reduced graphene oxide based on XRD graph pattern. Besides that, the width of the band gap energy in the overall of coconut shell charcoal were in the range of 0.07 – 0.67 eV which was still included in the range of reduced graphene oxide material. As we know that reduced graphene oxide is a semiconductor material, so it is strongly support that reduced graphene oxide can be made from an old coconut shell charcoal [7].

However, from the previous research it is not yet known how the shape, particle size and the distribution of the particle are in the form of powder and solution of an old coconut shell charcoal. So, this research focuses on the identification the shape, particle size and the distribution of the particle of rGO synthesized from the old coconut shell charcoal after getting an exfoliation and heating treatment using synchrotron SAXS and wide angel X-ray scattering (WAXS).

2. Experiment

The materials used in this study are an old coconut shell powder, and distilled water. The equipment used in this study are mortar, pestle mortar, crucible, sandpaper, furnace tube, drilling machine, ultrasonic cleaner, centrifuge, beaker glass, digital balance, cutter, spatula, aluminum foil, and rubber hoses. Sample preparation was done by cleaning the old coconut shell using sandpaper. It was burnt on fire until it turns into charcoal. The resulted and selected charcoal was then heated at oven overnight at temperature of 150^0C to remove water content. It was smoothed using the pestle mortar to form the coarse powder and using mortar to form the smooth powder and sieve using 500 mesh. Furthermore, it was heated using tube furnace at temperature 400°C for 3 hours in ambient air. Then, the aqueous solution of rGO was prepared by dispersing 10, 15 and 20 mg charcoal powder into 100 ml of distilled water by ultrasonic cleaner for 45 minutes and deposited by centrifuge for 15 minutes. Then, the precipitate in the solution was dried to be used in the form of powder. The samples were characterized using XRD, WAXS and SAXS to identify the structure, shape and particle size distribution.

3. Results and Discussion

SAXS measurements were carried out on beamline number 1.3 at Siam Photon Laboratory of Synchrotron Light Research Institute (SLRI), Nakhon Ratchasima, Thailand, to observe the particle size distribution, structural dimensions and shape of coconut shell charcoal which form the structure of rGO. The SAXS instrument provides nondestructive measurement and is able to measure the actual size of particles, distribution, shape, dimensions because the X-ray radiation from the instrument can be directly transmitted and scattered to the sample [8].

The intensity of the SAXS was generated by 9keV of X-Ray energy which is presented as a function of the scattering vector (q). Scattering vector is defined as $(4\pi/\lambda)\sin(\theta/2)$, where λ and θ are the wavelength of X-Ray and scattering angle of SAXS. The measurement used a CCD detector and 3,2 meter of sample to detector (SDD). The material standard that used during the measurement was Block Co-polymer (SEBS). Data result was analyzed by using the SAXSIT and SASfit software. The sample that used was in the form of powder and solution.

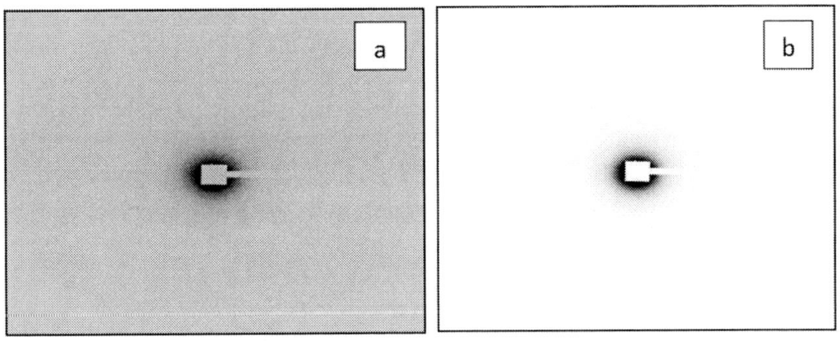

Figure 1. The SAXS pattern of rGO sample heated at 400°C in a) solution sample, b) powder sample

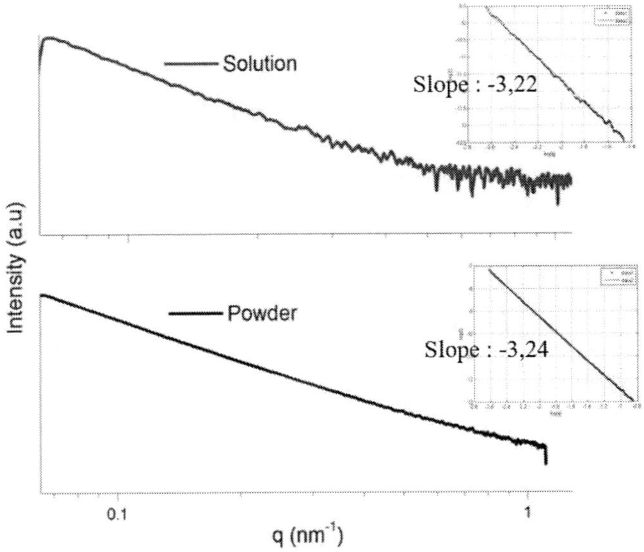

Figure 2. The SAXS graph and fitting slope of powder and solution

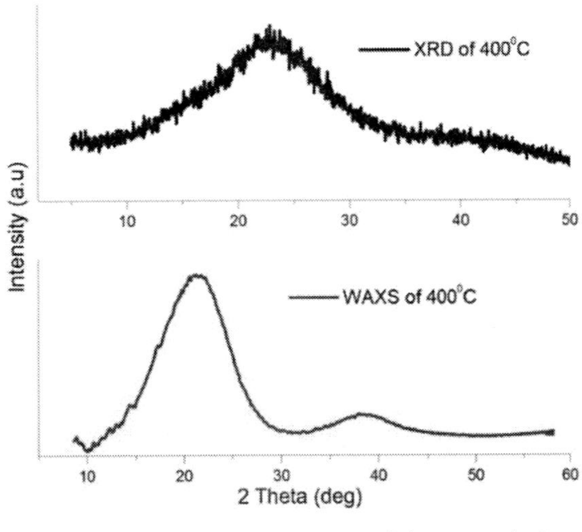

Figure 3. The XRD and WAXS patterns of the sample heated at 400^{0}C

The SAXS result in Figure1 showed that pattern of all type of samples have no black circular line in the middle of the pattern. Analysis of data reduction for the pattern was carried out by 4 steps: (i) read the pattern, (ii) align the pattern, (iii) subtract the background, and (iv) calculate SAXS profile I(q). Water was used as the background for the sample. The pattern indicates that the structure of the rGO is in the form of amorphous carbon structure and composed of random layers. These amorphous and random layers caused by the precursor material that used as a sample. An old coconut shell charcoal composed of cellulose, hemicellulose, lignin and a lot of metal substances from nature that form the amorphous structure.

The SAXS data for both type of samples, powder and solution, in Figure 2 show that the particle size of rGO in low range of momentum transfer (Q) is too big to be measured using SAXS. In the high range of Q, it shows the apparent absence of change in curvature which indicates that the small particles are typically seen on high Q range, having aggregate with a sphere shape and big particles. The big size of particles related to the exfoliation process during the experiment. Exfoliation process used to break down the van der waals bonding between the rGO layers to become thinner. This molecule bonding forces is weaker than covalent bonds such as the single or double bonding between carbon atoms that form a hexagonal structure. Energy vibration generated by the ultrasonic cleaner and the length of time (45 minutes) used in the process of exfoliation is not enough to make the oxygen, hydrogen and carbon bond between the rGO layers split together that produce particle size with extensive surface and thick layer. The average fitting slope in Figure 3 for powder and solution sample is -3.2, implying that the particles structure tends to be fractal (rough) surface ($3<D\leq4$), where this value revealed that rGO was a three-dimensional structure with a fractlly rough surface [9]. This fractal surface comes from the aggregation of some particles that is caused by low dispersion of the solution and also effected by exfoliation process. Furthermore, the slope value indicates that the dimension of the particles is in the form of 3D with a fractal surface because it still has a big size of particle of rGO. This simply exfoliation method still not enough to exfoliate the layers of rGO and change the particle dimension of rGO from 3D of sphere become 2D.

The WAXS measurement was conducted using a CCD detector, 89 cm distance of sample to detector (SDD), and the standard material of 4-Bromobenzoic Acid. The WAXS pattern in Figure 4 for all samples shows a similar pattern of the XRD data, consisting of two broad peaks which are located at $2\theta = \sim 23^0$ and $\sim 43^0$. In rGO, the carbon 002 peak is present at $2\theta = \sim 23^0$ and 100 peak is present at $2\theta = \sim 43^0$. The WAXS characterization shows that the exfoliation step has not affected the rGO phase formed. A shift that occurs at the angle of 43^0 due to difference in wavelength between WAXS of 1.38 $\AA$ and XRD of $\lambda = 1.54$ $\AA$ (Cu-K_α). Comparing to XRD with WAXS patterns, one may identify that there has a same phase in the sample as rGO. The broad peak corresponding to an amorphous structure is possibly a pile of random arrangement of thick parallel layers. The broad peak at $2\theta = \sim 23$ and 43^0 represent that the hkl of rGO are (002) and (100), respectively. The carbon layers are bound in the c-direction by weak van der waals forces [8].

Using Bragg equation, the inter-layer spacing and lattice parameter (a and c) of rGO can be measured. Comparing the calculation result of inter-layer spacing (d) of rGO from old coconut shell between XRD and WAXS, is indicated at $2\theta = \sim 23$ to be 4,03$\AA$, *at* $2\theta = \sim 43^0$ to be 2,10 $\AA$ by *XRD* and at $2\theta = \sim 23$ to be 3,61 $\AA$, *at* $2\theta = \sim 43^0$ to be 2,01 $\AA$ by WAXS. The lattice parameter at hkl (002) and (100) of WAXS (λ = 1,38 $\AA$) and XRD (λ = 1,54 $\AA$) are c = 7,232 $\AA$, a = 1,640 $\AA$ and c = 8,071 $\AA$, a = 1,673 $\AA$, respectively. Based on literature by Rajaura et al, most intense peak in the pristine graphite is appearing at $2\theta = 26{,}6°$ which is corresponding to (002) plane of graphite and inter-layer spacing is about 3,39 $\AA$. After chemical reduction by hydrogen, a new broad peak is appearing at $2\theta = 23{,}72°$ which corresponds to inter-layer spacing of 3,85 $\AA$. This is attributed to the removal of functional group and moisture of oxygen and hydrogen [10]. The similarity of d inter-layer spacing between rGO from old coconut shell and reference confirm that WAXS and XRD measurement form a rGO phase. Bigger value of inter-layer spacing and lattice parameter between measurement and common reference caused by the oxygen, hydrogen and impurity that bonded in hexagonal carbon structure of rGO. Inter-layer spacing and lattice parameter at hkl (002) and (100) between WAXS and XRD, shows that there is a small shift due to the differences of wavelength.

4. Conclusion

The structure of rGO in low Q range describes that the particle size is too big to be measured using SAXS, while in the high Q range, the apparent absence of change in curvature indicates that the small particles develop aggregate with a new shape and bigger particles. The average fitting slope of -3.2 implies that the particles structure tends to be fractal (rough) surface($3<D\leq4$), where this value revealed that rGO was a three-dimensional structure with a fractal rough surface due to the aggregation and exfoliation. The WAXS pattern for all samples were similar to the XRD data, featuring two peaks belonging to rGO phase. The similarity of d inter-layer spacing and lattice parameter between rGO from old coconut shell and reference confirm that WAXS and XRD measurement form a rGO phase. Inter-layer spacing and lattice parameter at hkl (002) and (100) between WAXS and XRD, shows that there is a small shift of *c*-axis due to the differences of wavelength.

Acknowledgement
The authors would appreciate partial support of this research by "Hibah Kompetensi" research grant, DRPM – Kemenristekdikti, 2016.

References

[1] Xu C, Shi X, Ji A, Shi L, Zhou C, Cui Y 2015 *Plos One* **10** 12
[2] Zhang C, Wei L, Xie X, Yuan C 2013 *Carbon* **62** 11
[3] Shalaby A, Nihtianova D, Markov P, Staneva A D, Iordanova R S, Dimitriev*et Y B.* 2015 *Bulgarian Chemical Communications* **47** 291
[4] Beaucage G, Kammler H K and S. E. Pratsinis S E 2004 *Journal of Applied Crystallography* **37** 523
[5] D.T. Britton D T, Odo E A, Gonfa G G, E O, Harting M 2009 *Journal of Applied Crystallography* **42** 448
[6] Nugraheni A Y, Nasrulloh M, Fandi A P, Astuti F, Darminto 2014 *Material Science Forum* **827** 285
[7] Sunaryono, Taufiq A, R, Putra E D P, Okazawa A, Watanabe I, Kojima N, Rugmai S, Soontaranon S, Zainuri M, Triwikantoro, Pratapa S, Darminto 2016 *NANO* **11** 1650027
[8] Chung D D L 2002 *Journal of Materials Science* 37, 1-15 (2002)
[9] Avnir D 1989 The Fractal Approach to Heterogeneous Chemistry: Surfaces, Colloids, Polymers. Chicester England Willey
[10] Rajaura R S, Srivastava S, Sharma V, Sharma P K, Lal C, Singh M, Palsania H S, Vijay Y K 2015 *Journal of Hydrogen Energy* **30** 11

IOP Conf. Series: Materials Science and Engineering **196** (2017) 012008 doi:10.1088/1757-899X/196/1/012008

Hematite from Natural Iron Stones as Microwave Absorbing Material on X-Band Frequency Ranges

Mochamad Zainuri

Departement of Physics, Faculty of Mathematics and Natural Sciences, Sepuluh Nopember Institute of Technology, Jl.Arief Rahman Hakim, Surabaya 60111 Indonesia

E-mail: zainuri@physics.its.ac.id

Abstract, This study has been investigated the effect of hematite as microwave absorbing materials (RAM) on X-Band frequency ranges. Hematite was succesfully processed by coprecipitation method and calcined at 500 °C for 5 hour. It was synthesized from natural iron stones from Tanah Laut, South Kalimantan, Indonesia. The products were characterized by X-ray diffraxtion (XRD), conductivity measurement, Vibrating Sample Magnetometer (VSM), and Vector Network Analyzer (VNA). The result was shown that hematite has conductivity value on (2.5–3).10^{-7} S/cm and be included as dielectric materials. The hysterisis curve was shown that hematite was a super paramagnetic materials. The product was mixed on paint with procentage 10% of total weight and coated on steel grade AH36 with spray methods. Then, the maximum of reflection loss on x – band's frequency range (8,2-12,4) GHz was -7 dB on frequency of 10.5 GHz. It mean that almost 50% electromagnetic energy was absorbed by hematite.

1. Introduction

RADAR is a technology that affort to detect the location, altitude, and direction of an object at resting or moving [1]. Detection of RADAR is based on Radar Cross Section (RCS), area on target that reflect electromagnetic waves (EM) [2]. Reflected waves was accepted by receiver and converted into data (location, velocity, and direction) of object. Along with the development of radar technology is also developed a system that can avoid radar detection, called stealth technology.

Stealth technology worked by manipulating the shape of the object so the reflection of waves can't be received by radar and using a material that can absorb and reduce the electromagnetic energy, called RAM. RAM reduces radar waves by mechanism of destructive superposition and impedance matching [3]. Energy absorption was obtained from polarization of electrical charges (dielectric loss), motion of magnetic domains on saturation magnetization (magnetic loss), motion of free electrons, and vibration of atoms. This processing depends from the characteristics of each material [4]. Based on electrical and magnetical properties, materials was divided into magnetic materials, dielectric materials, or dielectric and magnetic materials

Recently, the RAM has been used ferrite material that has a good magnetical and electrical properties [5]. Magnetite and Barium M-Hexaferrite (BaM) were ferrite material that has been applied as RAM. Both of them has a high reflection loss and means that almost of EM energy has been absorbed by material. Hematite was similar properties with magnetite but it was be acted as weak ferromagnetic at room temperature. So with this reason, hematite should have potencial to apply as RAM materials.

Magnetic nano materials have different properties of magnetic materials commonly where the magnetic nano materials will have a super paramagnetic properties. A material is called as nano magnetic if it has diameter size between on 1-100 nm. Super paramagnetic occurs on single domain of ferromagnetic and ferrimagnetic materials. It will acted as paramagnetic material when an external fields is applied on surrounding. The small size of particle made it becomes very reactive to an external magnetic field but the value of magnetization becomes zero when the field lost.

Indonesia has abundant natural iron stones that have a potency to be applied becomes a magnetic materials as RAM, such as hematite. Iron stones from Tanah Laut, South Kalimantan, have a high iron content about 98% so it can be changed becomes Hematite. This study synthesized hematite from iron stones and done characterization of the products. The goal of this study determined influence of hematite as RAM on X-Band's frequency.

2. Experimental

This study began an extraction process of natural iron stones to become iron powders. Then, it was cleaned by 96% amount of alcohol using ultrasonic cleaner to remove and reduce impuritas. Synthesis processing was done by co-precipitation method using HCl and NH_4OH solutions. First, iron powder was dissolved in hydrochloric acid solution and was heated to (60±5)°C for 20 min. The ammonium hydroxide solution was then slowly added to the iron (III) chloride solution under constant stirring until pH scale on 9. The final mixed solution was dried to (120±5)°C until turned into a dark powder. To produce the hematite, the dried precursor was calcined at 500°C for 5h. The phase of Hematite was characterized by X-rays Diffraction (XRD X'Pert PRO; CuKα radiation, λ = 1.54060 Å) with a step size 0.017°. The conductivity value was measured by two point probe methods and magnetic measurement was done by Vibrating Sample Magnetometer (OXFORD VSM 1.2H).

Coating process was began by mixed paint and hematite in a weight ratio of 1:9. Mixing process was done by mixer with velocity 800 rpm for 10 min. The final mixed solution was then coated on grade AH36 steel plate using a thinner and the thickness of coating was 3 mm. Microwave absorption on X-band frequency was measured by Vector Network Analyzer (VNA; VNA ADVANTEST type-3770).

3. Result and Discusion

The XRD for hematite powder, indicating only a single, essentially rhombohedral hematite phase, α-Fe_2O_3 is shown in Figure 1. The index *hkl* based on ICSD no. 201096. The absence of any other impurity phase confirms that Hematite (α-Fe_2O_3) was succesfully synthesized by coprecipation methods. The similar research was also done by Mufid [6], K. Supattarasakda [7], X. Yao [8], Radoslaw Przenioslo[9], etc.

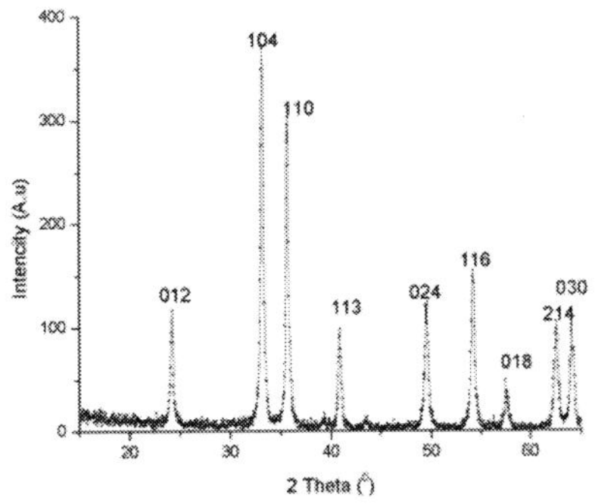

Figure 1. The XRD pattern of Hematite (α-Fe_2O_3)

The electrical conductivity for hematite sample is given in Figure 2. The plot are almost linier in the logaritmic scale indicating that conduction increases with increase in frequency. Based on the results, Hematite is a dielectric material that has electric conductivity between (1-0.35) 10^{-7} S/cm. Electric conductivity in dielectric material depended by charge polarization contribution. The polaron hopping mechanism can also result in electronic polarization contributing to low frequency dispersion [10]. Increasing frequency made localized electric charge carriers and electric conductivity increased.

The magnetization versus applied field was measured by VSM and was given in Figure 3. The saturation magnetization (M_s) and remanent magnetization (M_r) value from hematite were 3.2 emu/g and 1.2 emu/g. The coercivity field showed that it has value approach to zero. It means that hematite is super paramagnetic material. Super paramagnetic occurs at single domain of nano magnetic material, that has a diameter size on 1-100 nm. It will be paramagnetic material when an external magnetic field to be applied. Magnetic dipoles available inside paramagnetic materials rotate with the applied external magnetic field and become aligned with the magnetic field direction.

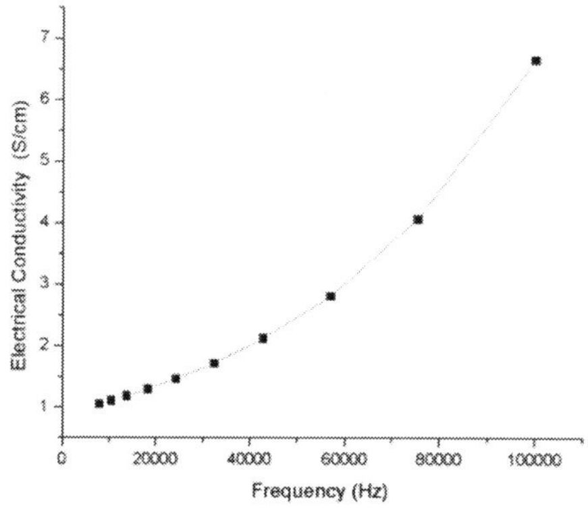

Figure 2. Graph between frequency and electrical conductivity of Hematite (α-Fe_2O_3)

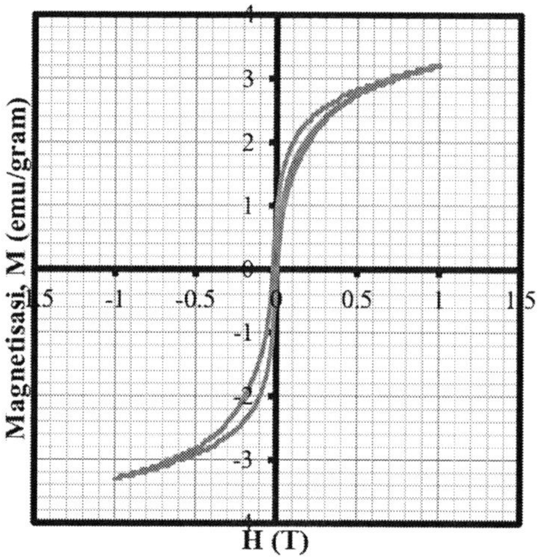

Figure 3. The hysteresis curve of Hematite (α-Fe_2O_3)

Vector Network Analyzer (VNA) was performed to determine the reflection loss of material, in this study used hematite. Figure 4 indicated that the maximum value of reflection loss by hematite was -7 dB on 10.5 Ghz. The other research determined that fifty percent of electromagnetic energy is absorbed at -3 dB. Then, it means that hematite can absorb more than fifty percent electromagnetic energy.

The absorption mechanism in hematite can be occurred at dielectric and magnetic material. The absorption electromagnetic energy by a dielectric material was done through the mechanism of polarization by the electric dipoles. Due to application of electric field, electric dipoles inside the dielectric material are aligned in the direction of the applied electric field. Electromagnetic energy was be converted to mechanical energy via dipole movement. Time required for electronic displacement was very short when compared to dipole rotation and thermal polarization. The other mechanism may be occurs if between the two dipoles have opposite direction so both of them was reduceeach other. The absorption by magnetic was done through rotation dipole magnetic. When an external magnetic field was applied, the overall dipole moment of the magnet will react according to the characteristics from the material. Magnetic dipoles in the paramagnetic material will attempt to rectify the direction of the dipole with the external magnetic field.

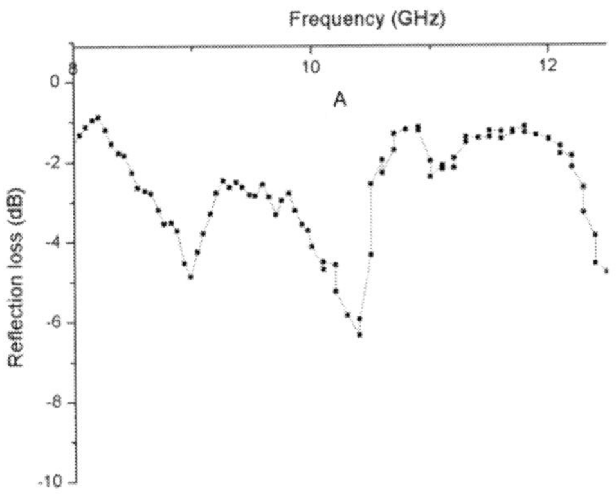

Figure 4. The Reflection Loss of Hematite (α-Fe_2O_3) on X-Band frequency

4. Conclusion and Suggestion

The results of this study can be write down conclusion, that are: Iron stones from Tanah Laut, South Kalimantan, Indonesia, can be synthesized into magnetic materials, such as Hematite. Hematite producted at 500°C is dielectric material that has conductivity value (1-0.35) 10^{-8} S/cm. Hematite has super paramagnetic properties with coerciyity value narrow to zero. Hematite can be applied as Radar Absorbing Material (RAM) on X-Band frequency.

References

[1] Lita A 2014 Thesis Institut Teknologi Sepuluh Nopember Surabaya
[2] Knott E F 2005 Radar Cross Section, Handbook of Radar,ed. Skolmik, M. I., Mc Graw Hills, USA, pp. 11.1-11.34.
[3] Gurer G 2011 Tesis, Middle East Technical University,Turki
[4] Radoslaw Przenioslo, Izabela Sosnowska, etc. Monoclinic deformation of the Crystal Latice of Hematite α-Fe_2O_3. Elsevier Ltd. 2014
[5] Lloyd D, Vainikka T, Ronkainen M and Kontturi K 2013*Electrochimica Acta* **109** 843–851
[6] Mufid A 2014 The Influence of Calcination Temperature on Quantitative Phase Hematite from Iron Stone Tanah Laut Kalimantan. Surabaya
[7] Supattarasakda K, Petcharoen K, Permpool T, Sirivat A, and Lerdwijitjarud W 2013 *Powder Technol.***249** 353
[8] Yao X, Tang C, Yuan G, Cui P, Xu X, and Liu Z 2011 *Electrochem. Commun.***13**1439
[9] Przenioslo R, Sosnowska I 2014 Monoclinic deformation of the Crystal Latice of Hematite α-Fe_2O_3. Elsevier Ltd.
[10] Shinde S S, Bhosale C H, Rajpure K Y 2011 *J. Alloys Compd.* **509** 3943

IOP Conf. Series: Materials Science and Engineering **196** (2017) 012009 doi:10.1088/1757-899X/196/1/012009

Synthesis of $Zn_{1-x}Cu_xO$ Nanoparticles by Coprecipitation and Their Structure and Electrical Property

Dyah Ayu Daratika[a], **Malik Anjelh Baqiya**[b], **Darminto**[c]

Advanced Materials Research Group, Department of Physics, Faculty of Mathematics and Natural Science, Institut Teknologi Sepuluh Nopember, Kampus ITS Sukolilo, Surabaya 60111, Indonesia

E-mail: darminto@physics.its.ac.id

Abstract. The $Zn_{1-x}Cu_xO$ (x = 0 - 6% wt) nanoparticles has been synthesized by coprecipitation method using zinc acetate dihydrate, and copper powder were employed with HCl and NH_4OH respectively as solvent and precipitating agents. The effect of Cu concentration on structural, optical, and electrical properties of $Zn_{1-x}Cu_xO$ nanoparticles were investigated. The diffraction patterns of XRD indicate that $Zn_{1-x}Cu_xO$ phase crystallized in the wurtzite structure having crystal size which was evaluated by using MAUD software, in the range of 28 - 79 nm. Electron microscope analysis shows the morphology of $Zn_{1-x}Cu_xO$ is nanowires, having finer grains with the increasing content of Cu. The Cu doping reduced the optical band gap energy from 3.10 eV to 2.80 eV, while the electrical conductivity increased from 1.18 x 10^{-8} to 24.25 x 10^{-8} S/cm. This result implies that Cu^+ or Cu^{2+} ions have substituted Zn^{2+} ions. However, doping of Cu more than 4% wt increase optical band gap which makes the electrical conductivity decrease. The electrical conductivity obtained from this study is significantly higher than that reported previously.

1. Introduction

Zinc oxide (ZnO) has emerged as a promising material for gas sensor, memory of optical and magnetic devices, piezoelectric tranducers, photodiodes, photodetectors, UV-light emitting diodes (UV-LEDs), biomedical, solar cell, and transparent conductive oxide [1]. ZnO is a semiconductor oxide group II-VI with 3.37 eV direct band gap, large excitation binding energy of 60 meV exciton, high electron mobility at room temperature, high optical transparency and good in mechanical, magnetic, optical, and electrical properties. In addition, ZnO is also a metal oxide material that is inexpensive, nontoxic, resistant to radiation, and has a high thermal and chemical stability.

In the next generation, transparent conductive oxide (TCO) will involve the p-n junction, thus the fabrication of metal oxide semiconductor requires n-type and p-type TCO. However, the development of high-quality TCO thin film limited to increasing the n-type conductivity. While, fabrication of the p-type TCO thin films still complicated, thus it remains a major challenge until now [2].The p-type TCO obtained by engineered the structure of ZnO with the addition of extrinsic doping with group IA and IB elements (Li, Na, K, Ag,Cu) in the place of Zn or by doping with group VA elements (N, P, As, Sb) in the place of oxygen to improve the electrical conductivity [3]. Among these doping elements, copper (Cu) is potential dopants to improve the electrical properties of ZnO, which has a low resistivity copper ~~is~~ ($1.67x10^{-6}$ Ω cm) [4]. The presence of copper (Cu) is used as a variable electrical conductance, thus it can control the electrical conductivity of ZnO. As a dopant element, copper has many physical and chemical properties that are similar to Zn, such as have the similar structure of the electron shells.

Synthesis of ZnO nanoparticles with addition of extrinsic doping can be done by using several methods such as auto-combustion, ball milling, sol-gel process, hydrothermal route, and co-precipitation method [4].One of the promising inorganic compounds synthesis method is coprecipitation, which is based on precipitation of more than one compound simultaneously when passing through the saturation point. In this research, $Zn_{1-x}Cu_xO$ (x = 0 – 6% wt) nanoparticles were synthesized by using coprecipitation method. The introduction of copper as dopant partially substituting Zn in ZnO powder and its effects will be studied in terms of structure, optical bandgap and electrical conductvity.

2. Experimental Procedure

The $Zn_{1-x}Cu_xO$ (x = 0 – 6% wt) nanoparticles has been synthesized by coprecipitation method using zinc acetate dehydrate [$Zn(CH_3COO)_2.2H_2O$], and copper powder (Cu) as starting materials, which were mixed and dissolved in an appropriate amount of 0.5 M HCl using magnetic stirrer at speed of 350 rpm.The NH_4OH of 0.5 M was gradually added to the solution by titration method to achieve pH of around 9. The precursor solution stirred using a magnetic stirrer at 85°C for 4 hours. The precipitate formed from the process was then filtered and washed using distilled water, and dried in an oven at 100 °C for 24 hours. The samples were then calcined in a furnace at 400°C for 3 hours.

The phase purity and crystal structure of samples were investigated by X-ray diffraction (XRD) using scanning angle 2θ ranging from 20°C to 70°C which operated at 40 kV and 30 mA with $CuK\alpha_1$ radiation at wavelength of 1.54 Å. The morphologies of $Zn_{1-x}Cu_xO$ nanoparticles were studied using scanning electron microscopy (SEM). UV-Vis spectrophotometer was used to measure the coefficient of optical transmittance, thus through a calculation the value of optical band gap was obtained.The electrical conductivity of the samples were examined by LCR meter.

3. Result of Discussion

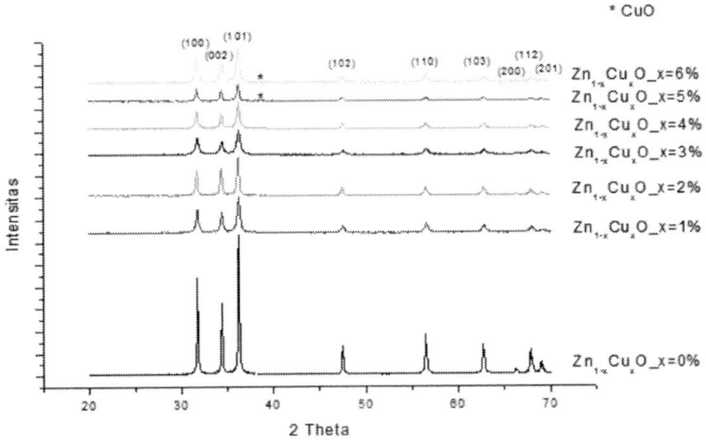

Figure 1. XRD patterns of the $Zn_{1-x}Cu_xO$

Fig.1 displays XRD patterns of $Zn_{1-x}Cu_xO$ (x = 0 – 6% wt) nanoparticles calcined at 400°C for 3 hours. Diffraction peaks have been observed to be sharp which implies high crystallinity of $Zn_{1-x}Cu_xO$ nanoparticles. The XRD data analyzed by software Match! was well matched with PDF (Powder Diffraction File) 00-079-0207 and has wurtzite crystal structure for all compositions. XRD results indicates that the Cu ions have substituted the position of Zn ions easily, due to the ionic radius of Cu^+/Cu^{2+} closed to that of Zn^{2+}. Theoretically, the copper with an atomic number of 29 has electron configurations that ends on d-orbital. Electron configuration of Cu is [Ar] $3d^9 4s^2$ or [Ar] $3d^{10} 4s^1$, so that both Cu^+ and Cu^{2+} ions have a probablility to present in the sample [2]. The presence of Cu^+ ions in the ZnO lattice will contribute as an extrinsic dopant to form acceptor states in the energy band structure of ZnO, thus change an n-type of $Zn_{1-x}Cu_xO$ to p-type semiconductor. Further, the presence of copper oxide phase has been detected at 5% and 6% wt Cu doped samples at diffraction angle (2θ) of 38.73°, though the CuO peak are very small, it shows an impurity phase in samples. Lattice parameters analyzed by refinement method using software Rietica, particle size analyzed using software MAUD (Material Analysis Using Diffraction), and volume cell that obtained from the analyzed of XRD data are tabulated in Tabel 1.

Table 1. Crystallite size, lattice parameters, and volume cell of $Zn_{1-x}Cu_xO$

Copper content (% wt)	Crystallite size (nm)	Lattice Parameter				
		a = b (Å)	c (Å)	α = β (⁰)	γ (⁰)	*Volume Cell*
0	79	3.254066	5.211942	90	120	47.795048
1	65	3.251919	5.208255	90	120	47.698223
2	55	3.250603	5.205092	90	120	47.630676
3	50	3.250454	5.205212	90	120	47.627419
4	37	3.250161	5.205280	90	120	47.619465
5	30	3.248422	5.202350	90	120	47.541725
6	28	3.248306	5.199845	90	120	47.515450

It can be seen in Table 1 that with the increasing content of copper, volume cell decreases for all samples. Lattice parameter decreases with the increasing copper content of the sample, due to replacing of Zn^{2+} ions by Cu^{2+} (0,73 Å) is smaller than that of Zn^{2+} (0,74 Å). Surface morphologies of pure ZnO and $Zn_{0.97}Cu_{0.03}O$, which shown in Fig. 2, revealed the formation of nanowires and having finer grains for sample with the increasing content of Cu.

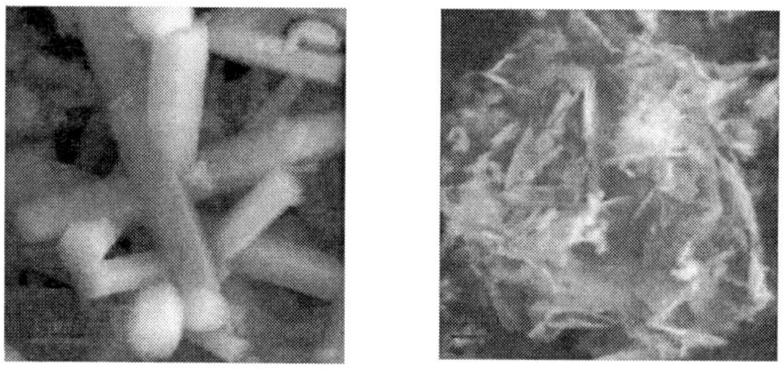

Figure 2. The morphologies of (a) pure ZnO, and (b) $Zn_{0,97}Cu_{0,03}O$ nanoparticles

UV-vis spectrophotometer was used to analyze the optical band gap of the $Zn_{1-x}Cu_xO$ as a semiconducting material. The transmission spectrum that has been obtained from measurements of UV-vis spectrophotometer was processed using Tauc Plot methods to obtain optical band gap. The Tauc relation to evaluate the optical band gap is shown on Eq. 1 [5].

$$(\alpha hv)^{1/n} = A(hv - E_g) \quad (1)$$

where A is a proportional constant, E_g is the energy band gap of semiconductor material, hv is the photon energy, and α is the optical absorption coefficient. The exponent n depends on the type of electron transition of the material, because ZnO has a direct band gap, so $n = ½$. The absorption coefficients were calculated by using Eq.2 :

$$\alpha = (-\ln T)/d \quad (2)$$

where T is transmittance of sample, and d is width of the cuvette. Optical band gap of the $Zn_{1-x}Cu_xO$ was obtained by plotting $(\alpha hv)^2$ vs. hv followed by extrapolating absorption edge by linear line in x-axis to find optical band gap [2].

The optical band gap energy and electrical conductivity of $Zn_{1-x}Cu_xO$ nanoparticles are preseted in Table 2. It is found that increasing of Cu concentration as dopant has reduced optical band gap energy from 3.10 eV to 2.80 eV, while the electrical conductivity increased from 1.18 x 10^{-8} to 24.25 x 10^{-8} S/cm. It implies that Cu^+or Cu^{2+} ions have substituted Zn^{2+} ions side. However, at concentration of Cu doping more than 4wt%, optical band gap energy increase which lead to decreasing of electrical conductivity.This may possibly due to the heterogeneity of phase on the sample. Previous research carried out by Sonawane et al. (2008) [6] described that the Cu-doped ZnO nanoparticles prepared with the same method have much lower electrical conductivity in the range of 1.6 x 10^{-10} to 1 x 10^{-9}S/cm with monotonic increase in the increasing contain of Cu.

Tabel 2. The optical band gap energy and electrical conductivity of $Zn_{1-x}Cu_xO$ nanoparticles

Copper content (x)	Optical band gap energy	Electrical conductivity (x 10^{-8})
0 %wt	3.05 eV	1.18 S/cm
1 %wt	2.90 eV	1.25 S/cm
2 %wt	2.88 eV	1.57 S/cm
3 %wt	2.85 eV	3.38 S/cm
4 %wt	2.80 eV	24.25 S/cm
5 %wt	3.06 eV	1.03S/cm
6 %wt	3.10 eV	0.65 S/cm

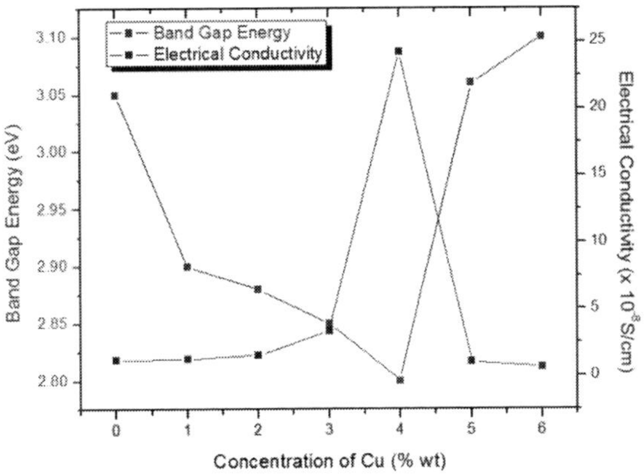

Figure 3. The optical band gap energy and electrical conductivity of $Zn_{1-x}Cu_xO$ nanoparticles

4. Conclusion

The $Zn_{1-x}Cu_xO$ nanoparticles has been successfully syntesized using coprecipitation method. Cu doping has led to the decreasing lattice parameters and forming finer grains of ZnO crystal. The presence of Cu until 4% wt has reduced the optical band gap energy, while the electrical conductivity increased by the presence of Cu^{+} or Cu^{2+} ions to create an acceptor energy level. For the whole samples, the significantly improved electrical properties have been obtained in this study compared to the previous report.

Acknowledgment

This research was partially supported by “Hibah KLN”, Kemenristekdikti, Indonesia, 2017.

References

[1] Al-Ghamdi A A, Al-Hartomy O A, El Okr M, Nawar A M, El-Gazzar S, El-Tantawy F, Yakuphanoglu F 2014 *Spectrochim. Acta. A. Mol. Biomol. Spectrosc.* **131** 512
[2] Yuan H, Xu M, Huang Q Z 2014 *J. Alloys Compd.* **616** 401
[3] Zaharescu M, Mihaiu S, Toader A, Atkinson I, Calderon-Moreno J, Anastasescu M, Nicolesc, M, Duta M, Gartner M, Vojisavljevic K, Malic B, Ivanov V A, Zaretskaya E P 2014 *Thin Solid Films* **571** 727
[4] Singhal S, Kaur J, Namgyal T, Sharma R 2012 *Phys. B Condens. Matter* **407** 1223
[5] Muthukumaran S, Gopalakrishnan R 2012 *Opt. Mater* **34** 1946
[6] Sonawane Y S, Kanade K G, Kale B B, Aiyer R C 2008 *Mater. Res. Bull.* **43** 2719

IOP Conf. Series: Materials Science and Engineering **196** (2017) 012010 doi:10.1088/1757-899X/196/1/012010

Electrical Properties of $La_{0.67}Sr_{0.33}Mn_{0.8}Ni_{0.2}O_3$ Synthesized by Sol Gel Method

B Kurniawan[1] , U Widyaiswari[1], I Fauziyah[1] and A Imaduddin[2]

[1] Department of Physics, Universitas Indonesia, Depok 16424, Indonesia
[2] Research Center of Metallurgy and Material, LIPI, Puspiptek, Serpong 15314, Indonesia

E-mail: bkuru07@gmail.com

Abstract. The previous study of lanthanum manganites showed that physical properties of $LaMnO_3$ compound has changed when Sr doped the La site. $La_{0.67}Sr_{0.33}MnO_3$ compound give the highest $T_C = 370$ K and showed metal insulator transition. It happens as a result of changed in Mn^{3+}/Mn^{4+} ratio. Its indicated that Mn has the most important rule to the physical properties changed. In order to shift Tc of the compound to the near room temperature, Ni doped was given to Mn site of LSMO. In this paper, 20% Mn site was doped by Ni. $La_{0.67}Sr_{0.33}Mn_{0.8}Ni_{0.2}O_3$ compound synthesized by sol gel method and characterized using XRD. Then, single phase compound resisitivity was measured to see the electrical properties this compound. The XRD result showed that sample have single phase with rhombohedral structure and R-3c space group. Resistivity as a temperature function graph does not show resistivity peak and metal insulator transition. The resistivity measurement showed that below 300 K, resistivity inscreased with decreased of temperature that indicates insulator behavior. It caused by decreased of double exchange interaction and increase of superexchange interaction. From this result, we can conclude that Ni doped decreased electrical properties of the sample.

1. Introduction

Lanthanum manganites $La_{1-x}A_xMnO_3$ (A = Sr, Ba, Ca) have been studied and become subject of research interest because of their magnetic and electrical properties which can be varied and applied in magnetic random access memory, and magnetic sensor [1,2]. Doping in La site would change structure and electrical properties of material. The previous study of lanthanum manganites showed that physical properties of $LaMnO_3$ compound has changed when Sr doped the La site. $La_{0.67}Sr_{0.33}MnO_3$ compound give the highest $T_C = 370$ K and showed metal insulator transition [3]. It happens as a result of changed in Mn^{3+}/Mn^{4+} ratio. Its indicated that Mn has the most important rule to the physical properties changed.

In order to shift Tc of the compound to the near room temperature, Ni doped was given to Mn site of LSMO. In this paper, 20% Mn site was doped by Ni. $La_{0.67}Sr_{0.33}Mn_{0.8}Ni_{0.2}O_3$ compound was synthesized by sol gel method. Sample was characterized using XRD and the resistivity was measured using cryogenic magnetometer.

2. Experimental Details

$La_{0.67}Sr_{0.33}Mn_{0.8}Ni_{0.2}O_3$ sample was synthesized using a sol gel method. Suitable proportions of La_2O_3, $Sr(NO_3)_2$, $Mn(NO_3)_2.4H_2O$, and $Ni(NO_3)_2.6H_2O$ were used as starting materials. Each starting material solved in aquabidest, while La_2O_3 precursor was solved in diluted HNO_3. The resulting solution of each starting material mixed and stirred at 80°C. Ammonia solution was added to the mixed solution to adjust pH until reached 5,6. Then, the solution was stirred and dried it at 120°C for 2 h until sol gel formed. The gel then calcined at 500°C, grinded, and re-calcined at 850°C. The re-calcined sample was pressed into pellet and sintered at 1200°C for 2 h. The sample then characterized using XRD, while the resistivity was measured using cryogenic magnetometer.

3. Result and Discussion

3.1. Structure Characterization

Structure of sample is characterized using X-Ray Diffraction (XRD), and the result shows in figure 1. The figure 1 shows that sample has single phase with no impurity.

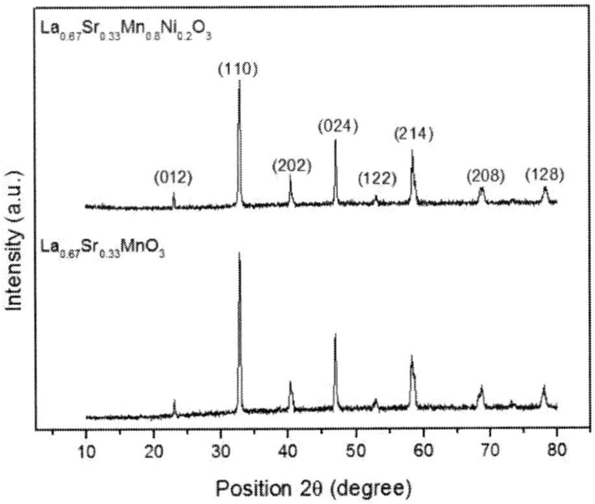

Figure 1. XRD pattern of $La_{0.67}Sr_{0.33}MnO_3$ sample [4] and $La_{0.67}Sr_{0.33}Mn_{0.8}Ni_{0.2}O_3$ sample.

The result of XRD characterization is analyzed using Rietveld refinement analysis method. The Rietveld refinement results show that sample has Rhombohedral structure with R-3c space group. Refinement result of $La_{0.67}Sr_{0.33}Mn_{0.8}Ni_{0.2}O_3$ sample compared with $La_{0.67}Sr_{0.33}MnO_3$ sample [4] and summarized in table 1.

Table 1. Data obtained from X-Ray diffraction refinement with Rietveld analysis of $La_{0.67}Sr_{0.33}MnO_3$ sample [4] and $La_{0.67}Sr_{0.33}Mn_{0.8}Ni_{0.2}O_3$ sample.

Parameters	$La_{0.67}Sr_{0.33}MnO_3$	$La_{0.67}Sr_{0.33}Mn_{0.8}Ni_{0.2}O_3$
a (Å)	5.5005	5.4767
a (Å)	13.368	13.316
Volume (Å^3)	350.26	345.90
Crystallite size (Å)	535.6	572.8

3.2. Resistivity

The result of resistivity measurement show in figure 2. It show that sample has insulator behavior in temperature range 100-300 K with resistivity value range 0.01-0.34 Ω.m. The insulator behavior is indicated by increasing of resistivity with temperature decreased. The result does not show resistivity peak as shown in $La_{0.67}Sr_{0.33}MnO_3$ sample and denote higher resistivity compared with $La_{0.67}Sr_{0.33}MnO_3$ sample resistivity value range [4]. When Ni doped Mn site, there are more localization of electron,

IOP Conf. Series: Materials Science and Engineering **196** (2017) 012010 doi:10.1088/1757-899X/196/1/012010

because the ratio of Mn^{3+}/Mn^{4+} will be decreased and suppress the Double Exchange (DE) interaction [5].

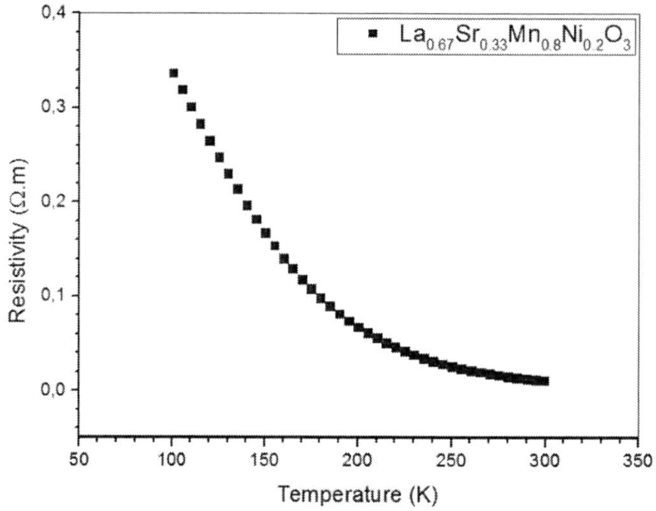

Figure 2. Resistivity of $La_{0.67}Sr_{0.33}Mn_{0.8}Ni_{0.2}O_3$ sample in respect of temperature at zero magnetic field.

Resisitivity of this sample could be analyzed using Small Polaron Hoping (SPH) model with equation (1) [5,6,7,8].

$$\rho = \rho_\alpha T \exp\left(\frac{E_A}{k_B T}\right) \tag{1}$$

Where E_A is the sum of the activation energy required to create free carriers and the energy that activates carriers for hoping [5], k_B is Boltzmann's constant with value $8.617 \times 10^{-5} eV.K^{-1}$, and ρ_α is residual resistivity [6,7,8].

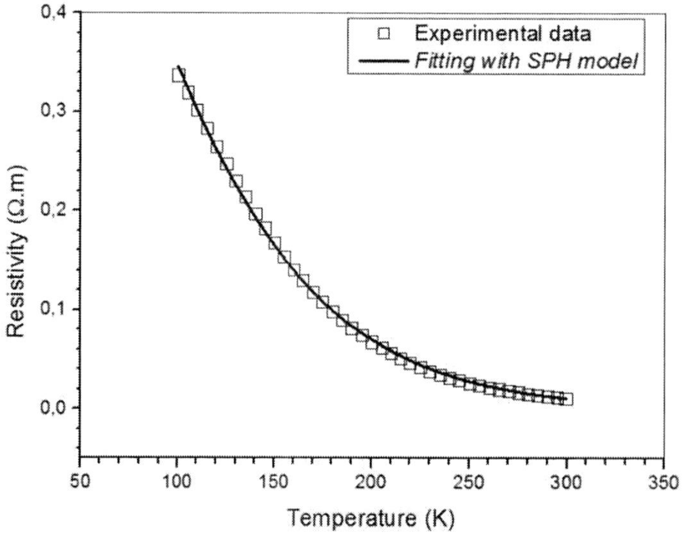

Figure 3. Theoretical fit of insulator state resistivity data for $La_{0.67}Sr_{0.33}Mn_{0.8}Ni_{0.2}O_3$ using SPH model.

When Ni doped Mn site, DE interaction will be reduced due to decrease of Mn^{3+}-O^{2-}-Mn^{4+} bond. In the insulator state as shown by $La_{0.67}Sr_{0.33}Mn_{0.8}Ni_{0.2}O_3$ sample, many electrons are localized, due to the decrease of Mn^{3+}/Mn^{4+} ratio and the coupling between electrons and phonons that form polarons. When given temperature increases, the energy supplied is also getting bigger, and more polaron activated, so that the resistivity decreases. It explain the insulator behavior that showed by decreases of resistivity when temperature increased.

Value of obtained fitting parameters of $La_{0.67}Sr_{0.33}Mn_{0.8}Ni_{0.2}O_3$ sample are showed in table 2. Quality of fitting is evaluated by the value of squared linear correlation coefficients (R^2) [6]. We obtained R^2 value above 99.9%, and it means that the fitting that has been done has a good quality.

Table 2. Best-fitting parameters obtained from equation (1) of $La_{0.67}Sr_{0.33}Mn_{0.8}Ni_{0.2}O_3$ sample

Parameters	Value
$\rho_\alpha(\Omega.m)$	0.034
$E_A(\times 10^{-7}\, meV)$	1.973
R^2 (%)	99.93

4. Conclusion

$La_{0.67}Sr_{0.33}Mn_{0.8}Ni_{0.2}O_3$ sample was synthesized by sol gel method. XRD result showed that sample has single phase with no peak of impurity with rombhohedral structure and R-3c space group. The resistivity measurement showed that sample has insulator behavior with resistivity value range 0.01-0.34 Ω.m in temperature range 100-300 K. Analysis of resistivity found that small polaron hoping (SPH) model fit the experimental data with R^2 value 99.93%.

Acknowledgments

We sincerely thank to Universitas Indonesia for funding this research through Publikasi Internasional Terindeks untuk Tugas Akhir (PITTA) grant with number of contract 1999/UN2.R12/HKP.05.00/2016.

References

[1] Salamon M B and Jaime M 2001 *Rev. Mod. Phys.* **73** 583
[2] Cao D, Zhang Y, Dong W, Yang J, Bai W, Chen Y, Wang G, Dong X and Tang X 2015 *Ceram. Int.* **41** 381
[3] Jonker G H and Van Santen J H 1950 *Physica* **XVI** 337
[4] Widyaiswari U, Kurniawan B, Imaduddin A and Saptari S A 2016 *J. Phys.: Conf. Ser.* **776**
[5] Ahmed A M and Mohamed A E A 2015 *Rare Metals* **35** 551
[6] Dhahri A, Jemmali M, Dhahri E and Hlil E K 2015 *Dalton Trans.* **44** 5620
[7] Manjunatha S O, Rao A, Lin T, Chang C and Kuo Y 2015 *J. Alloy Compd.* **619** 303
[8] Manjunatha S O, Rao A and Okram G S 2015 *J. Alloy Compd.* **640** 154

IOP Conf. Series: Materials Science and Engineering **196** (2017) 012011 doi:10.1088/1757-899X/196/1/012011

Protected Vacuum Annealing Effect on Single Crystals of $Pr_{1-x}LaCe_xCuO_4$ in the Overdoped Regime

M A Baqiya[1], B Triono[1], Darminto[1], T Adachi[2], A Takahashi[3], T Konno[3], M Watanabe[3], T Prombood[3], Y Koike[3]

[1] Department of Physics, Institut Teknologi Sepuluh Nopember (ITS), ITS Campus, Sukolilo, Surabaya 60111, Indonesia

[2] Department of Engineering and Applied Sciences, Sophia University, 7-1 Kioi-cho, Chiyoda-ku, Tokyo 102-8554, Japan

[3] Department of Applied Physics, Graduate School of Engineering, Tohoku University, 6-6-05 Aoba, Aramaki, Aoba-ku, Sendai 980-8579, Japan

E-mail : malikabits@physics.its.ac.id

Abstract. We report the effect of protected vacuum annealing on the superconducting (SC) and magnetic properties of $Pr_{1-x}LaCe_xCuO_4$ (PLCCO) single crystals in the overdoped regime (x = 0.13 – 0.20). The powders and single crystals of PLCCO with all x were synthesized by the solid state reaction and grown by the travelling solvent floating zone (TSFZ) method, respectively. The powder x-ray diffraction (XRD) results analyzed by the Rietveld method have shown that there is a systematic decrease in the c-axis length with increasing x, suggesting the proper substitution of Ce for Pr in overdoped PLCCO up to x = 0.20. The so-called two-step annealing performed in the overdoped regime was essential for increasing the SC volume fraction estimated from the magnetic susceptibility measurements. Furthermore, the temperature dependence of the normal-state magnetic susceptibility for reduced crystals of PLCCO with all x has exhibited a Curie-Weiss-like behavior regardless of various steps of the protected vacuum annealing process.

1. Introduction

There are still open questions concerning the role of the oxygen reduction in the electron-doped high-T_c superconductivity. In the electron-doped cuprates with the so-called T'-structure, the superconducting (SC) regime in the generic phase diagram is quite narrow compared with the hole-doped ones [1, 2]. However, the recent investigation in thin films of T'-$(La,RE)_2CuO_4$ (RE = Sm, Eu, Gd, Tb, Lu, and Y) has revealed that they show Ce-free superconductivity [3]. This phenomenon is related to the role of the reduction annealing process for the occurrence of superconductivity in the T'-cuprates. The superconductivity in the T'-cuprates is very sensitive to the amount of excess oxygen at the apical site (O(3)) above or below the Cu site. The presence of the excess oxygen suppresses superconductivity

To achieve the optimum critical temperature (T_c) for the electron-doped T'-cuprates, it is necessary to find an optimum oxygen reduction process, as performed in T'-$Pr_{0.88}LaCe_{0.12}CuO_{4+\delta}$ single crystals [4]. An improved reduction technique has been used by Brinkmann *et al.* [5] to establish the superconductivity of $Pr_{2-x}Ce_xCuO_4$ over a wide doping-range between x = 0.04 and 0.17. They protected the single crystal surface with the polycrystalline pellets on the top and bottom having the same composition in order to overcome the destruction of the surface. Then, by employing a similar improved reduction annealing, named the protected vacuum annealing, we have obtained SC crystals with the T'-structure of $Pr_{1.3-x}La_{0.7}Ce_xCuO_{4+\delta}$ (x = 0.10) [6]. The difference from the technique by Brinkmann *et al.* [5] is that we have covered the crystal with polycrystalline powders of the same composition during the vacuum annealing process. Then, a so-called two-step annealing process has been developed by Krockenberger *et al.* [7] to achieve superconductivity of Ce-free thin films of Pr_2CuO_4. They claimed that the second-step annealing, performed at a lower temperature, provided "healing" of the defective CuO_2 plane by the oxygen rearrangement from the apical site to the regular in-plane site. The current paper reports the effect of protected vacuum annealing combined with the two-step annealing process on the SC and magnetic properties of single crystals of $Pr_{1-x}LaCe_xCuO_4$ (PLCCO) with x = 0.13 – 0.20.

2. Experimental

The polycrystalline powders and single crystals of PLCCO with x = 0.13 – 0.25 were synthesized by the solid sate reaction and the traveling solvent floating zone (TSFZ) method, respectively. The synthesized powders and the crystals prepared were characterized by the powder x-ray diffraction (XRD) to determine the crystal structure and the lattice constants. The quality and the composition of the crystals were checked to be good by the x-ray back-Laue photography and the inductively coupled plasma analysis, respectively. The reduction annealing was performed by combining the protected vacuum annealing and the two-step annealing process for all PLCCO crystals as described in Refs. [6] and [7], respectively. In order to determine T_c of the reduced samples, magnetic susceptibility measurements were performed as a function of temperature using a SC quantum interference device (SQUID) magnetometer.

3. Results and Discussion

XRD patterns displayed in Figure 1(a) indicate good quality of the crystals. Figure 1(b) shows the Ce-concentration dependence of the c-axis length for as-grown and reduced crystals of PLCCO. The c-axis length was estimated by analyzing the (00l) peak position of the powder x-ray diffraction pattern. It is found that the c-axis length decreases with increasing Ce-concentration up to x ~ 0.20, which is consistent with the former report [8]. This is because trivalent rare earth ions, in this case Pr^{3+}, are replaced by Ce^{4+} ions with a smaller ionic radius. For the a-axis length, it seems to increase slightly with increasing Ce-concentration. This can be understood as being due to the stretching of the Cu-O bonding owing to the increase in the number of electrons in the antibonding orbital. With increasing x above x ~ 0.20, the c-axis length tends to be saturated, meaning the solubility limit at x ~ 0.20.

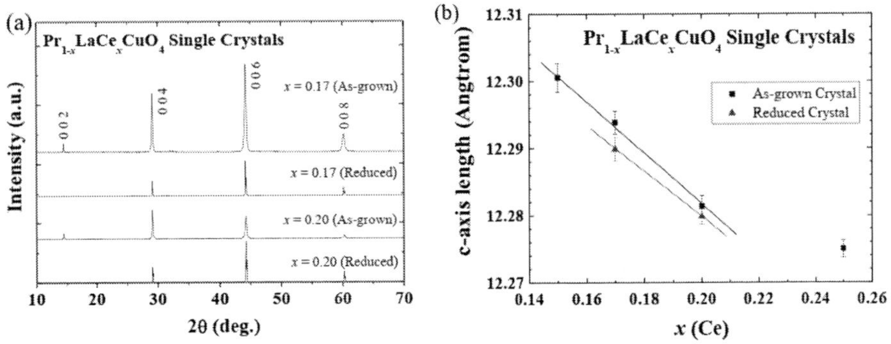

Figure 1. (a) XRD patterns and (b) Ce-concentration dependence of the c-axis length for both as-grown and reduced crystals of $Pr_{1-x}LaCe_xCuO_4$ (PLCCO). PLCCO with x = 0.17 was obtained by the two-step protected vacuum annealing, whereas PLCCO with x = 0.20 was obtained by the one-step protected vacuum annealing.

IOP Conf. Series: Materials Science and Engineering **196** (2017) 012011 doi:10.1088/1757-899X/196/1/012011

Through the reduction annealing, it is found that the c-axis length decreases, indicating a sufficient oxygen removal. In the as-grown crystals, there exists excess oxygen at some apical sites. The presence of the excess oxygen expands the c-axis lattice parameter. By removing the excess oxygen, the c-axis length is reduced. In fact, the oxygen removal occurs not only from the apical site but also from the CuO_2 plane being important for the appearance of superconductivity. Therefore, based on the results of PCCO thin films [7] and $Pr_{1.3-x}La_{0.7}Ce_xCuO_4$ single crystals [6], it is necessary to perform the two-step annealing for Ce-free and underdoped PLCCO crystals as well as PLCCO crystals in the overdoped regimes to recover the oxygen deficiency in the CuO_2 plane.

In this report, several annealing conditions were carried out to obtain the optimum T_c for reduced PLCCO crystals of $x = 0.17$ and 0.20. Figures 2(a) and 2(b) display the temperature dependence of the magnetic susceptibility for PLCCO with $x = 0.17$ and 0.20, respectively. It is found for $x = 0.17$ that the excess oxygen cannot be effectively removed at a low annealing-temperature. At a high annealing-temperature, however, the decomposition of the sample occurred. Therefore, the two-step protected annealing was necessary up to $x = 0.17$. The optimum T_c of $x = 0.17$ was obtained by the two-step protected vacuum annealing on the condition of the first annealing at 900 °C for 12 h and the second annealing at 500 °C for 12 h. Through this annealing, a sharp SC transition was observed at $T_c = 5.2$ K.

In PLCCO with $x = 0.20$ shown in Figure 2(b), behaviors different from those of $x = 0.17$ were observed. The optimum T_c was achieved only by the one-step annealing at a low temperature of 800 °C for 24 h. Both a high annealing-temperature and a long annealing-time caused the decrease in the SC volume fraction due to the less-reduction and over-reduction, respectively. Moreover, the observed Meissner value of $x = 0.20$ at 800 °C for 24 h is quite small. T_c in this case was lower than the measured lowest temperature 2 K in the present SQUID measurements. It is about 1/100 of Meissner signal of the reduced crystal of PLCCO with $x = 0.17$. Thus, the observed Meissner signal of $x = 0.20$ at 800 °C for 24 h is due to filamentary superconductivity. Accordingly, the reduced crystal of PLCCO with $x = 0.20$ was considered to be almost non-SC. It has been found that the SC transition becomes broad with increasing x above $x = 0.17$, which is consistent with the former result [9], indicating the increasing hardness to obtain chemically homogeneous samples.

Figure 3 displays the temperature dependence of the normal-state magnetic susceptibility in 1 T (H // c-axis) for reduced crystals of PLCCO with $x = 0.13 - 0.20$. A hump is observed around 100 K for all x, probably related to the Pr^{3+} moments included in the crystals [10]. At high temperatures above ~150 K, the temperature dependence of the magnetic susceptibility exhibits a Curie-Weiss-like behavior for all x. This result is similar to that in the hole-doped cuprates of $La_{2-x}Sr_xCuO_4$ (LSCO) in the overdoped regime [11], where the Curie-Weiss-like behavior of the magnetic susceptibility at high temperatures has been understood to be due to magnetic moments generally induced in the overdoped regime of hole-doped cuprates [12]. Additionally, it has been found that the behavior of the magnetic

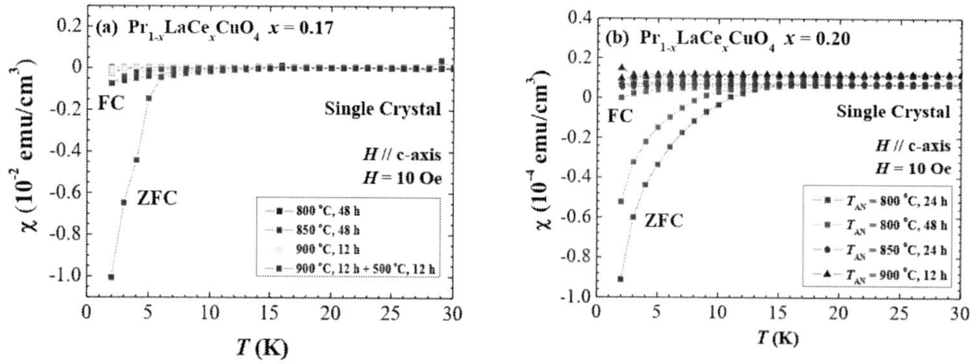

Figure 2. Temperature dependence of the magnetic susceptibility χ for $Pr_{1-x}LaCe_xCuO_4$ single crystals with (a) $x = 0.17$ and (b) $x = 0.20$ on various annealing conditions on both zero-field cooling (ZFC) and field cooling (FC).

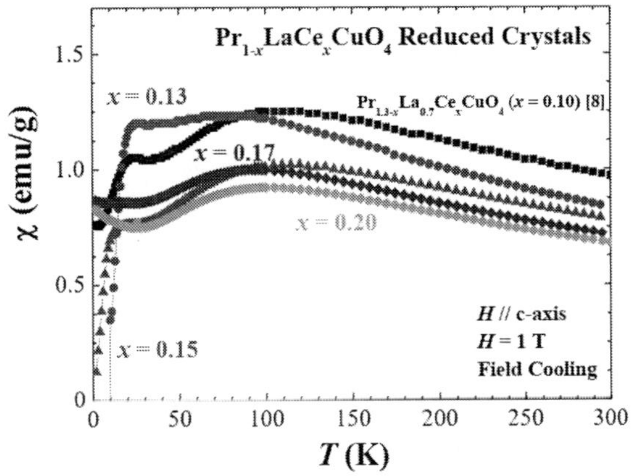

Figure 3. Temperature dependence of the normal-state magnetic susceptibility χ for reduced crystals of $Pr_{1-x}LaCe_xCuO_4$ with $x = 0.13 - 0.20$, together with $Pr_{1.3-x}La_{0.7}Ce_xCuO_4$ ($x = 0.10$) [6].

susceptibility is independent of various steps of the reduction annealing process in the overdoped regime of T'-PLCCO.

4. Summary

The effect of protected vacuum annealing on the SC and magnetic properties of PLCCO single crystals in the overdoped regime has been investigated. It has been found that a combination technique between the protected vacuum annealing and the two-step annealing process is very essential for achieving the optimum SC transition and increasing the SC volume fraction for both underdoped and overdoped T'-PLCCO. The normal-state magnetic susceptibility has been found to behave a Curie-Weiss-like dependence at high temperatures which is a general characteristic for both hole- and electron-doped cuprates in the overdoped regimes. Additionally, the magnetic behavior at high temperatures has been found to be independent of various steps of the reduction annealing process in the overdoped regime.

References

[1] Chu C W, Deng L Z and Lv B 2015 *Physica C* **514** 290-313
[2] Fournier P 2015 *Physica C* **514** 314-338
[3] Tsukada A, Krockenberger Y, Noda M, Yamamoto H, Manske D, Alff L and Naito M 2005 *Solid State Commun* **133** 427-431
[4] Wang Y L, Huang Y, Shan L, Li S L, Dai P, Ren C and Wen H H 2009 *Phys. Rev. B* **80** 094513
[5] Brinkmann M, Rex T, Bach H and Westerholt K 1995 *Phys. Rev. Lett.* **74** 4927-4930
[6] Adachi T, Mori Y, Takahashi A, Kato M, Nishizaki T, Sasaki T, Kobayashi N and Koike Y 2013 *J. Phys. Soc. Jpn* **82** 063713
[7] Krockenberger Y, Irie H, Matsumoto O, Yamagami K, Mitsuhashi M, Tsukada A, Naito M and Yamamoto H 2013 *Sci. Rep* **3** 2235
[8] Naito M and Hepp M 2001 *Physica C* **357–360** 333-336
[9] Fujita M, Matsuda M, Lee S H, Nakagawa M and Yamada K 2008 *Phys. Rev. Lett* **101** 107003
[10] Risdiana, Adachi T, Oki N, Koike Y, Suzuki T, Watanabe I 2010 *Phys. Rev. B* **82** 014506
[11] Wakimoto S, Birgeneau R J, Kagedan A, Kim H, Swainson I, Yamada K and Zhang H 2005 *Phys. Rev. B* **72** 064521
[12] Le Bras G, Konstantinovic Z, Colson D, Forget A, Carton J P, Ayache C, Jean F, Collin G and Dumont Y 2002 *Phys. Rev. B* **66** 174517

Structural and Magnetic Properties of n-type Superconductor $Eu_{2-x}Ce_xCuO_{4+\alpha-\delta}$

Risdiana, M Saputri, M F Sobari, A I Hanifah, W A Somantri and T Saragi

Department of Physics, Padjadjaran University, Jl. Raya Bandung-Sumedang km. 21 Jatinangor, Sumedang, 45363 Indonesia

E-mail: risdiana@phys.unpad.ac.id

Abstract. The n-type superconducting materials of $Eu_{2-x}Ce_xCuO_{4+\alpha-\delta}$ (ECCO) with $x = 0.11$, 0.13, 0.15, 0.16 have been successfully prepared by solid state reaction method, in order to study their structural and magnetic properties of ECCO. The results of X-ray diffraction characterization showed that main peaks of T' tetragonal structure were observed in all samples. Magnetic properties are investigated by DC magnetic-susceptibility measurements that carried out at low temperatures down to 2 K using a standard SQUID magnetometer in a magnetic field of 5 Oe on field cooling. It is found that the highest T_c of 15 K is observed for sample with $x = 0.15$ and decreased for sample with x = 0.16. No trace of superconductivity is observed for $x = 0.11$ and 0.13.

1. Introduction

First material of high-T_c superconductors (HTSC) discovered in 1986 was single-layer cuprates called 214 cuprates. These 214 cuprates consist of two doping systems, namely, hole-doped (p-type) and electron-doped (n-type) superconductor. The parent compound of these cuprates is R_2CuO_4 (R=La,Eu,Nd,Pr). Superconductivity appears by randomly substituting some Sr atoms for La and forming $La_{2-x}Sr_xCuO_4$. When a La^{3+} ion is replaced by Sr^{2+}, the CuO_2 plane loses one electron so that one mobile hole remains in the CuO_2 planes, leading to the formation of a hole-doped system. On the other hand, when a Eu^{3+} is replaced by Ce^{4+}, the CuO_2 plane gets an excess electron, leading to the formation of an electron-doped system of $Eu_{2-x}Ce_xCuO_4$. Lots of studies both in experiment and theories have been carried out in hole-doped system [1-5]. The schematic phase diagram of hole-doped system of $La_{2-x}Sr_xCuO_4$ has been well established. The undoped system is an antiferromagnetic insulator. On charge doping, the Neel temperature drops rapidly and the long-range antiferromagnetic order disappears for $x \geq 0.02$. A superconducting phase appears in a range of x between 0.05 and 0.27.

In electron-doped system, because of difficulties in sample preparations, some informations about physical properties are very limited. In these systems, superconducting samples are obtained not only after charge doping, but also after the heat treatment of as-grown samples in a reducing atmosphere [6]. This is the reason for the difficulty in preparing superconducting samples in electron-doped system. In the phase diagram of the electron-doped system, the antiferromagnetic regime survives in a wide range of electron concentration. It may be due to the fact of electrons doped into the Cu-3d state without giving rise to any magnetic frustration [7]. The highest T_c value in electron-doped cuprates is over 5 times lower than that in the hole-doped ones. For $Nd_{2-x}Ce_xCuO_4$ and $Pr_{2-x}Ce_xCuO_4$, the highest

T_c was observed at x = 0.15. Superconductivity for both systems was observed in very limited Ce-concentration range in between 0.13 and 0.17 atmosphere [6].

In this paper, we studied structural and magnetic properties of electron-doped superconducting cuprates of $Eu_{2-x}Ce_xCuO_{4+\alpha-\delta}$ (ECCO) with x = 0.11, 0.13, 0.15, 0.16.

2. Experimental

Polycrystalline samples of $Eu_{2-x}Ce_xCuO_{4+\alpha-\delta}$ with x = 0.11, 0.13, 0.15, 0.16 were prepared by ordinary solid-state reaction method with some details procedure are described in our previous reports [8,9]. As-grown samples of ECCO were post-annealed in flowing Ar gas of high purity at various temperatures from 900 to 920 °C for 8 – 20 h, in order to remove the excess oxygen (α) at the apical site. The reduced oxygen content (δ) was estimated from the weight change before and after annealing. From our previous experiments, the best δ values getting high T_c superconductors are obtained when the value of δ is nearly equal to the value of α [9-risdiana PRB-PLCCO]. However, we cannot control the value of δ well. For the present study, we obtained set samples with δ values from 0.027 to 0.044. Structural properties were determined by powder X-ray diffraction (XRD). DC magnetic-susceptibility (χ) measurements were carried out at low temperatures down to 2 K using a standard SQUID magnetometer in a magnetic field of 5 Oe on field cooling.

3. Results and Discussion

Figure 1 shows XRD pattern of $Eu_{2-x}Ce_xCuO_{4+\alpha-\delta}$ with x = 0.11, 0.13, 0.15, 0.16 and δ values of 0.027, 0.033, 0.044, and 0.039, respectively. All main peaks of T' tetragonal structure were observed in all samples. The values of lattice parameters of ***a*** and ***c*** are summarized in Tabel 1. The value of ***a*** increased ranging from 3.89930 Å to 3.90342 Å with increasing the Ce content (x). On the other hand, value of ***c*** slightly decreased with increasing x. This tendency indicated that substituting some Ce atoms with ion radii ~ 1.034 Å for Eu with ion radii ~ 0.947 Å, expands in-plane squares of tetragonal unit cell.

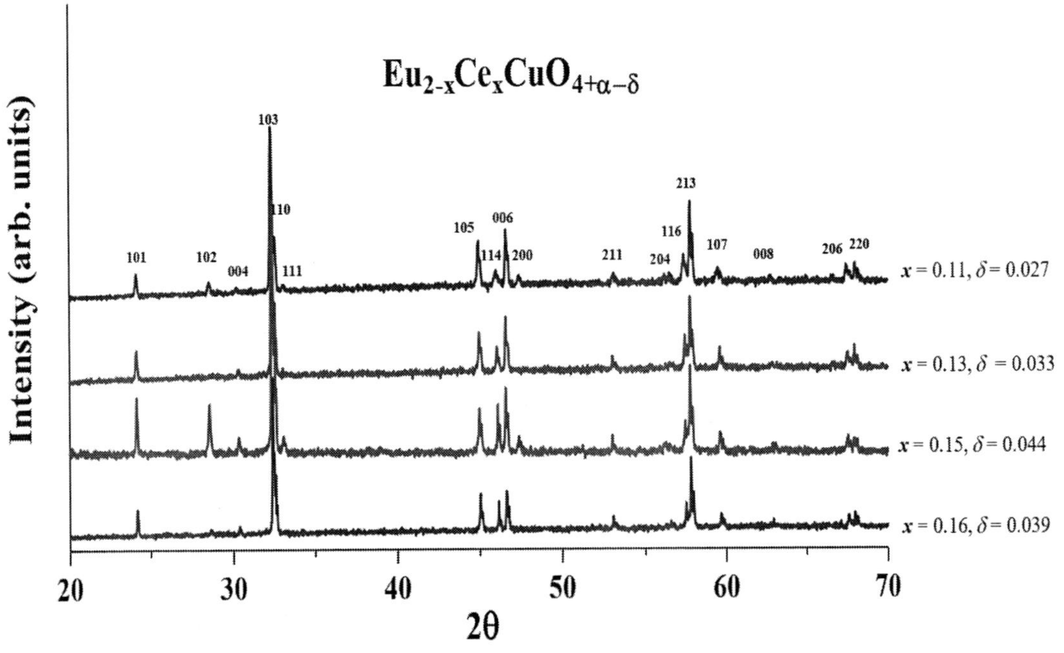

Figure 1. XRD pattern of $Eu_{2-x}Ce_xCuO_{4+\alpha-\delta}$ with x = 0.11, 0.13, 0.15, 0.16 and δ values from 0.027 to 0.044.

Table 1. Lattice parameters of $Eu_{2-x}Ce_xCuO_{4+\alpha-\delta}$.

x	a (Å)	c (Å)
0.11	3.89930	11.84332
0.13	3.90106	11.83311
0.15	3.90129	11.82682
0.16	3.90342	11.80504

Figure 2 (a) shows Temperature dependence of DC magnetic-susceptibility (χ) on field cooling at 5 Oe for ECCO with x = 0.11, 0.13, 0.15, 0.16. For x = 0.11, 0.13, paramagnetic behavior is observed with no trace of superconductivity. For x = 0.15, diamagnetic behavior with χ smaller than 0 is observed starting from about 15 K. This temperature can be defined as T_c onset for this sample. For x = 0.16, T_c onset decreases to be around 13 K as shown in Figure 2 (b). The volume fraction of the superconducting state, V_{SC}, has been estimated from the absolute value of χ at 2 K. We determined the value of χ at 2 K of x = 0.15 sample as corresponding to 100 % of V_{SC}. Then, V_{SC} for each sample is estimated as each value of χ at 2 K divided by that of x = 0.15. V_{SC} decreases markedly from 100 % to below 61 % with increasing x. V_{SC} is completely zero when the concentration of Ce is smaller than 0.13. From these results, it is found that optimum doped in ECCO samples appear at x = 0.15 and δ ~0.044. Decreasing of T_c value with increasing x above 0.15 was analogously found at overdoped condition in hole-doped system of $La_{2-x}Sr_xCuO_4$.

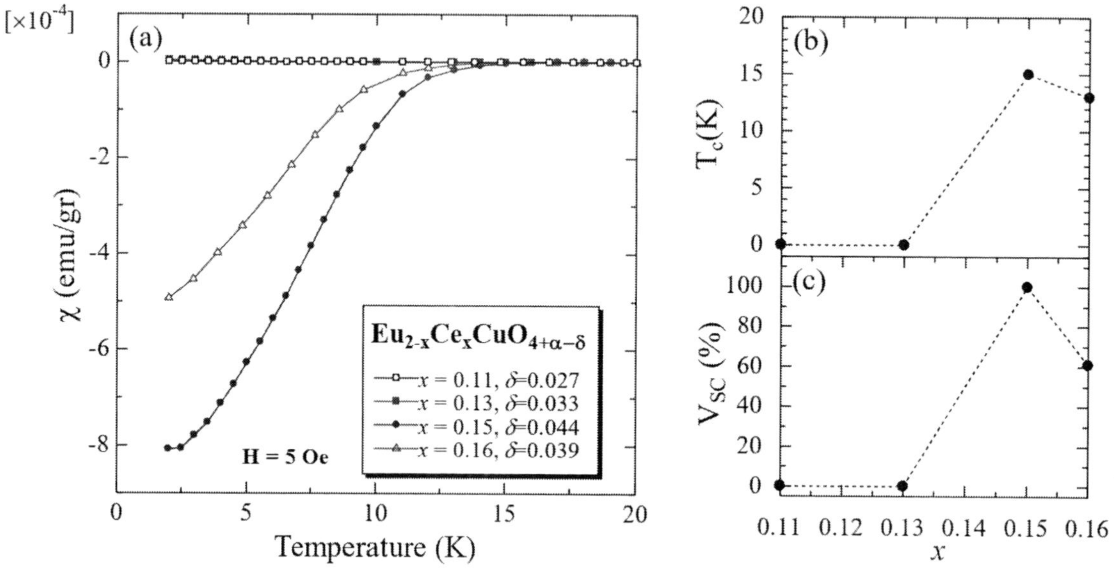

Figure 2. Temperature dependence of DC magnetic-susceptibility (χ) on field cooling at 5 Oe for $Eu_{2-x}Ce_xCuO_{4+\alpha-\delta}$ with x = 0.11, 0.13, 0.15, 0.16 (a), the Ce content (x) dependence of T_c (b) and the volume fraction of the superconducting state, V_{SC} (c) with various δ values from 0.027 to 0.044.

Comparing with other electron-doped superconducting cuprates such as $Nd_{2-x}Ce_xCuO_4$ (NCCO) and $Pr_{1-x}LaCe_xCuO_4$ (PLCCO) [6,10], the highest T_c value of ECCO is lower than that of NCCO and PLCCO samples. It might be due to the size of tetragonal unit cell of samples. From our present data, ECCO with T_c ~ 15 K has unit cell with a ~ 3.90129 Å and c

~ 11.82682 Å. It was reported that NCCO with T_c ~ 24 K has unit cell with *a* ~ 3.94 Å and *c* ~ 12.10 Å [6], while PLCCO with T_c ~ 26 K has unit cell with *a* ~ 4.0 Å and *c* ~ 12.3 Å [10].

4. Conclusion

The n-type superconducting materials of $Eu_{2-x}Ce_xCuO_{4+\alpha-\delta}$ (ECCO) with x = 0.11, 0.13, 0.15, 0.16 have been prepared by solid state reaction method, in order to study their structural and magnetic properties of ECCO. The results of X-ray diffraction characterization showed that main peaks of T' tetragonal structure were observed in all samples. DC-susceptibility measurements show superconductivity with T_c onset about 15 K for sample with x = 0.15 and decrease for sample with x = 0.16. No trace of superconductivity is observed for x = 0.11 and 0.13. The highest T_c value of ECCO is lower than that of NCCO and PLCCO samples that might be due to the smellest size of tetragonal unit cell of samples.

Acknowledgments
We would like to thank Y. Koike, T. Adachi, M. Kato, T. Noji, T. Kawamata and H. Sato for their technical support in the x-ray diffraction and magnetic-susceptibility measurements at Tohoku University, Japan. These works were supported by research grant of International Research Collaboration and Scientific Publication 2016 (No: 431/UN6.3.1/PL/2016).

References

[1] Suzuki K, Adachi T, Tanabe Y, Koike Y, Kawamata T, Risdiana, Suzuki T, Watanabe I 2010 *Phys. Rev.* B **82** 054519
[2] Risdiana, Tanabe Y, Omori K, Suzuki T, Watanabe I, Koda A, Higemoto W, Koike Y 2008 *Phys. Rev.* B **77** 054516
[3] Uzumaki T, Hashimoto K, Kamehara N 1992, *Physica* C **202** 175-187
[4] Koike Y, Adachi T, Oki N, Risdiana, Yamazaki M, Kawamata T, Noji T, Kudo K, Kobayashi N, Watanabe I, Nagamine K 2005 *Physica* C **426-431** 189
[5] Tarascon J M, Greene L H, Barboux P, McKinnon W R, Hull G W, Orlando T P, Delin K A, Foner S, McNiff Jr E J 1987 *Phys. Rev.* B **36** 8393
[6] Takagi H, Uchida S, Tokura Y 1989 *Phys. Rev.* B **62** 1197
[7] Tranquada M, Heald S M, Moodenbaugh A R, Liang F, Croft M 1989 *Nature* **337** 720
[8] Risdiana, Adachi T, Oki N, Koike Y, Suzuki T, Watanabe I 2010 *Phys. Rev.* B **82** 014506
[9] Risdiana, Safriani L, Somantri W A, Saragi T, Adachi T, Kawasaki I, Watanabe I, Koike Y 2014 *Advanced Materials Research* **896** 354
[10] Isawa K, Nagano M, Fujita M, Yamada K 2002 *Physica* C **378-381** 142

Enhancement of Photo- and Sonodegradation of Methylene Blue using Fe_3O_4/ZrO_2 by Adding Two Different Graphene Material

Yogi Kristianto[1,2], Ardiansyah Taufik[1,2], and Rosari Saleh[1,2]
[1] Departemen Fisika, Fakultas MIPA, Universitas Indonesia, Kampus UI Depok, Depok 16424, Indonesia

[2] Integrated Laboratory of Energy and Environment, Fakultas MIPA, Universitas Indonesia, Kampus UI Depok, Depok 16424, Indonesia

E-mail: rosari.saleh@ui.ac.id

Abstract. In this study, magnetite Fe_3O_4/ZrO_2/nanographene platelets (NGP) and Fe_3O_4/ZrO/graphene composonsites were prepared successfully by depositing the Fe_3O_4/ZrO_2 composites onto the surface of NGP and graphene using co-precipitation method as an effort to improve photo- and sonocatalytic performance. The structure and magnetic properties of the as-prepared samples were characterized by XRD and TGA. The catalytic activity of Fe_3O_4/ZrO_2/NGP and Fe_3O_4/ZrO_2/graphene composite towards methylene blue (MB) from the aqueous solution were investigated under UV light and ultrasound irradiation, separately. It was found that the incorporation of NGP and graphene in Fe_3O_4/ZrO_2 could greatly improve photo- and sonocatalytic performance. The incorporation of graphene with the higher surface area than NGP exhibits the best catalytic performance. In addition, scavengers of reactive species on the degradation activity were investigated.

1. Introduction

Due to widely used in industries, dye, such as methylene blue (MB), has become one of the most dangerous organic pollutant, because they are hardly degraded and has been suspected as mutagen and carcinogen [1,2]. Therefore, scientists proposed some techniques to degrade dyes from aqueous solutions, such as coagulation, electrochemical treatment and chemical oxidation. Compared to the other techniques, advanced oxidation process (AOPs), such as photo- and sonocatalysis, using metal oxide semiconductor (MOS) have shown excellent removal of various contaminant dye [3].

Among the various of MOS, zirconium dioxide (ZrO_2) has been widely studied as a catalyst to support wastewater treatment due to their several reasons, such as high stability, non-toxicity, low cost, and environmentally-friendly [4,5]. Unfortunately, single metal oxide ZrO_2 has limitations such as high recombination rate, indicating low catalytic performance.

Recently, graphene as a two-dimensional material provides several unique properties such as high specific surface area, good electron acceptor and excellent charge carrier mobility [6-8]. Based on these features, depositing MOS into the graphene materials can effectively enhance separation of photogenerated electron-hole pairs, prevent their recombination rate, and increase the number of active sites, which is mean could improve catalytic performance. Another way to improve catalytic performance by depositing MOS with magnetite material (Fe_3O_4) because it could facilitate separation of the electron-hole pairs effectively [9]

In these regards, we tried to depositing Fe_3O_4/ZrO_2 composites onto the surface of NGP and graphene via co-precipitation method. The photo- and sono- catalytic of both prepared samples were tested for the degradation of MB as an organic pollutant model. The aim of this work is to evaluate the influence of adding as-prepared composites on the graphene materials for degrade MB.

2. Experimental

Materials: All the chemical reagents were of an analytical grade and were used without further purification. Iron(II) sulfate heptahydrate ($FeSO_4.7H_2O$, 99%), zirconium(IV) chloride ($ZrCl_4$) and sodium hydroxide (NaOH) were purchased from Merck. NGP (S.A. 30 m^2/g) and graphene (S.A. 400 m^2/g) were purchased from Angstron Materials.

Preparation of samples: Step 1. The Fe_3O_4, ZrO_2 nanoparticles were synthesized using the same method as described previously [10,11]. Step 2. The Fe_3O_4/ZrO_2 composites were synthesized using ultrasonic-assisted method. Briefly, ZrO_2 nanoparticles were mixed with Fe_3O_4 in aqueous and ethanol mixture. Then, the mixed solution was given ultrasonic for 2 h, followed by centrifugation to obtain the product on its precipitated. The obtained product was dried in vacuum at 80°C for 12 h to obtain Fe_3O_4/ZrO_2 composites. Step 3. To coupled NGP and graphene with Fe_3O_4/ZrO_2 composites same method as described in our previous publication was used [12].

Characterization: X-ray Diffraction (XRD) performed on a Rigaku Miniflex 600 diffractometer with Cu K-α radiation (λ =1,5406 Å) was used to obtain the crystal structure of samples and the thermal stability of samples was investigated using Thermal gravimetric analysis (TGA).

Photo- and sono- catalytic activity: MB was used as a model organic pollutant. The photo- and sono- catalytic activity of samples was investigated via degradation of an MB aqueous solution with a concentration of 20 mg/L at room temperature with UV light irradiation and ultrasound (US) irradiation, respectively. The catalyst was dissolved into MB solution. The solution was continuously stirred in the dark for 30 min to ensure that an equilibrium of adsorption/desorption. Subsequently, the solution was exposed to UV light and US irradiation, respectively. A small amount of the water sample was collected through centrifugation every 15 min to measure the MB concentration with UV-vis spectrophotometer. Two lamp of 40 W UV-C was used as UV light source and commercially ultrasonic bath which operated at a fixed frequency and power of 40 kHz and 150 W was used as an ultrasound source.

Scavenger experiment: In order to determine the species involved in photo- and sono- catalytic activities, different radical scavengers were added to the solutions of MB. Di-ammonium oxalate, sodium sulfate, and tertbutyl alcohol (TBA) were used as scavengers for holes, electrons, and hydroxyl radical, respectively. The same measurements as described above were used.

3. Result and Discussion

Figure 1 (a) shows the XRD patterns of all prepared samples. As seen, XRD patterns of NGP and graphene are quite different. NGP has a sharp peak at 26.49° indicating graphitic-like structure while graphene has a broad peak around 23° indicating amorphous structures [13]. The observed diffraction peaks at 2θ = 30.18°, 35.03°, 50.36°, and 59.94° correspond to the [111], [200], [220], and [311] plane, respectively, which can be readily assigned to the tetragonal phase of ZrO_2. Meanwhile, peak at 2θ = 30.14°, 35.49°, 43.28°, 53,76°, 57.20° and 62.83° which shows each plane [220], [311], [400], [422], [511] and [440] of cubic spinel phase. The XRD patterns of FZ/NGP and FZ/graphene composites show the presence of cubic spinel phase of Fe_3O_4 and tetragonal phase of ZrO_2, followed by the addition of graphitic-like structure from NGP for FZ/NGP. However, no typical diffraction peaks of graphene are observed within the FZ/graphene composites, probably due to the low content (10wt%) and relatively weak diffraction intensity of graphene. Furthermore, there were no characteristic peaks from other crystalline impurities that were detected by XRD, suggesting that the as-prepared samples have good purity.

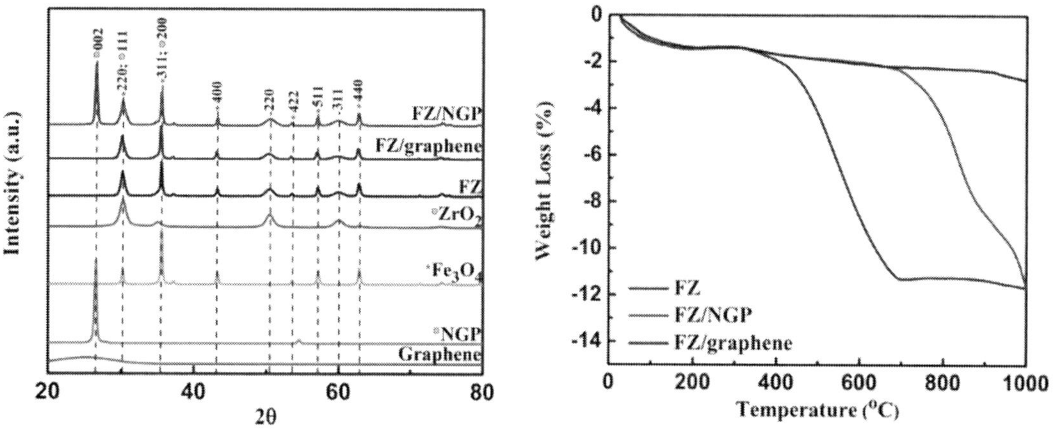

(a) (b)

Figure 1. (a) XRD patterns of all prepared samples. (b) TGA plot of FZ, FZ/graphene and FZ/NGP.

Figure 1 (b) shows the typical TGA plot of FZ, FZ/graphene and FZ/NGP composite. As shown in fig. 1 (b), TGA plot of FZ exhibit one decomposition region while FZ/graphene and FZ/NGP exhibits two decomposition region to significantly weight loss upon the increase in temperature. The first one occurring in the temperature around 100°C, which can be attributed to the loss of absorbed water on the surface of FZ, FZ/graphene and FZ/NGP samples. The next one in the temperature around 430°C and 670°C for FZ/graphene and FZ/NGP, respectively. It indicated the combustion process of graphene and NGP in the FZ/graphene and FZ/NGP composites, respectively, which proves the presence of graphene and NGP.

The photo- and sono- catalytic activity of the prepared samples is shown in fig. 2 (a). The obtained results revealed that the presence of graphene materials in FZ composites could improve greatly catalytic performance. The incorporation of graphene with higher specific surface area (S.A.) than NGP area exhibit the best catalytic performance in both photo- and sono- catalytic. The reason is that

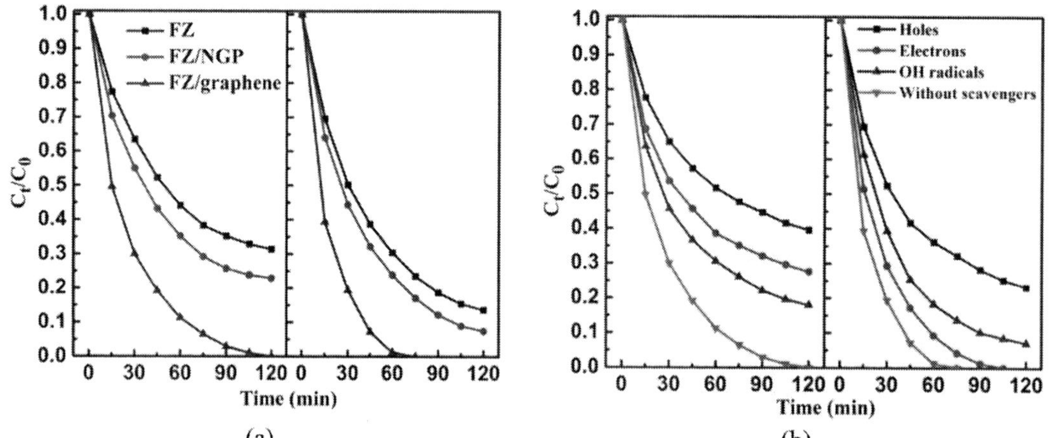

(a) (b)

Figure 2. (a) Degradation of methylene blue on photocatalytic activities (left) and sonocatalytic activities (right). (b) The effect of main active species on FZ/graphene: photo- (left) and sonocatalytic (right).

increasing surface area of samples may also relate to increasing the opportunity for contact between the catalyst and light, thereby increase the rate of oxidation process [14]. Moreover, supportive properties of NGP and graphene could trap electron to hinder recombination of electron-hole pairs, which also increase degradation rate [15]. In the sonocatalytic activity, ultrasound irradiation could disaggregate catalyst particles, thus increasing surface area [16]. That's one reason why sonocatalytic could bring better degradation efficiency than photocatalytic.

The effect of scavengers was performed to study the main active species in the photo- and sono-catalytic process. The control experiment was carried out by adding several trapping agents, Di-ammonium oxalate, sodium sulfate, and TBA as holes, electrons and hydroxyl radicals scavenger, respectively. From fig. 2 (b) we can see that, in photocatalytic the main active species follows: hole > electron > hydroxyl radicals, while in sonocatalytic the main active species follows: hole > hydroxyl radicals > electron. The different trend of the main active species between photo- and sono- catalytic might be because the different mechanism on them. In sonocatalytic, there is a process called hot spot which can pyrolysis water molecules into hydroxyl radicals, thus increasing the population of hydroxyl radicals as strong oxidation agent [17].

4. Conclusions

In this study, FZ/NGP and FZ/graphene composites were successfully synthesized by depositing the FZ onto the surface of NGP and graphene which confirmed by XRD and TGA measurements. The results revealed that incorporation of NGP and graphene could enhance the photo- and sono- catalytic performance. The sample with an incorporation of graphene exhibits better catalytic performance compared to NGP in both photo- and sono- catalytic activity.

5. References

[1] Ambashta R D and Sillanpaa M 2010 *Journal of Hazardous Materials* **180** 38-49.
[2] Lee S M and Ong S T 2014 *APCBEE Proc.* **9** 165-169.
[3] Guimaraes J R, Maniero M G and De Araujo R N 2012 *Journal of Environmental Management* **110** 33-39.
[4] Guo G Y and Chen Y L 2005 *Journal of Solid State Chemistry* **178** 1675-1682.
[5] Tahmasebpour, Babaluo A A and Razavi Aghjeh M K 2008 *Journal of European Ceramic Society* **28**, 773-778.
[6] Zhang W, Dong F, Xiong T and Zhang Q 2014 *Ceramics International* **40** 9003-9008.
[7] Du A J, Sanvito S, Li Z, Wang D W, Jiao Y and Liao T 2012 *Journal of American Chemical Society* **134** 4393-4397.
[8] Han, Wang P and Dong S 2012 *Nanoscale* **4** 5814-5825.
[9] Wu W, Jiang C Z and Roy V A L 2015 *Nanoscale* **7** 38-58.
[10] Taufik A, Susanto I K and Saleh R 2015 *Materials Science Forum* **827** 37-42.
[11] Kristianto Y, Taufik A and Saleh R 2016 *Journal of Physics: Conference Series* **776** 012040
[12] Kristianto Y, Taufik A and Saleh R 2017 *AIP: Conference Proceedings* **1788** 10.1063/1.4968379
[13] Dong L, Chen W, Deng N and Zheng C 2016 *Chemical Engineering Journal* **S1385-8947** 31101–9 (accepted manuscript).
[14] Hu J, Ma J, Wang L and Huang H 2014 *Journal of alloys and compounds* **583** 539-545.
[15] Wu H, Fan J, Liu E, Hu X, Ma Y, Fan X, Li Y and Tang C 2015 *Journal of alloys and compounds* **623** 298-303.
[16] Silva A M T, Nouli E, Carmo-Apolinario A C, Xekoukoulotakis N.P and Mantzavinos D 2007 *Catalysis Today* **124** 232-239.
[17] Jyothi K P,Yesodharan S and Yesodharan E P 2014 *Ultrasonic chemistry* **1** 1787-1796.

IOP Conf. Series: Materials Science and Engineering **196** (2017) 012014 doi:10.1088/1757-899X/196/1/012014

Fabrication of Semi-quasi Solid DSSC using Spiro Material as Hole Transport Material

L Safriani, W P Primawati, C Mulyana, T Susilawati, and A Aprilia

Physics Department, Faculty of Mathematic and Natural Sciences, Universitas Padjadjaran, Jl. Raya Bandung Sumedang Km. 21, Jatinangor, Sumedang 45363, Indonesia

E-mail : lusi.safriani@phys.unpad.ac.id

Abstract. Dye Sensitized Solar Cells (DSSC) has been emerging a promising development in recent years. DSSC is a low-cost solar cell belonging to the third generation of solar cells. However, the conversion efficiency of DSSC is still far behind compared to silicon based solar cells. To produce long stability of DSSC, the used of solid state electrolyte is recommended instead of liquid electrolyte, though solid state DSSC also has problem relating to a lack of pore-filling hole transport material into mesoporous TiO_2. In this work an attempt to improve performance of DSSC has been done by adding hole transport material into mesoporous TiO_2 layer and optimizing fabrication method. In the first part of the work, we used low T_g material spiro-TAD and spiro-TPD as hole transport material with mosalyte and hybrid polymer as gel electrolyte to obtain a semi-quasi solid DSSC. In the second part, we modified fabrication method by annealing process before spin-coated spiro material into dye-coated TiO_2 substrate. Current–voltage measurement of semi-quasi solid DSSC was performed using halogen lamp. We found that the used of spiro-TPD as hole transport give the best power conversion efficiency $\eta = 2.03\%$ of semi-quasi solid DSSC.

1. Introduction

Recently, the development of dye-sensitized solar cells (DSSC) attracts many researchers due to its achievement of DSSC performance that has reached efficiency as high as 18.7% [1] with some modification on its electrode. There are three main components of DSSC i.e. a dye-sensitized photoanode, an electrolyte and counter electrode [2]. Fine tuning any of these components have resulted in increasing of DSSC's efficiencies [3]. A dye-sensitized photoanode contain of a semiconductor material which always been TiO_2 (anatase) mesoporous to hold dye. Dye itself is a material act as photo-sensitizer which is adsorbed on the semiconductor surface. On this layer, an excitation of charge transfer dye results in injection of an electron into conduction band of semiconductor layer.

Normally an electrolyte is a liquid electrolyte based on the I_3^-/I^- redox couple [4]. However, the use of liquid electrolyte shows some disadvantages such as sealing problems and long-term durability, solvent evaporation and leakage of volatile organic solvent [4,5]. One alternatives to liquid electrolyte is quasi-solid electrolyte, ionic liquids (e.g. 1-propargyl-3- methylimidazolium iodide, bis(imidazolium) iodides and 1-ethyl-1-methylpyrrolidinium) and polymer gel (e.g.

poly(ethyleneoxide), poly(vinylidinefluoride) and polyvinyl acetate) containing redox couples are commonly used as quasi-solid electrolytes to overcome leakage problems [6,7,8].

One of factors that could inhibit the efficiency of DSSC is related to charge transport especially hole transport in porous TiO_2. The incomplete pore-filling of hole transport materials into porous TiO_2, leads to poor dye-regeneration and reduces device efficiency [9]. A good candidate material for a transparent hole transport material is small molecules so it will be easy to penetrate into porous TiO_2 layer. Organic material based on spiro material has been used as hole transport material in DSSC [10] due to its charge transport properties [11]. In this research, combination of gel electrolyte and spiro material was used to form semi-quasi solid DSSC to overcome the solvent evaporation problem and increase device performance.

2. Experimental methods

Fabrication of DSSC based on combination of gel electrolyte and spiro material was divided into three steps. Firstly, photoanode was prepared using screen printing on FTO substrate. One layer of Ti Nanooxide T/SP was deposited FTO substrate and subsequently followed by Ti Nanooxide MC/SP and Ti Nanooxide T/SP. It was then gradually annealed until 500°C for 30 minutes on hotplate to produce mesoporous TiO_2 layer. Secondly, photoanode was immersed on Ruthenium dye (535-bisTBA; purchased from Solaronix) and chenodeoxolydacid (1:10). After overnight, FTO/TiO_2/Ru-dye was rinsed by acetonitrile several times to remove residue of dye molecules. Hole transport material (HTM) based on spiro material dissolve in chlorobenzene was spin-coated on FTO/TiO_2/Ru-dye. Two kind of spiro material were used in this experiment, 2,2′,7,7′-tetra- kis(diphenylamino)-9,9′-spirobifluorene (spiro-TAD) and 2,2′,7,7′-tetra(m-tolyl-phenylamino)-9,9′-spirobi- fluorene (spiro-TPD) which purchased from Lumtec. Drilled platinum coated FTO substrate (Pt/FTO) was used as counter electrode and stack together with working electrode FTO/TiO_2/Ru-dye separated by hot melt surylin film having thickness 25 μm. Finally, gel electrolyte which mixed between co-polymer TMSPMA and ionic liquid electrolyte mosalyte (purchased from Solaronix) was injected to the hole of Pt/FTO side and then sealed with transparent tape to avoid leaking. To investigated performance of device, current voltage (J-V) characteristics was measured using source lamp with power input 36.5 mW/cm^2 integrated with Yokogawa GS 200 DC voltage-current source and Yokogawa digital multimeter 7555.

3. Results and Discussions

Absorption spectrum of TiO_2 layer combined with Ru-dye and spiro material is presented in figure 1. It is clearly shown that TiO_2/Ru-dye/spiro material layer has higher absorption compared to TiO_2/Ru-dye layer which due to the addition of both spiro-TPD and spiro-TAD. An absorption at around 500 nm (visible region) is correlated to Ru-dye absorption.

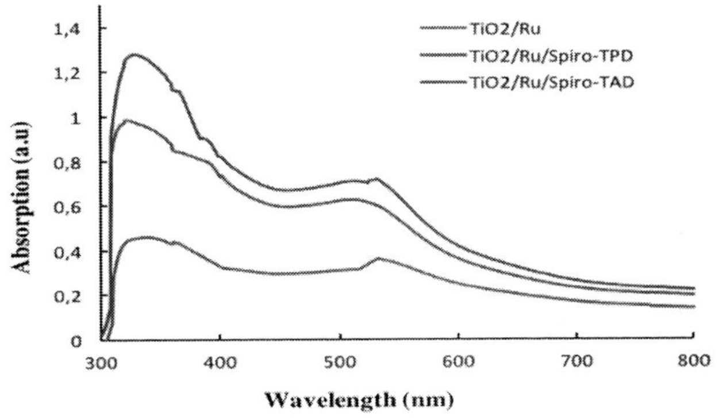

Figure 1. Absorption spectrum of TiO_2/Ru-dye and TiO_2/Ru-dye/spiro material

Current density-voltage characteristics of DSSC with structure FTO/TiO_2/Ru-dye/spiro material (HTM)/gel electrolyte/Pt-FTO was carried out under irradiated of halogen lamp with 0.3 mW/cm2 as power input and shown di figure 2. All parameters such as short circuit current density (J_{sc}), open circuit voltage (V_{oc}), Fill Factor (FF) and efficiency (η) which observed and calculated were presented in table 1.

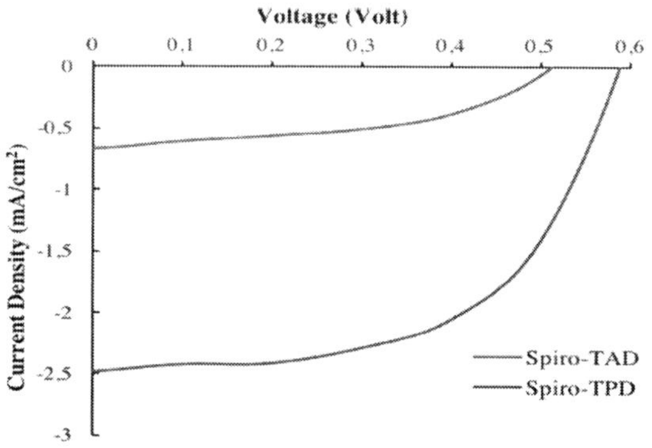

Figure 2. J-V curve of DSSC with different spiro material as HTM

Table 1. Parameters of DSSC with different spiro material as hole transport material (HTM)

Spiro material (HTM)	J_{SC} (mA/cm^2)	V_{OC} (Volt)	FF (%)	η (%)
Spiro-TAD	0.67	0.51	45	0.24
Spiro-TPD	2.48	0.59	56	1.86

Energy conversion efficiency η obtained for semi-quasi solid DSSC with spiro-TAD and spiro-TPD as HTM are 0.24% and 1.86% respectively. A lower efficiency of DSSC with spiro-TAD as HTM correlated with its energy level which lower than that of Ru-dye. This could inhibit charge transfer from Ru-dye to highest occupied molecular orbital (HOMO) of spiro-TAD. On the other hand, spiro-TAD has a higher HOMO level than RU-dye to facilitate charge transfer and increase the efficiency.

In order to enhance performance of DSSC with structure FTO/TiO_2/Ru-dye/spiro material (HTM)/ gel electrolyte/Pt-FTO, we did two kind modification in fabrication method. In previous result, mesoporous TiO_2 layer was resulted from coating of three times Ti Nanooxide, but this cause a thick mesoporous TiO_2 layer which could prevent a penetration of Ru-dye. To optimize the penetration process of Ru-dye into mesoporous TiO_2 layer, Ti Nanooxide was deposited on FTO substrate twice and followed by annealing process as described in previous section. We also did annealing process before spin coating of spiro material into dye-coated TiO_2 substrate to remove air bubbles which trapped inside mesoporous TiO_2 layer. Current density-voltage characteristics and parameters device produced from optimized fabrication methods is presented in figure 3 and table 2 respectively.

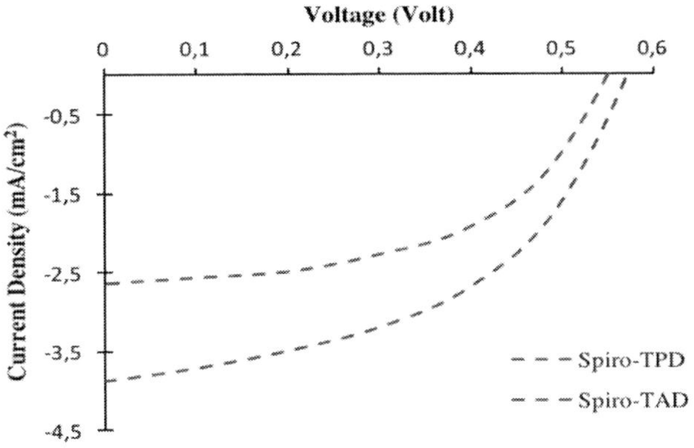

Figure 3. J-V curve of DSSC with different spiro material as HTM produced from optimized fabrication methods

Table 2. Parameters of DSSC with different spiro material as hole transport material (HTM) produced from optimized fabrication methods

Spiro material (HTM)	J_{SC} (mA/cm^2)	V_{OC} (Volt)	FF (%)	η (%)
Spiro-TAD	2.64	0.55	33	0.97
Spiro-TPD	3.88	0.57	48	2.03

Based on current density-voltage characteristic, it is clearly shown that after optimized fabrication method device, efficiency η is increased both for device with spiro-TAD and spiro-TPD as HTM. Twice coating process resulted thinner mesoporous TiO_2 layer so that Ru-dye penetrated in all area of mesoporous TiO_2 layer and prevented empty space of Ru-dye. Annealing process before spin-coated of spiro material also give contribution to broaden contact area between Ru-dye and spiro material to ease charge transfer from Ru-dye to spiro material. Therefore, the combination between the used of HTM based on spiro material and modification process that have been done could be one of the ways to increase power conversion efficiency of semi-quasi solid DSSC.

4. Conclusions

Semi-quasi solid DSSC have been fabricated using gel electrolyte and spiro material as hole transport material. In order to enhance device performance, the optimized fabrication methods were done by making thinner mesoporous TiO_2 layer to ease Ru-dye penetration into TiO_2 layer, and by anneal the TiO2/Ru-dye layer before spin-coated of spiro material to broaden contact area between Ru-dye and spiro material. The highest power conversion efficiency, $\eta = 2.03\%$, is resulted from device which fabricated with optimized methods and used spiro-TPD as hole transport material.

5. Acknowledgement

Authors gratefully acknowledges the financial support from Universitas Padjadjaran under research PUPT (Program Unggulan Perguruan Tinggi) with contract No. 431/UN6.3.1/PL/2016.

References

[1] Jiazang C, Bo L, Jianfeng Z, Jianghong Z, Zhenping Z. 2012 *J. Phys. Chem. C* **116** 14848
[2] Ali N, Hussain A, Ahmed R, Wang M.K, Zhao C, Ul Haq B and Fu Y Q 2016 *Renewable and Sustainable Energy Reviews* **59** 726
[3] Obotowo I N, Obot I B and Ekpe U J 2016 *Journal of Molecular Structure* **1122** 80

[4] Bastianini M, Vivani R, Nocchetti M, Costenaro D, Bisio C, Oswald F, Meyer T B and Marchese L 2014 *Solar Energy* **107** 692
[5] Ye M, Wen X, Wang M, Iocozzia J, Zhang N, Lin C and Lin Z 2015 *Materials Today* **18** 155
[6] Bai S, Bu C, Tai Q, Liang L, Liu Y, You S, Yu Z, Guo S and Zhao X 2013 *ACS Appl. Mater. Interfaces* **5** 3356
[7] Achari M B, Elumalai V, Vlachopoulos N, Safdari M, Gao J, Gardnera J M and Kloo L 2013 *Phys. Chem. Chem. Phys.* **15** 17419
[8] Wang L, Zhang H, Wang C and Ma T 2013 *ACS Sustain. Chem. Eng.* **1** 205
[9] Yuan W, Zhao H, Baker and G L 2014 *Organic Electronics* **15** 3362
[10] Saragi T P I, Spehr T, Siebert A, Fuhrmann-Lieker T and Salbeck J 2007 *Chem. Rev.* **107** 1011
[11] Saragi T P I, Fuhrmann-Lieker T and Salbeck J 2006 *Adv. Funct. Mater.* **16** 966

The Microstructure Analysis of Barium M- Hexaferrite Particles Coated by Pani Conducting Material with In Situ Polymerization Process

M. Zainuri and L Amalia

Departement of Physics, Faculty of Mathematics and Natural Sciences, Sepuluh Nopember Institute of Technology Jl.Arief Rahman Hakim, Surabaya 60111 Indonesia.

Email: zainuri@physics.its.ac.id

Abstract, Barium M-Hexaferrite (BaM) was synthesized by coprecipitation method and doped with Zn. Polyaniline (PANI) was synthesized by chemically and doped DBSA. The composite of PANI/BaM was synthesized by in situ polymerization method. The phase identification of the sample was performed by XRD, FTIR and SEM. Based on XRD data, the phase composition of BaM and hematite are 85.52 % and 14.48%. The characteristic peaks of PANI occur at 3435, 1637, 1473, 1298, 1127, 1009, and 799 cm^{-1}. The characteristic metal oxide stretching peaks of BaM occurs at 575 and 437 cm^{-1}. There is no phase changing in PANI/BaM composite. Based on SEM photography, the shape of BaM is hexagonal. The particle size of BaM powder ranges from 400-700 nm. The qualitative interfacial bonding between PANI and BaM particles are conducted very well and the both materials have good wettability.

1. Introduction

Conducting polymer has attracted much attention in recent years because of its properties. The material has high conductivity that can be arranged its dopant concentration level [1]. Among the conducting polymers, polyaniline (PANI) is polymer that mostly using in experiments because it is easy to synthesis and good environmental stability [2]. It has good stability in air, high electrical conductivity, and can be used to the protection of electromagnetic wave [3]. The other applications of PANI are in electrochemical displays, sensors, catalysis, redox capacitors, secondary battery [4]. The present of dopant, such as Dodecyl Benzene Sulphonic Acid (DBSA), it can increase the electrical conductivity of PANI [5]. DBSA plays role as dopant and surfactant [6].

Barium M-Hexaferrite (BaM) is hexagonal molecular structure which use to permanent magnet, magnetic data recording, and microwave absorber. It is because high anisotropic magnetocrystalline, high magnetization, good chemical stability, and good corrosion protection, where is magnetic and electrical properties can be arranged [7]. The magnetization of Barium Hexagonal ferrite can be further improved by ion Zn^{2+} as dopant [8-9]. BaM can be synthesized by iron sand, natural material. The compounds of natural iron oxide (Fe_3O_4 and Fe_2O_3) were found in a great amount in iron sand formed. To increase the technology value of iron sand in Indonesia, it is changed to microwave absorber [10]. The composite of PANI/BaM is a good combination to Radar Absorber Material (RAM) [4].

According [11] the elecromagnetic functionalized polyaniline/$BaFe_{12}O_{19}$ composite was successfully synthesized by in situ polymerization process. In this work, the composite of PANI/BaM was synthesized by in situ polymerization, where BaM was obtained by coprecipitation and PANI as conducting material was prepared by chemical process. The samples were characterized by various experimental measurement such as X-Ray diffraction, FTIR spectroscopy, and SEM. The microstructure of PANI/BaM composite was investigated.

2. Experimental

2.1 Materials

Fe_3O_4 is obtained from iron sand (Tulungagung, Indonesia), Barium Carbonate ($BaCO_3$) (ACS reagent grade, MP Biomedicals, LLC), Zn powder (pro analys, Merck), Aceton (pro analys, Merck), Hydrochloric Acid (HCl) (pro analys, Merck), Ammonium Persulfate (pro analys, Merck), Dodecyl Benzene Sulfonic Acid (DBSA) (70% wt solution in 2-propanol, pro analys, Aldrich).

2.2 Preparation of Barium M-Hexaferrite

BaM was synthesized by coprecipitation method. Fe_3O_4, Zn and $BaCO_3$ were dissolved in HCl. That solution was blended by magnetic stirrer for 45 minutes and added NaOH to make precipitate. The precipitate was washed with distilled water and dried at 100°C. Then BaM precursor was calcinated at 1000°C for 5 hours.

2.3 Preparation of Polyaniline

PANI was synthesized by chemical process. Aniline and DBSA were dissolved in distilled water for 30 minutes then slowly added APS solution. The mixture was allowed to react under continous stirring at 0-10°C for 8 hours. PANI was washed and filtered with aceton and distilled water till the filtrate becomes colorless. Then it was dried in room temperature to get PANI powder.

2.4 Preparation of Composite

PANI/BaM composite was synthesized by in situ polymerization. Aniline and DBSA were dissolved in distilled water for 30 minutes then mixed BaM and stirred till homogeous. APS solution was slowly added dropwise to the above mixture and allowed to react under continous stirring at 0-10°C for 8 hours. The composite was washed and filtered with aceton and distilled water till the filtrate becomes colorless. Then it was dried in room temperature to get composite powder.

2.5 Characterization

The phase identification of the sample was performed with XRD using Philips X'Pert MPD with Cu Kα (λ=1.54056 Å). The XRD data was analyzed by using Rietica, a Rietveld-based refinement software to determine phase composition. The diffraction was recorded from 15° to 75°. FTIR measurement of the sample was recorded on PerkinElmer spectroscopy at 400-4000 cm^{-1} to analyze the bonding formed. The morphology of sample was performed by SEM EVO MA 10 at an accelerating voltage of 20kV.

IOP Conf. Series: Materials Science and Engineering **196** (2017) 012015 doi:10.1088/1757-899X/196/1/012015

3. Result and Discussion

3.1 In situ polymerization
PANI/BaM composite was synthesized by in situ polymerization of aniline in the presence of BaM particles. During the polymerization process, color changing occurs [12]. The color changing shows that chemical reaction between initiator and aniline monomer is happened. The polymerization in the acidic environment causes the surface of the BaM is positively charged. Therefore, the adsorption of amount anions may happen and compensate the positive charges on BaM surface. The while, the specific adsorption of these anions on BaM may also occur [11]. It happens when aniline monomers are changed to cationic anilinium ions in acidic environment. The interaction between anions adsorbed on BaM surface and the ions of cationic anilinium appear. The interaction is electrostatic forces between BaM particles and PANI chains [13]. The aniline monomers are electrostatic attracted to BaM surface then polymerized. It shows that polyaniline is formed and precipitated in mixture which change into dark green. The color of emeraldine salt is dark green. Emeraldine salt is a form of conducting polyaniline.

3.2 Phase analysis
XRD patterns of the prepared PANI, BaM and PANI/BaM composite have shown in Figure 1. It is confirmed that polyaniline which contains the characteristics of PANI at 2θ = 20.5° and 25.5°. It is similar to other experiment result [11] that PANi peaks at 2θ = 20.4° and 25.4°. PANi is amorphous polymer material. For BaM, it is confirmed that $BaFe_{12}O_{19}$ (identified by collection number 00-039-1433) and α-Fe_2O_3 (identified by collection number 00-033-0664). $BaFe_{11.7}Z_{0.3}O_{19}$ is identified as the major phase with a small amount of residual α-Fe_2O_3. The phase composition of $BaFe_{11.7}Z_{0.3}O_{19}$ and α-Fe_2O_3 is determined by Rietica software with Rietveld refinement. The weight percentage of phase composition are 85.52 % $BaFe_{11.7}Z_{0.3}O_{19}$ and 14.48% α-Fe_2O_3. The value of the GoF (goodness-of-fit) is 1.68. It shows the refinement is accepted. In BaM case, the input of parameters for refinement is hexagonal structure, a=b= 5.8932 Å and c= 23.1877 Å. The PANI/BaM composite presents the characteristics BaM peaks and amorphous PANI. The main peaks of BaM still stand in PANI/BaM composite and the amorphous phase of PANI also take places. The peaks of BaM in composite are weaker than the peaks in BaM. It is showed on the decreasing of intensity of BaM peaks.

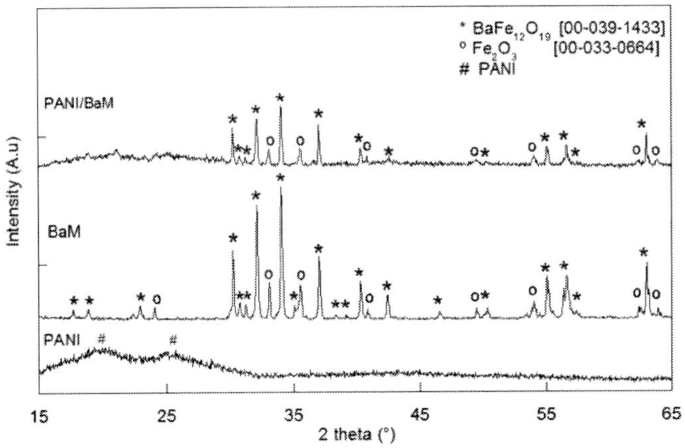

Figure 1. XRD patterns of PANI, BaM, and PANI/BaM composite

3.3 Spectroscopy FTIR

The FTIR spectroscopy of PANI, BaM dan PANI/BaM composite have shown in Figure 2. The characteristic peaks of PANI occur at 3435, 1637, 1473, 1298, 1127, 1009, and 799 cm^{-1}. The peaks at 3435, 1637, 1473, and 1298 cm^{-1} are attributed to the characteristic N-H stretching, C=N stretching, C=C stretching of quinoid and C-N stretching. The peak at 1127 cm^{-1} is assigned N=Q=N vibration mode. The peaks at 1009 and 799 cm^{-1} show S=O stretching of DBSA and benzenoid rings. The characteristic metal oxide stretching peaks of BaM occur at 575 and 437 cm^{-1}. The metal oxide stretching peaks in [14] occur at 600 and 460 cm^{-1}. PANI/BaM composite presents the characteristics PANI peaks and metal oxide peak of BaM. The main peaks of PANI present in PANI/BaM composite and the metal oxide peak of BaM may also be seen. This FTIR measurement completely supports XRD data that composite material does not occur the phase changing. The interactions between conducting material and BaM particles happen because of secondary bonding such as electrostatic bonding and the polarity of both materials.

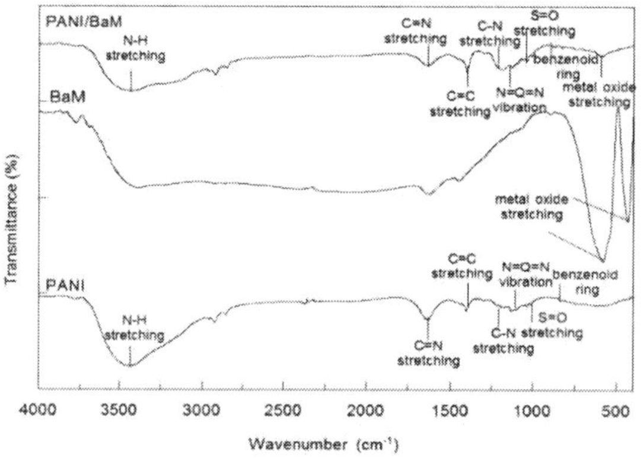

Figure 2. FTIR spectroscopy of PANI, BaM, and PANI/BaM composite

3.4 SEM Morphology

SEM images of BaM and PANI/BaM composite have shown in Figure 3. It can be seen that the shape of BaM is hexagonal. The particle size of BaM powder ranges from 400-700 nm. In Figure 3a the particles are polydisperse, and some of them form multi particle aggregates because of the magneto-dipole interparticle interactions. The distribution of BaM particles is dispersed homogenously. After coating with PANI (Figure 3b), BaM particles stamped in conducting material, PANI. The BaM particles are attracted to PANI each other so making composite formed by in situ polymerization method. From the composite microstructure in Figure 3b, the qualitative interfacial bonding between PANI and BaM particles are conducted very well. Based on the microstructure of PANI/BaM composite, the both materials have good wettability because PANI coated BaM particles.

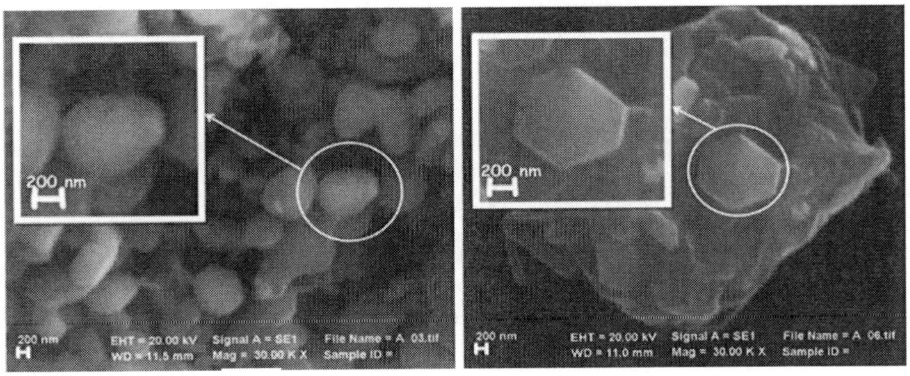

a b

Figure 3. The SEM images of (a) BaM particles, and (b) PANI/BaM composite

4. Conclusion

PANI/BaM composite was successfully synthesized by in situ polymerization of aniline in the presence of BaM particles. From the XRD data and FTIR measurement, it can be conclude that there is no the phase changing in composite PANI/BaM. It is secondary bonding between PANI and BaM particles. Based on SEM photograph, the shape of BaM is hexagonal. The particle size of BaM powder ranges from 400-700 nm.. The qualitative interfacial bonding between PANI and BaM particles are conducted very well. The PANI and BAM particles have good wettability

References

[1] Wan M 2008 Tsinghua University Press, Beijing
[2] Diarmid M A G 2001*Reviews of Modern Physics* **73** 701
[3] Li X, Wang G, Li X 2005 *Surf. Coatings Techno.* **197** 56
[4] Ting T H, Wu K H 2010 *Journal of Magnetism and Magnetic Materials* **322** 2160
[5] Tsotra P, Friedrich K 2004 *Synthetic Metals* **143** 237
[6] Han Y G, Kusunose T, and Sekino T 2009 *Synthetic Metals* **159** 123
[7] Paul K B M 2007 Physica B **388**
[8] Ramli I 2012 Thesis Department of Physics, Institute of Technology Sepuluh Nopember Surabaya
[9] Rosyidah K C 2013 Final Project. Department of Physics, Institute of Technology Sepuluh Nopember Surabaya
[10] Mashuri 2012 Dissertation. Department of Physics, Institute of Technology Sepuluh Nopember Surabaya
[11] Jiang J, Lun-Hong A, Da-Bin Q, Liu H and Li L C 2009 *Synthetic Metals* **159** 695–699
[12] Pradhana I G B 2009 Final Project. Department of Physics, Institute of Technology Sepuluh Nopember Surabaya
[13] Li Y, Zhang H, Liu Y, Wen Q, and Li J 2008 *Nanotechnology* **19** 16
[14] Yang C C, Gung Y J, Hung W C, Ting T H, and Wu K H 2010 *Composites Science and Technology* **70** 466

IOP Conf. Series: Materials Science and Engineering **196** (2017) 012016 doi:10.1088/1757-899X/196/1/012016

Synthesis and Characterization Materials M-Barium Hexaferrite Doping Ions Co-Mn Nano Particle

Susilawati, A Doyan, Sahlam

[1]Magister Science, Post Graduate of Mataram University, Mataram, Indonesia

E-mail: susilawatihambali@yahoo.co.id

Abstract. This research has been success in the synthesis of M-Barium hexaferrite (BaM) doping Co-Mn ions using coprecipitation method are expected to be applied as a base material in the coating RADAR. M-Barium hexaferrite (BaM) are $BaFe_{12-2x}Co_xMn_xO_{19}$ synthesized with various concentrations (x = 0.0, 0.1, 0.2, 0.3) and the calcinations temperature (T = 400°C, 600°C, 800°C).The materials characterization using a X-Ray Diffraction (XRD), Transmission Electron Microscope (TEM), Inductance Capacitance and resistance (LCR) meter, and Vibrating Sample Magnetometer (VSM) Instruments. The measurement results using XRD shows the material has a hexagonal crystalline structure. The measurement results using a TEM show a sample of nano crystal materials with grain diameters up to 40 nm and spacing of the crystal lattice. The measurement results using a LCR-meter shows the electric conductivity of 1.15 x 10^{-6} S/cm to BaM without doping, 3.75 x 10^{-6} S/cm to 0.1 doping concentration calcination temperature of 400 °C, and 1,23 x 10^{-5} S/cm to 0.3 doping concentration calcination temperature of 800 °C, thus including semiconductor materials. The magnetic properties of materials using a VSM test results show the value of coercivity of 0.1 T; remanence value of 0.06 emu/ g; and the saturation value of 0.42 emu/g. The results above show BaM Co-Mn metal doping potential as anti-radar material.

1. Introduction

The mechanism is not detecting an object by radar to be based on two aspects that is the places designed with the geometry of angular radar absorbing structure (RAS), so that the reflection of electromagnetic waves cannot be recaptured by the receiver, and the airframe is covered by absorbent material radar wave, radar absorbing materials (RAM) Barium in the form of M-Barium hexaferrite $BaFe_{12}O_{19}$ [1]. A material can absorb radar when it interacts with the electric field of the wave, so many developed materials that are magnetic. The materials that have the ability to absorb radar include M-barium hexaferrite $BaFe_{12}O_{19}$ [2], because of the nature of magnetocrystalline anisotropy is high, it is necessary to engineering material properties such as conductivity, coercivity, saturation magnetization, magnetization remanence, and the Curie temperature to fit the needs by giving Fe ion doping with other elements such as Co, Zn, Ni, Mn, Ti, and others up to a certain amount. Giving doping is not expected to cause changes in the structure of the base material $BaFe_{12}0_{19}$ considering the dimensions of the doped material and almost the same doping [3]. Research is growing rapidly with a variety of methods to obtain barium M-Barium hexaferrite (BaM) in nano size, because the particle size largely determines the magnetic characteristics of the BaM. The method used is the sol-gel, crystallization gas, ball milling, coprecipitation, aerosol, precipitation of hydrothermal and mechanical integration. The coprecipitation is one method of synthesis of inorganic compounds based on the deposition of more than one substance together when passing through the point of saturation [4]. In this study synthesized material M-barium hexaferrite ($BaFe_{12}O_{19}$), Co-Mn ions doping using coprecipitation method with variation of dopant concentration and calcination temperature until $BaFe_{12-x}Co_xMn_xO_{19}$ generated. The characterization is performed to determine the structure, electrical properties and magnetic properties of materials $BaFe_{12-x}Co_xMn_xO_{19}$ formed.

2. Materials and Methods

$BaFe_{12-2x}Co_xMn_xO_{19}$ synthesized using coprecipitation method. The basic material used is a powdered barium carbonate ($BaCO_3$) with Mr = 197.34 g/mol, Iron (III) chloride hexahydrate ($FeCl_3.6H_2O$) with Mr =270.38 g/mol, Cobalt (II) chloride hexahydrate ($CoCl_2.6H_2O$) with Mr = 201.95 g/mol, and Manganese (II) chloride tetrahydrate ($MnCl_2.4H_2O$) with Mr = 197.96 g/mol is a product of Merck KGaA mixed with appropriate stoichiometric ratio. $BaFe_{12-2x}Co_xMn_xO_{19}$ synthesized with various concentrations (x = 0.0, 0.1, 0.2, 0.3) and the calcinations temperature (T = 400°C, 600°C, 800°C). Synthesis $BaFe_{12-2x}Co_xMn_xO_{19}$ with x = 0.3 is done in several stages, first dissolving 22.2856 g $FeCl_3.6H_2O$ in 1.552 HCl using a magnetic stirrer for 30 minutes without heating, dissolving 1.7732 g $BaCO_3$ added 0.6556 ml of HCl 37% with Mr = 12.063 using a hot plate and a magnetic stirrer at 70°C for 2 hours dissolve powder $CoCl_2.6H_2O$ 0.6414 and 0.5335 g MnCl2.4H2O in 0, 1619 ml H_2O. The second stage is the third mixed solution while stirring with a magnetic stirrer for 45 minutes. The third is dripped slowly 11.7815 ml NH_4OH with Mr = 35.06 last in the solution so that a precipitate is formed is then filtered and washed with distilled water until the normal pH (pH = 7).

The last stage in the sludge drying oven with a temperature of 80° C for 4 hours and then pulverized to form a fine powder. Samples were then characterized using XRD, TEM, LCR meters, and VSM instruments. The crystal structure of the products was characterized by X-ray Diffraction (XRD) brand Rigaku SmartLab operated at a voltage of 40 kV and current of 30 mA with a target Cu (λ 1.541862 Å) were recorded in the region of 2θ. Transmission Electron Microscope (TEM) brand Tecnai G2 SN-D6892 M operated at a voltage of 120 kV - 2000 kV was used for characterizing the size of nano particles. The electrical properties of the material was measured by using LCR-meter Yokogawa brand Hioki 3522-50. The magnetic properties of the materials measurement by vibrating sample magnetometers (VSM) OXFORD type VSM1.2H.

3. Results and Discussion

Synthesis of barium M-hexaferrite Co and Mn metal doping using dopant variations coprecipitation with x = 0.0 ; 0.1 ; 0.2 ; 0.3 and calcination temperature variations of 400, 600, and 800 °C. Samples with doping concentrations of 0.1 and 0.2 before calcination appear darker than the sample with a doping concentration of 0.3, this can happen because of possible formation of salt $MnCl_3$ (black) is not stable, reaction of $MnCl_2$ with hydrochloric acid (HCl) at low temperature, and will break down again at temperatures above 40 °C [5], until after calcination 400 °C reddish brown color of the sample becomes brighter than the sample with a doping concentration of 0.3. Calcination gave very big influence on the synthesis process materials such as in the color of sample.

Figure 1 shows the diffraction pattern of the sample with the dopant concentration x = 3 and the calcinating temperature of 800 °C. Based on the results of XRD, the phase that is formed is a Manganese-Iron-Oxide hexagonal structure with identity $a = b =$ 5,04 Å and $c =$ 13,76 Å where $\alpha = \beta = 90^o$ and $\gamma = 120^o$; $V = 302{,}7$ A^3, in theory [6], barium M-hexaferrite own identity and where and refers to data crystallographic information file (CIF). These results indicate that the addition of the dopant concentration can regenerate identity and does not change the basic structure of barium M-hexaferrite, in accordance with the statement [7], that the addition of dopants with a small concentration not to alter the basic structure of BaM. Crystal size becomes finer with increasing calcination temperature. Components of the solid ingredients and spread evenly with statistical parameters: $\chi^2 = 1.0016$; S = 1.0008 ; $R_{wp} = 9.49$; and the position of the particle layer (d_{hkl}) constituent materials exhibit repetition with a regular pattern, indicating the material is formed consisting of layers of the crystal, where the size and the distance between the crystal layer is viewed using TEM test equipment.

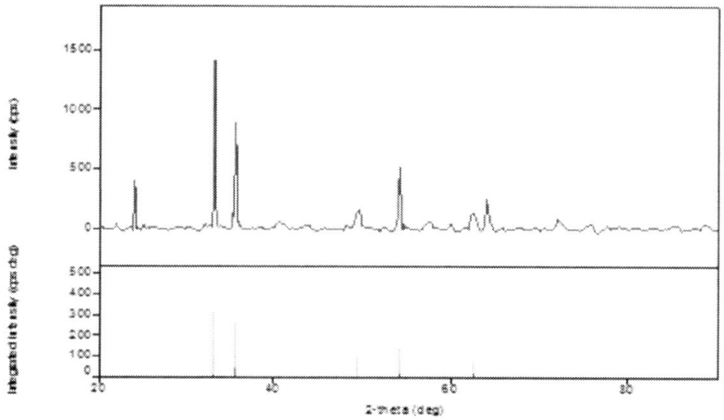

Figure 1. XRD diffraction patterns for material $BaFe_{11,4}Co_{0,3}Mn_{0,3}O_{19}$ and calcination temperature of 800 °C

Tests using a TEM are performed to determine the size and composition of the particles formed in the sample. The structure of the building blocks of the sample can be seen in more detail for TEM can measure up to on the order of nanoparticles. The observation using a TEM as shown in Figure 2(a,b).

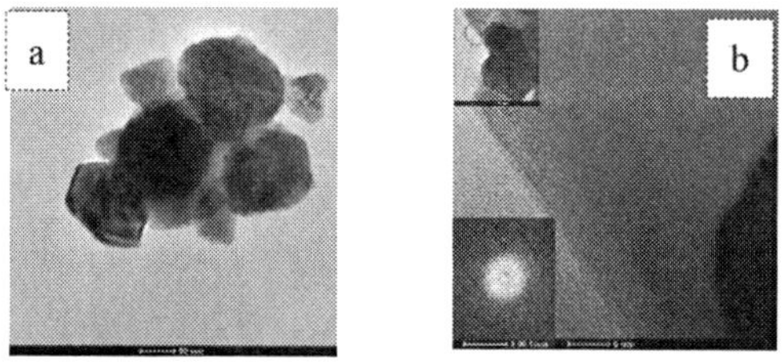

Figure 2(a,b). The observation using TEM , (a = picture crystal sizes on the order of 50 nm, b = image layer and the crystal lattice spacing on the order of 5 nm).

Figure 2a is a result of observation using a TEM with a magnification of up to 50 nm in order to determine the crystal grain size, by comparing measuring scale with the scale of the observable obtained crystal grain size vary, the diameter of the smallest and greatest diameter. In Figure 2b, the upper left corner is the result of observation of crystal layers on the order of 50 nm part circled and enlarged up to the order of 5 nm, revealing layers of crystal neat, ensuring constituent particles sampled BaM is the layers of the crystal, reinforcing the results of XRD before. Crystals formed pattern tends to elongate color to the gray- transparent, consistent with research conducted of the order of 10 nm observation [8]. The bottom left corner is the observation by the order of 5 nm to determine the distance between the crystal lattice, by comparing the measuring scale with the scale of the observable, the distance between the crystals obtained. Based on the above, the results and analysis of samples BaM in this study is a nanoparticle materials. Another study using sol-gel method to get the particle size of BaM by 38, 34, and 33 nm, 50-100 nm [9], a method of self-propagating obtain a particle size of 40 nm [10].

Table 1 shows the measurement results using a LCR meter to measure the electrical properties of materials.

Table 1. The results of testing the electrical properties of the metal doping samples BaM Co-Mn

Sample	Co-Mn 0,2 400 °C	Co-Mn 0,3 800°C	Nodoping
Diameter (cm)	1.17	1.17	1.17
Thick (cm)	0.3	0.304	0.394
$Z_{average}$ (Ω)	7.44E+04	2.30E+04	3.19E+05
A (cm^2)	1.075132	1.075132	1.075132
σ (S/cm)	3.75E-06	1.23E-05	1.15E-06

The average impedance of materials $3{,}19\,x\,10^5\ \Omega$ (without doping); $7{,}44\,x\,10^4\ \Omega$ ($x = 0{,}2;\ T = 400\,^oC$), and $2{,}30\,x\,10^4\ \Omega$ ($x = 0{,}3;\ T = 800\,^oC$), so that the electrical conductivity of the material if it is calculated using the formula $\sigma = \frac{1}{R} = \frac{1}{Z}$, obtained as σ values in Table 1. of $1{,}15\,x\,10^{-6}$ S/cm (without doping) ; $3{,}75\,x\,10^{-6}$ S/cm ($x = 0{,}2;\ T = 400\,^oC$); and $1{,}23\,x\,10^{-5}$ S/cm ($x = 0{,}3; T = 800\,^oC$). Based on these data, the dopant concentration and calcination temperature affect the conductivity properties of materials and samples that form a semiconductor material for conductivity value in the range of up to S/cm [11]. Another study by the sol-gel method to get conductivity value of 10.03 x 10^{-4} S/cm to BaM doping Mg [12], and without doping of 1 x 10^{-5} to 6 x 10^{-5} S/cm [13]. This research resulted in a better semiconductor material on doping concentration and calcination temperature of 800 °C ($BaFe_{11,4}Co_{0,3}Mn_{0,3}O_{19}$) with a conductivity of $1{,}23\,x\,10^{-5}$ S/cm.

Figure 3 shows the results of measurement of magnetic properties of materials $BaFe_{11,4}Co_{0,3}Mn_{0,3}O_{19}$.

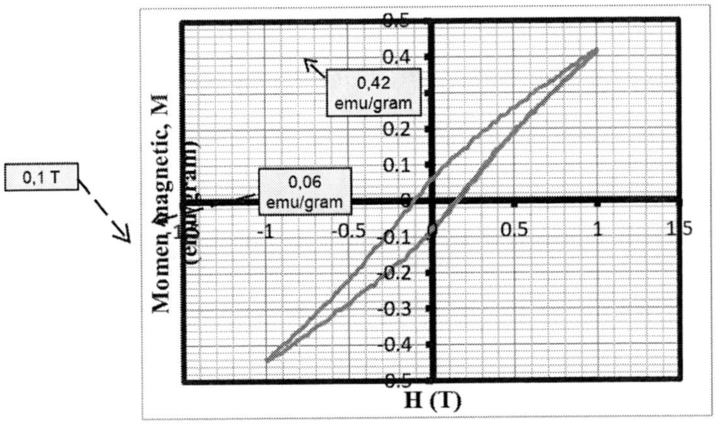

Figure 3. Hysteresis curve samples $BaFe_{11,4}Co_{0,3}Mn_{0,3}O_{19}$

The value of coercivity (Hc) of 0.1 T, saturation (Ms) of 0.42 emu/g and remanence (Mr) of 0.06 emu/g. In theory, Barium hexaferrite has a coercivity of 6700 Oe, Curie temperature of 450 °C, the saturation magnetization of 78 emu/g. Previous research, produce BaM coercivity value of 163 mT [14], 0.32 T, with a solid reaction method acquire 98.22 kA/m to 27.4 Mn dopant to dopant Co [15], BaM doping Mn-Ti obtain 557 Oe [16]. With the sol-gel method, getting 30.304 kA / m for doping Co-Ni and 42.472 kA/m for doping Co-Sr [20], getting 1181.07 Oe for doping La and Co, and get coercivity coefficient of 0.1104 for doping Co-Zn [17]. This research obtains remanence of 0.06 emu /g, the other study, by the reaction of a solid gain of 7.58 emu/g for doping Mn-Ti; through the sol-gel method acquire 29.22 and 26.12 emu/g for doping La and Co. Magnetic saturation in this study get 0.42 emu/g, more research with the sol-gel method to get Ms = 43.67 emu/g for doping Co-Zn, Ms = 32.07 emu/g for doping Mn-Cu-Sr, with a solid reaction method to obtain Ms = 0.35 and 0.38 Mn doping to doping Co [18]. Based on the resulting magnetic values, the BaM material doping metals Co and Mn for x = 0.3 and the calcination temperature of 800 °C with coprecipitation methods have successfully lowered the hard magnetic properties of BaM into a soft magnetic, so the potential to be applied as a base material absorbing microwaves.

4. Conclusion

Synthesis of BaM with Co and Mn metal doping with coprecipitation method has resulted $BaFe_{12-2x}Co_xMn_xO_{19}$ at the doping concentration x = 0.3 and the calcination temperature of 800 °C to form $BaFe_{11,4}Co_{0,3}Mn_{0,3}O_{19}$. BaM produced is a hexagonal crystal structure of nanoparticles of materials with the identity of $a = b = 5{,}04$ Å; $c = 13{,}76$ Å and $\alpha = \beta = 90^0$; $\gamma = 120^0$. Electrical properties $BaFe_{11,4}Co_{0,3}Mn_{0,3}O_{19}$ in the range of a semiconductor with a conductivity value $1{,}23\ x\ 10^{-5}$ S/cm. The material of BaM to be soft magnetic field values coercivity (Hc) of 0.1 T; magnetic remanence (Mr) of 0.06 emu/g; and the value of magnetic saturation (Ms) of 0.42 emu/g. BaM produced can be applied as a base for anti- radar material.

Acknowledgement

Thanks to the Government of Indonesia through Kemenristekdikti to grant Prime Research Universities (PUPT) in 2013 to 2016 on behalf of Susilawati, Ph.D. Thank you to all staff and laboratory analytical chemistry Mataram University and laboratory physics Department LIPI Indonesia and to all those who have helped to realize this article.

References

[1] Af'idah N, Indahnia E and Darminto 2011 Paper presented at the National Graduate Seminar XI – ITS Surabaya
[2] Feng Y B, Qiu T, Shen C Y 2007 *Journal of Magnetism and Magnetic Materials* ***318***
[3] Ahmeda Okashab, Kershi 2008 *Journal of Magnetism and Magnetic Materials* ***320***
[4] Susilawati, Doyan A, Taufik M, Wahyudi 2011 Research Report featured colleges, University of Mataram
[5] chem-is-try.org., (2008). (Online): http://www.chem-is-try.org/tabel periodik/kobal/, Accessed October 6, (2014)
[6] G B Teh, N Saravanan, D A Jefferson 2007 *Materials of Chemistry and Phy*sics **105** 253
[7] Kaur T and Srivastava A K 2013 *International Journal of Research in Mechanical Engineering & Technology* **3** 171-173
[8] Swamy P M P and Basavaraja S 2011 Barium ferrite nanoparticles prepared by self-propagating low-temperature combustion method and its characterization. *Indian Academy of Sciences* **34** 1319–1323.
[9] Said L M 2013 Personality Analysis of Electrical Conductivity in Some Kind of Material with Potential Methods Fall . *Journal Teknosains* **7** 66-77
[10] Chauhan C C, Jotania R B, Jotania K R 2012 *International Journal of Advanced Engineering Research and Studies.* I. 25-27
[11] Parween N 2012 Thesis. National Institute of Technology, Rourkela
[12] Hasanah E, Budi A S, Adi W A and Suguhartono I 2012 Paper presented at the National Seminar of Applied Physics III University of Airlangga, Surabaya , September 15.
[13] Sulistyo, Marhaendrajaya I and Priyono 2012 *Periodic Journal of Physics* **2** 63
[14] Priyono and Ahyani M 2010 MIPA UI .
[15] Rosler S, Wartewig P and Langbein H 2003 *Cryst. Res. Technol* **38** 927-934.
[16] Aylin G *et.al.* 2012 *Material and Techology* **46** 305
[17] Saidah I N and Zainuri M 2012 *Journal of Science and Arts of ITS* **1** 41
[18] Wisnu A A and Manaf A 2012 *Journal of Basic and Aplied Scientific Research* University of Indonesia. Jakarta.

Effect of Cerium Doped on the Poly(3-(Trimethoxysilyl)propyl methacrylate) Characteristic as Corrosion Protection Material of Carbon Steel

D N Rochmah, N Syakir, T Susilawati, S Suryaningsih, and Fitrilawati*

Department of Physics, University of Padjadjaran, Jalan Raya Bandung-Sumedang KM 21, Sumedang 45363, West Java, Indonesia

E-mail: fitrilawati@phys.unpad.ac.id

Abstract. The hybrid polymer precursor was synthesized from monomer of 3-(trimethoxysilyl) propyl methacrylate (TMSPMA) using sol-gel method and doped with inhibitor of Cerium Nitrate Hexahydrate with a concentration of 0.2%. The synthesized material was coated on a carbon steel surface by solution casting technique and followed by a photopolymerisation process. Corrosion tests were performed by using Electrochemical Impedance Spectroscopy (EIS) in 3.5% NaCl at the critical temperature of 75°C. Result of EIS data and their fitting analysis using an equivalent circuit model shows that a coating of poly(TMSPMA)-Cerium on the surface of carbon steel form a layer of protection and caused increasing of impedance value significantly. The impedance is higher compared to the carbon steel that coated with poly(TMSPMA) only.

1. Introduction

Coating technique has been widely used for protection corrosion of metal surfaces. The coating materials for that purpose need to have good mechanical strength and hydrophobicity [1]. The application of hybrid polymer as coating material to protect steel and other metals from corrosion attack have been reported [2,3]. Besides, inhibitor is also widely used for corrosion protection [2, 4]. Lanthanides have been identified as good corrosion inhibitor since they have low toxicity and can be used as cathodic inhibitor [4]. Amongs the lanthanide, Cerium is widely used as an inhibitor for corrosion protection in industrial application [5].

We synthesized coating material using hybrid polymer precursor of 3–(trimethoxysylil) propyl methacrylate (TMSPMA) and doped with Cerium inhibitor for corrosion protection of carbon steel, and measured the material performance using Electrochemical Impedance Spectroscopy (EIS).

2. Experiments

The precursor of hybrid polymer were polymerized from monomer TMSPMA using sol-gel method as described previously [6, 7]. The inhibitor of Cerium Nitrate Hexahydrate was doped into the polymer precursor with a concentration of 0.2% in order to obtain poly(TMSPMA)-Ce. Before coated with the poly(TMSPMA)-Ce precursor, the carbon steel (API 5L X65) surfaces were polished and cleaned with acetone and dried on air. The synthesized poly(TMSPMA)-Ce precursors then dissolved with ethanol and added with 1% Irgacure-819 as an initiator. The polymer precursors were coated on the surface of carbon steel, then prebaked and photopolymerized using UV light. Beside carbon steel/poly(TMSPMA)-

Ce, we also prepared carbon steel coated with the poly(TMSPMA) named as carbon steel/ poly(TMSPMA) and carbon steel covered with Cerium then followed by poly(TMSPMA) that named as carbon steel/Ce/poly(TMSPMA).

The corrosion tests were done using EIS measurements that performed in the electrochemical cell and equipped with temperature controlled apparatus. The electrochmeical cell consists of three electrodes that are carbon steel as a working electrode, platinum as an auxiliary electrode, and calomel as reference electrode. The EIS measurements were conducted in 3.5% NaCl electrolyte at a critical temperature of carbon steel corrosion of 75°C [6]. Electrochemical cells that has been assembled then placed on a magnetic stirrer and connected to a water bath and a computer pre-installed software Gamry Reference 3000. The EIS characterization was done by varying the frequency range of 20 mHz to 100 kHz. The potentiostat voltage was varied on a range of -10 mV to +10 mV with a sweep rate of 0.2 mV/s, and AC voltage of 10 mV.

3. Results and Discussion

The synthesized precursors of hybrid polymer poly(TMSPMA) has yield of 89% and apparent as transparent gel-like. Addition of Cerium Nitrate Hexahydrate with a concentration of 0.2% does not change the color of the hybrid polymer precursor.

EIS characterization performed on carbon steel that has been coated with poly(TMSPMA) and poly(TMSPMA) which was blended with Cerium inhibitor are meant to measure an improvement passivation character of the poly(TMSPMA) that combined with Cerium. This characterization generate Nyquist curve that shows the value of the impedance of the electrodes and Bode curve showing the relationship between frequency and phase. The EIS data in form of Nyquist curves, Bode curves, and Bode phase curves are shown in figure 1 and figure 2, respectively.

The Nyquist of Carbon steel/poly(TMSPMA) layer has a semi-circle shape, that behaves like passive protection since the poly(TMSPMA) is an isolator. The Nyquist curve of Carbon steel/Ce/poly(TMSPMA) has a same pattern as Carbon steel/poly(TMSPMA)-Ce, that has two semi-circular shape and shows capacitive loop. Both of them show characteristic of Cerium that performed inhibition-like protection as indicating by downward looping curve that has correlation with inductive characteristic. Capacitive loop shows the charge transfer in the corrosion process, whereas inductive loop suggesting possible adsorption-desorption process of the inhibitor molecule. An adsorption will occur when an inhibitor molecule is absorbed and bound at carbon steel surface, and desorption will occur when inhibitor apart from the surface [8].

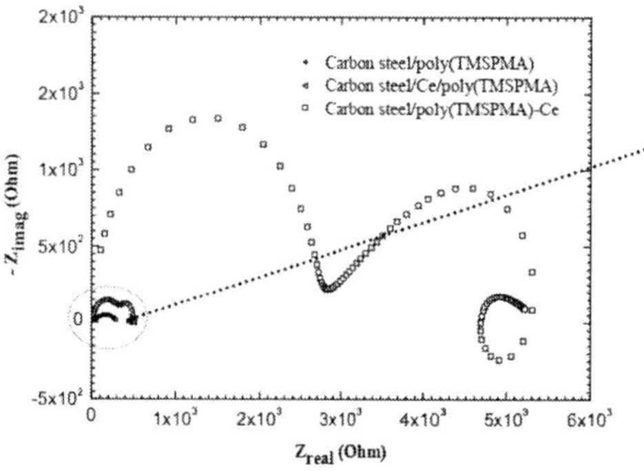

(a)

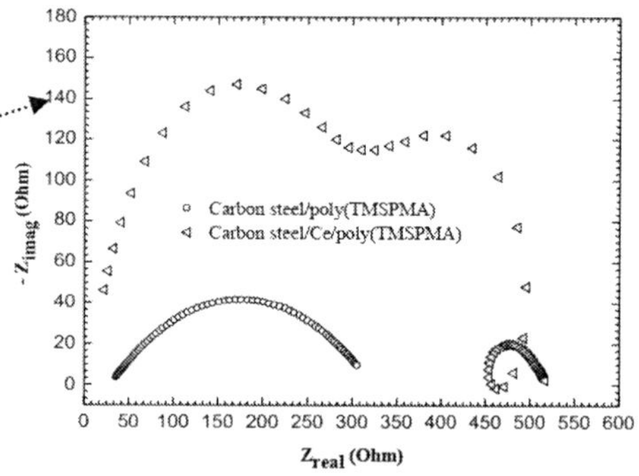

(b)

Figure 1. EIS Nyquist curves of the protected carbon steel (a), and its zooming (b)

Figure 2(a) show modulus impedances and phase of all of coated carbon steels in the measurement range of EIS scanning frequency. The total impedances of carbon steel have significantly increased with addition of Cerium on the hybrid polymer. High modulus impedance value indicates higher resistance value of the coated carbon steel. The Bode curves that have higher impedance values at a low frequency indicate a passivity of the carbon steel surface due to coating materials. The phase curve in figure 2(b) shows difference phase response at higher frequency.

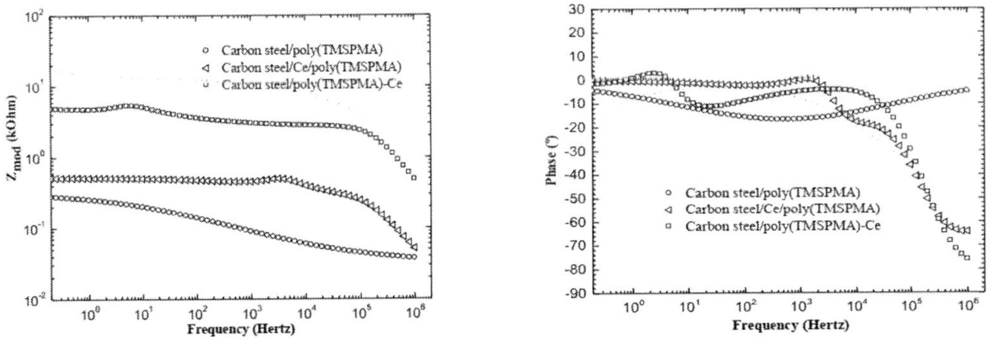

(a) (b)

Figure 2. EIS Bode-plot of the protected carbon steel, (a) Modulus impedance, and (b) Phase

Fitting the data is the first step to be able to analyze characteristics of corrosion protection resulting from EIS measurements. We employ data fitting using equivalent circuit modeling that utilizes Gamry Echem Analyst software which is integrated with the equipment Reference Gamry-3000. The initial fitting was done using a series of double resonance, which makes the RC circuit made of two series. However double resonance circuit seem not able to follow the data points. The model of equivalent circuit should represent protection characteristics as passive protection or as inhibit protection. In an inhibit protection, an inductive characteristic should be added in the equivalent circuit model. A number of semicircle curve in the Nyquist curve indicates the number of resonating equivalent circuit. The Nyquist curve of Carbon steel/poly(TMSPMA) is not an ideal semicircular because the curve experienced 'depression' with center below the Z_{real} axis. It shows the behavior frequency dispersion associated with the irregularities of the surface of the electrode [8]. Therefore to do a fitting of this data element, additional Constant Phase Element (CPE) is required to facilitate the fitting process. Elements of the capacitor to the second loop is replaced with CPE to improve performance of fitting. In addition, the inductor elements are added as well because there are data points form an arch. Based on these consideration, we propose equivalent circuit as shown in figure 3.

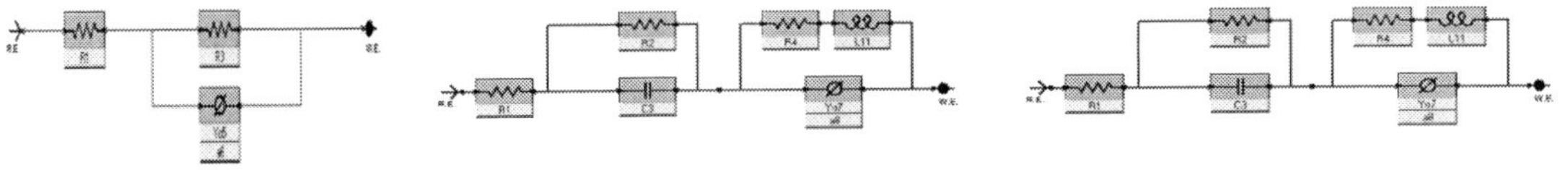

(a) (b) (c)

Figure 3. Equivalent circuit model for data fitting, (a) Carbon steel/poly(TMSPMA), (b) Carbon steel/Ce/poly(TMSPMA), and (c) Carbon steel/poly(TMSPMA)-Ce.

Fitting on the Nyquist curve will automatically provide fitting of Bode curve and Bode phase curve. Bode curves can be used to analyze changes of impedance magnitude and phase of the system in response to changes of the input signal frequency. The fitting results of each component value in the equivalent circuit model are listed in table 1. The resistance value of the Carbon steel/poly(TMSPMA)-

Ce is higher than the Carbon steel/poly(TMSPMA) and the Carbon steel/Ce/poly(TMSPMA). This is due to a protective layer of blend poly(TMSPMA)-Ce. The capacitance value of the carbon steel/poly(TMSPMA)-Ce is lower than the Carbon steel/Ce/poly(TMSPMA). It implies that the smaller the capacitance value, the better protection. Overall, the characterization of electrodes with different coatings have been able to reveal the characteristics of poly(TMSPMA) and Cerium inhibitors as a coating material to minimize corrosion of the carbon steel. Only required optimization of the carbon coating on steel to increase resistance to corrosion. Also deposition of Cerium on the surface of carbon steel forms a passive layer since Cerium has more negative potential than steel, resulting in a cathodic protection on the carbon steel. Adding of 0.2% corrosion inhibitor of Cerium improved film's corrosion performance and barrier properties compared to samples without inhibitor.

Table 1. The parameter values of EIS data fitting results

Carbon steel /poly(TMSPMA)	**Carbon steel /Ce/poly(TMSPMA)**	**Carbon steel /poly(TMSPMA)-Ce**
R1(ohm) = 290.5	R1(ohm) = 1.243×10^{-6}	R1(ohm) = 5.326×10^{-8}
R3(ohm) = 31.29	R2(ohm) = 230.0	R2(ohm) = 2585.0
Yo5(S*s^a)= 6.259×10^{-4}	R4(ohm) = 285.0	R4(ohm) = 2805.0
a6 = 0.352	C3(farad) = 1.631×10^{-9}	C3(farad) = 3.461×10^{-10}
	Yo7(S*s^a)= 3.590×10^{-5}	Yo7(S*s^a)= 9.570×10^{-5}
	a8 = 0.501	a8 = 0.350
	L11(henry)= 4.470×10^{-2}	L11(henry)= 150.0

4. Conclusions

We have successfully prepared corrosion protection materials of hybrid polymer precursors from monomer of TMSPMA doped with Cerium inhibitor. Addition of Cerium inhibitor into the hybrid polymer precursors does not change the physical condition of precursors that is still clear gel and viscous. The result of corrosion test use EIS experiment showed that coating material protection corrosion poly(TMSPMA)-Cerium on the carbon steel increased impedance value significantly.

Acknowledgment

This project was partly supported by Hibah Pengembangan Kapasitas Riset Dosen Universitas Padjadjaran No. 1094/ UN6.3.1/PL/2016

References

[1] Chou T P, Chandrasekaran C, Limmer S J, Seraji S, Wu Y, Forbess M J, Nguyen C and Cao G Z 2001 *Journal of Non-Crystalline Solids* **290** 153
[2] Suzana A F, Ferreira E A, Benedetti A V, Carvalho H W P, Santilli C V, Pulcinelli S H 2016 *Surface & Coatings Technology* **299** 71
[3] Coana T, Barrosob G S, Machadoa R A F, de Souzac F S, Spinellic A and Motzb G 2015 *Progress in Organic Coatings* **89** 220–230
[4] Zand R Z, Verbeken K and Adriaens A 2013 *Int. J. Electrochem. Sci.* **8** 548
[5] Ashraf P M and Edwin L, 2004 *India Journal of Chemical Technology* **11** 672
[6] Melati H A, Suratno W, Hidayat R, Syakir N, Fitrilawati 2011 Proceeding of the 2nd International Seminar on Chemistry (UNPAD) 32
[7] Pitriana P, Syakir N, Hidayat R and Fitrilawati, 2011 *Jurnal Material dan Energi Indonesia* **1** 167
[8] Sandler S R, Karo W, Bonesteel J-A, Pearce E M 1998 *Polymer Synthesis and Characterization: a Laboratory Manual* (Academic Press)

Structural and Magnetic Characterization of Mn and Fe–doped ZnO Nanopowders

Putra Dewangga Candra Seta[a], Pelangi Eka Yunita[b], Malik Anjelh Baqiya[c], Darminto[d]

Advanced Materials Research Group, Department of Physics, Faculty of Mathematics and Natural Science, Institut Teknologi Sepuluh Nopember, Kampus ITS Sukolilo, Surabaya 60111, Indonesia

E-mail: putradewangga99@gmail.com

Abstract. The $Zn_{1-x}A_xO$ (A = Mn, Fe and x = 0 - 0.03) nanopowders, have successfully been synthesized by coprecipitation method. The starting materials of dihydrate zinc acetate, iron and mangan powders were employed using HCl and NH_4OH as solvent and precipitating agents respectively. The solution was stirred at 48°C for 4h to induce a precipitation and the resulted precipitate was then heated at 100 °C for 24h in air followed by annealing at 400 °C for 3h. Based on the XRD spectra, the doped samples with x up to 0.02 form the single wurtzite phase of ZnO with crystal size around 20-60 nm, where the dopants, both Mn and Fe, have partially substituted Zn. Meanwhile, samples with x = 0.03 were obesrved to contain impurity phases. The magnetization for the magnetic field up to 1 tesla vizualizes a monotonically increased weak-ferromagnetism for Fe doping, while exhibits from diamagnetic, paramagnetic to weak-ferromagnetic behaviors for Mn doping. This magnetic interaction becomes complicated by the presence of magnetic impurities due to Mn dopant.

1. Introduction

Dilute magnetic semiconductors (DMSs) have gained a considerable attention from scientific community because of their wondering optical and magnetic properties which have applications in 'Spin electronics' also known as spintronics [1]. After these predictions about the dilute magnetic semiconductors, simulations of Sato and Katayama-Yoshida [2] showed that Ferromagnetic could also be achieved in V, Cr, Fe, Co, and Ni doped ZnO. Over last decade, most of the research efforts were devoted to Mn and Co doped ZnO systems [3]. Very few studies were reported for Fe doped ZnO ;however, reports showed inconsistency in magnetic behavior of Fe doped ZnO. Zinc ionic can substitution with mangan ionic which give ferromagnetism behaviour [4].

In this research, a series of The $Zn_{1-x}A_xO$ (A = Mn, Fe and x = 0 - 0.03) nanopowders, have successfully been synthesized by coprecipitation method. The influence of Fe concentration and Mn concentration on the structural and magnetic properties of the samples was presented and discussed. Considering their essential importance for applications in spin-optoelectronic devices in the future, the optical properties of the films were also studied.

2. Experimental

The study was conducted by using a powder material Dyhidrat Zinc Acetate, Iron powders, Mangan powders, solvents HCl, and NH4OH precipitating agent. The synthesis process by mixing Zinc Acetat Dyhidrat powders and (Mn, Fe) powders with HCl and NH_4OH. The Solution was stirred for 4h at 84

IOP Conf. Series: Materials Science and Engineering **196** (2017) 012018 doi:10.1088/1757-899X/196/1/012018

°C and then heated in air for 24h at 100 °C and annealed for 3h at 400 °C. $Zn_{1-x}A_xO$ (A = Mn, Fe) nanoparticle synthesis performed with the variation of (Mn, Fe) dopant (x = 0,00, 0,01, 0,02, 0,03) to the solution precursor. Characterization testing performed using X-ray diffraction (XRD), *Vibrating Sample Magnetometer* (VSM), and Scanning Electron Microscopy (SEM). XRD data analysis performed using the Match software, Rietica, and Material Analysis Using Diffraction (MAUD). The crystal structure of the $Zn_{1-x}A_xO$ (A = Mn, Fe) nanoparticle was determined by XRD. Magnetic measurements were carried out on an vibrating sample magnetometer (VSM) with the magnetic field applied parallel to the sample.

3. Result of Discussion

Figure 1 shows the X-ray diffraction patterns of the $Zn_{1-x}Fe_xO$ and $Zn_{1-x}Mn_xO$.

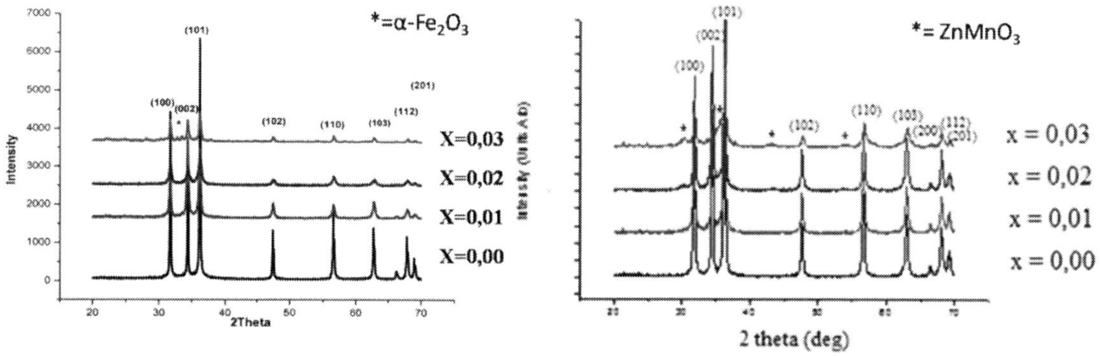

Figure 1. XRD spectra of $Zn1-xFexO$ and Zn1-xMnxO.

All of the diffraction peaks can be indexed to the wurtzite structure of ZnO (space group P63mc). With increasing Fe concentration x in the $Zn_{1-x}Fe_xO$, the clear shifts of the (002) peaks to higher angles are observed. This phenomenon indicates a reduction in the lattice constant with Fe doping, which is expected as smaller Fe ions are incorporated into the Zn sites of the ZnO lattice without changing the wurtzite structure. Also, the (002) peak exhibits an increased line broadening as x increases, suggesting a significant lattice-defect formation by Fe doping. In order to properly substitute Zn^{2+} ionic sites with maintaining charge neutrality of the sample, Fe ions need to have a valence of +2. It is noted that the ionic radius of Fe^{2+} is larger than that of Zn^{2+} by about 5%, where as that of Fe^{3+} is smaller by about 10% [5]. The lattice parameter should not decrease with enhancing x if there is Fe^{2+} that replaces the site of Zn^{2+}. Therefore, it can be inferred that Fe^{2+} and Fe^{3+} ions coexist in the $Zn_{1-x}Fe_xO$, Fe^{2+} and Fe^{3+} ions are expected to distort the lattice structure for holding charge neutrality. The higher concentrations of Fe may provide more chances for Fe to substitute Zn^{2+} sites, resulting in the lattice constants further shrink with the increase of the Fe concentration x. Considering that both the valence state and the ionic radius are the major factors determining the solubility of the dopant [5], high Fe solubility (x=0.02) for the present $Zn_{1-x}Fe_xO$ samples is attributable to the existence of the Fe^{2+} ions with high density. They do not show any secondary phases such as FeO, Fe_3O_4 and $ZnFe_2O_4$ in the detection of our diffractometer. This may be attributed to the limitation of XRD characterization that small amount of impurities cannot be detected. However, for highly doped sample (x=0.03) we observe the appearance of a spinel α-Fe_2O_3 phase are marked with *. α-Fe_2O_3 exhibits weak-ferromagnetism. The analysis of diffraction data was carried out using Rietica software for calculate lattice parameter, and MAUD software for calculate crystal size. The cause of shiftin (002 and 101) reflection peak is because of variation in lattice spacing a, c and crystal size as shown in Table 1.

The XRD results also indicate that the Mn^{2+} ions systematically substituted for the Zn^{2+} ions in the sample without changing the wurtzite structure, as shown in the inset of Figure 1. The most intense diffraction peak (101) is clearly evident with a slight shift into low angular scale, whereas the corresponding intensity decreases compared with undoped ZnO. Furthermore, the intensity of the diffraction peaks decreases, and the width broadens due to the formation of smaller grain diameters as a result of an increase in disorder on Mn^{2+} doping. But, high Mn solubility (x=0.03) for the present $Zn_{1-x}Mn_xO$ samples is attributable to the existence of $ZnMnO_3$ phase [6]. In our experiment, Fe and Mn dopant influence the crystalline size of ZnO, because the creation of new nucleation centers from the dopant atoms is retard for the growth of ZnO crystals. So that crystal size of ZnO with dopant (Fe or Mn) less than ZnO undopped. This is probably connected with due to the difference of ionic radius between dopant (Fe or Mn) and Zn^{2+}.

Table 1. Lattice parameters and crystal size for $Zn_{1-x}Fe_xO$ and $Zn_{1-x}Mn_xO$

Latice Parameter	Dopant	x=0	x=0.01	x=0.02	x=0.03
a=b (Å)	Mn	3.2559	3.2494	3.2509	3.2449
	Fe		3.2514	3.2506	3.2524
c (Å)	Mn	5.2149	5.1998	5.2056	5.1863
	Fe		5.2054	5.2054	5.2099
Crystal size (nm)	Mn	61.2	51.4	47.7	ZnO = 25.2 and $ZnMnO_3$= 90.7
	Fe		32.8	38.1	43.5

a) b) c)

Figure 2. SEM images of : (a) ZnO , (b) $Zn_{0.99}Fe_{0.01}O$, and (c) $Zn_{0.99}Mn_{0.01}O$.

The surface morphology of undoped (a), Mn doped (b) and Fe doped (c) is shown in Fig. 2(a)–(c). it is seen that the average size of the particles is in sub-micrometre size. Pure ZnO showed rod-like structures and as dopant (Mn and Fe) content increased, rod-like structures transformed into sheets- and plates-like. Further increase of dopant (Mn and Fe) causes a transformation from sheets and plates to spheres with finer grains. The defects caused by dopants may be responsible for deformation of structures, and hence different morphology.

Magnetic characterizations for $Zn_{1-x}Fe_xO$ and $Zn_{1-x}Mn_xO$ were done by using VSM at room temperature. Fig. 3a shows the hysteresis graph for $Zn_{1-x}Fe_xO$. It has been observed that our samples did not show full saturation; this effect appears due to the ambiguous nature of $Zn_{1-x}Fe_xO$ phase which contributes in both ferromagnetic and paramagnetic behaviors. It has also been shown in previous studies that the $ZnFe_2O_4$ phase exhibits anti-ferromagnetic and α-Fe_2O_3 phase exhibits weak-ferromagnetism behavior which may suppress the magnetic signal [7]. However, the increase/decrease in the lattice parameter with increasing doping percentage indicates that it is not only due to Fe^{2+} ions and Fe^{3+} ions substitution in the ZnO lattice, but also due to any secondary phases such as FeO, α-Fe_2O_3, Fe_3O_4 and $ZnFe_2O_4$ [8]. The magnetization for the magnetic field up to 1 tesla vizualizes a monotonically increased weak-ferromagnetism for Fe doping.

In Fig. 3b, when ZnO was doped by Mn, the magnetization curve generated was different for each composition of dopant. On the curve with x = 0.01, it can be seen that the curve indicates a nature of diamagnetic. A few of Mn dopants will occupy some Zn position in this doping state. When x = 0.02, a curve featuring a superparamagnetic behavior is observed. Further, for x = 0.03, the ZnO behaves as weak ferromagnetic. This magnetic interaction becomes more complicated by the presence of magnetic impurities due to Mn dopant [9].

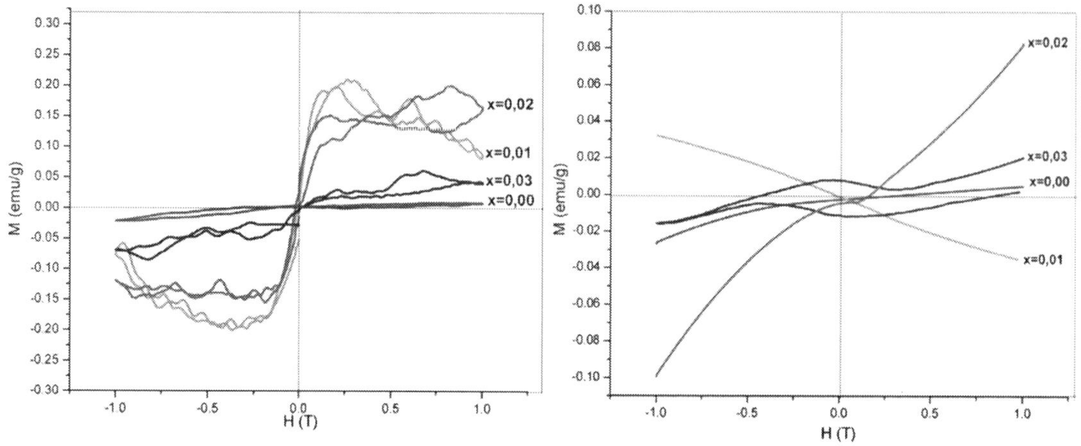

(a) (b)

Figure 3. The magnetization curve of : (a) $Zn_{1-x}Fe_xO$, and (b) $Zn_{1-x}Mn_xO$.

4. Conclusion

A series of $Zn_{1-x}Fe_xO$ and $Zn_{1-x}Mn_xO$ with the atomic fraction, *x*, in the range of 0–0.03 were synthesized by coprecipitation method. A systematic study was done on the structural and magnetic properties as a function of dopant concentration. XRD results indicated that there are not any secondary phases and Fe^{2+} as well as Fe^{3+} substituted for Zn^{2+} of ZnO host, while Mn^{2+} ions systematically substituted for the Zn^{2+} ions in the sample without changing the wurtzite structure for Mn doping. The magnetization for the magnetic field up to 1 tesla vizualizes a monotonically increased weak-ferromagnetism for Fe doping, while exhibits from diamagnetic, paramagnetic to weak-ferromagnetic behaviors for Mn doping. This magnetic interaction becomes complicated by the presence of magnetic impurities due to Mn dopant.

Acknowledgment

This research was partially supported by "Hibah KLN", Kemenristekdikti, Indonesia, 2017.

References

[1] T Dietl, H Ohno, F Matsukura, J Cibert, D Ferrand *Science* 2000 **287** 1019.
[2] K Sato, H Katayama Yoshida 2000 *Jpn.J.Appl.Phys.* **39** L555
[3] S W Jung, S J An, G C Yi 2002 *Appl.Phys.Lett.* **80** 4561
[4] Kim K J, Y R Park 2002 *J. Apll. Phys.* **8** 1420
[5] Z Jin, T Fukumura, M Kawasaki, K Ando, H Saito and T Sekiguchi 2001 *Appl. Phys. Lett.* **78** 3824
[6] Fauzana, B Z Azmi, M G M Sabri, W R Wan Abdullah and M Hashim 2013 *Sains Malaysiana* **42** 1139-1144
[7] Y Q Wang, S L Yuan, L Liu, P Li, X X Lan, Z M Tian, J H He, S Y Yin 2008 *J. Magn. Magn. Mater.* **320** 1423
[8] S Karamata, R S Rawata, P Leea, T L Tana, R V Ramanujanc 2014 Structural, elemental, optical and magnetic study of Fe doped ZnO and impurity phase formation, Progress in Natural Science: Materials International **24** 142–149
[9] Omri, j El Ghoul, O M Lemine, M Bououdina, B Zhang, L El Mir 2013 *Journal of Superlattices and Microstuctures* **60** 139-147

IOP Conf. Series: Materials Science and Engineering **196** (2017) 012019 doi:10.1088/1757-899X/196/1/012019

Microstructure and Magnetic Properties of Optimally Annealed $Ni_{43}Mn_{41}Co_{5}Sn_{11}$ Heusler Alloy

Nastiti Elwindari[1], Budhy Kurniawan[1], Candra Kurniawan[2], Azwar Manaf[1]

[1]Physics Department, Faculty of Science, Universitas Indonesia, Depok 16424, Indonesia.

[2] Research Center for Physics LIPI, Puspitek Office Area, Tangerang Selatan, Banten, Indonesia

E-mail : azwar@ui.ac.id

Abstract. In this work, synthesis and characterization of a polycrystalline $Ni_{43}Mn_{41}Co_{5}Sn_{11}$ (NMCS) alloy are reported. Alloy preparation was conducted by melting the constituent components of the designated alloy under an inert Argon (Ar) atmosphere in a vacuum mini arc-melting furnace. Microstructure observation to the as-cast and annealed ingots showing dendritic structure in the as-cast sample. Series of annealing treatment to the sample at 1173 K have changed dendrites progressively in the homogeneous structure after 24 hours annealing time. The annealed sample consisted of a NiMnCoSn main phase with 99.3 % volume fraction. Hence, the 24 hours annealed ingot is a single phase alloy. The curie temperature of the annealed NMCS alloys was found in the range 348~351 K. Loop hysteresis evaluation of the annealed ingots showed that ingot which annealed for 12 hours showed the largest magnetization value of 57.96 emu/g.

1. Introduction

Heusler alloy is one of rare-earth-free material based Ni-Mn-X (X=Sn, In, Sb) which have been attracting serious attention among researchers, especially Heusler alloys of low temperature magnetic transition and those of good intrinsic property. In addition to the intrinsic properties like Curie temperature, total magnetization, entropy change (ΔS_M) are the primarily important factor, the preparation process which strongly determine the microstructure of alloys and magnetic ordering[1] also important to obtain the large value of refrigeration capacity (RC) and entropy change at the same time.

The arc-melting furnace is one of the established tools from which alloys can be made through melting process and successive solidification. Despite arc melting furnace is a more straightforward tool for the production of alloys; however, an additional treatment like a long time annealing the as-cast ingot at a sufficiently high temperature is still required in order promote grain homogenization toward homogeneous single phase alloys. The arc-melting technique in many cases has proven useful, in alloys fabrication including that of Heusler alloys, with improved Magneto caloric effect (MCE) properties [2,3] despite more modern technique like rapid solidification by melt-spinning technique, which effectively substantially shortened annealing stage is readily available.

It is mostly accepted that ingots which obtained from arc melting method contained inhomogeneous grains. Thus, the solidified ingot needs to be annealed for a long period of time to facilitate the atomic diffusion in order to obtain a uniform phase. Ghosh and Aksoy *et al.*[4,5] have studied the influence of annealing time on MCE of Ni-Mn-Sn bulk samples at 1173 K from 0 h to 24 h and a maximum magnetic entropy change could be found when annealing time is close to 24 h. In this research work, we studied the effect of annealing treatments to the microstructure and magnetic properties of $Ni_{43}Mn_{41}Co_5Sn_{11}$ alloy. The magnetic properties of the annealed ingots were evaluated to determine the total magnetization and Curie temperature.

2. Experiment Method

We used Ni, Mn, Co, and Sn powders with nominal purity better than 99.9%. All the starting powders with were mixed, as the feedstock for the preparation of alloy with a designated $Ni_{43}Mn_{41}Co_5Sn_{11}$ composition. The mixed powders were blended, hand grinded in a mortar and then poured into a cylindrical die. The green compact was obtained after pressured by an uniaxial pressure and sintered at 1173 K for 1 hour for densification. The dense sample was then melted using a mini vacuum arc melting furnace for several times to ensure a homogeneous element distribution throughout the sample. The cast alloy was inserted into a quartz tube under a flowing argon gas during annealing treatment at 1173 K for annealing times 0, 6, 12, and 24 hours to obtain more homogenous samples. The microstructure changes of as-cast NMCS and the effects annealing timewere obtained by scanning electron microscopy (SEM) using a JEOL JSM-6510LA microscope equipped with energy dispersive spectrometry (EDS)

3. Result and Discussion

Plot of the x-ray diffraction (XRD) traces of series annealed NMCS alloy samples compared with that of as-cast sample is shown in Figure 1. Phase analysis based on diffraction peaks revealed the presence of crystalline $Ni_{43}Mn_{41}Co_5Sn_{11}$ phase as the main phase in the alloys which is in accordance with database in Inorganic Crystal Structure Database (ICSD). Some additional phases were also identified. Intensity increment of three primary peaks after annealed treatment were observed at diffracted positions around 40°-45°, 60°-65°, and 75°-80° arise from crystal with the orientation hkl index [220], [400] and [422]. It happened because the main phase NMCS alloy before anneal has not been present in the phase of fully crystalline, but yet still presence in the form of dendrites, so further treatment was needed in order to reach a proportion of optimal concentration diffusion. This was demonstrated by plot of 24 hours annealed sample showing almost single phase alloy.

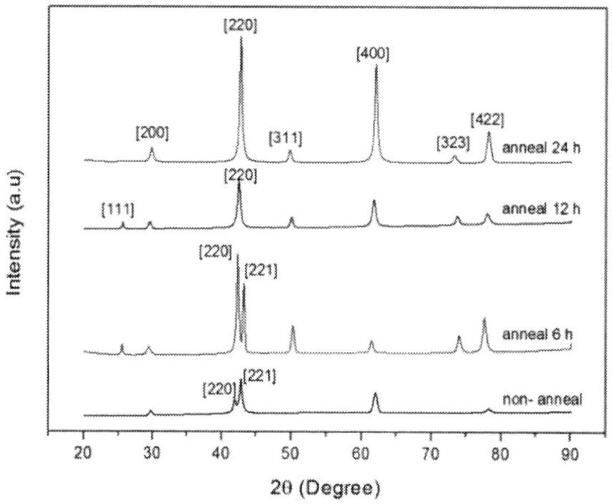

Figure 1. Plot of difraction trace for annealed $Ni_{43}Mn_{41}Co_5Sn_{11}$ samples compared with that of as-cast.

Table 1. Volume fraction of as-cast and annealed $Ni_{43}Mn_{41}Co_5Sn_{11}$alloy.

Phase	Volume Fraction (%)			
	As-cast	Annealed 6 h	Annealed 12 h	Annealed 24 h
NiMnCoSn	68.7	48.6	45.3	99.3
Ni_2MnSn	-	49.4	52.6	-
Ni_3Mn	3.0	2.0	2.2	0.7
Ni_3Sn	16.9	-	-	-
Mn_2Sn	11.4	-	-	-
Total	100	100	100	100

The peaks indicated that alloy phase are growing during the annealing time. In the peak plane [220] shows the crystal structure of austenite with cubic L_{21} Heusler structure. The peaks of annealed sample are mixed with a small amount of some extra phases, most likely Mn-rich phase around 25~35 and 60~65 degree. This result indicates that material peaks were eliminated for further process. Longer annealing time, obtain a uniform phase. Because of the longer atom diffusion distance in a bulk sample is need longer annealing time to obtain the sample homogeneous [1,6]. As the increase annealing time, the stress can gradually release and at the same time, the grain growth can also reduce friction.

Table 1 summarizes result of volume fraction of phases in alloys which determined by quantitative XRD analysis of XRD data in Figure 1. It indicates a successful annealing treatment at 1173 K for 24 hours to transform as-cast sample of multi-phase dendrite structure to almost single phase alloy with NiMnCoSn phase as the main phase.

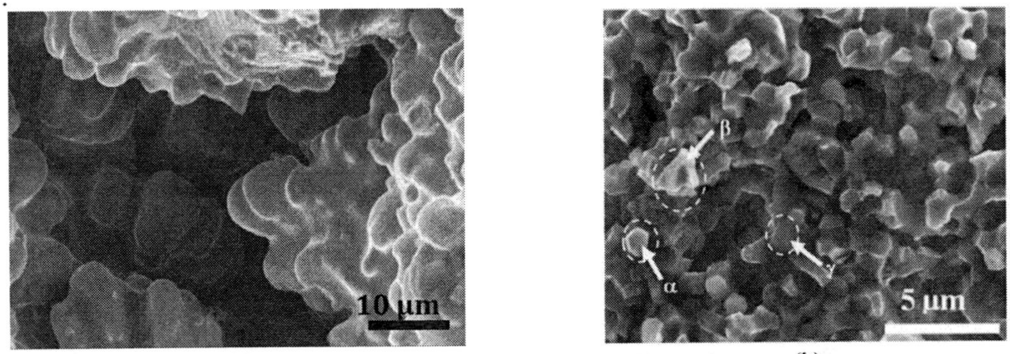

(a) (b)

Figure 2. (a) The original as-cast alloy contains randomly equiaxed dendritic grains (b) Microstructure of annealed alloy showing the grains with a mean size of 1.5 µm after 12 h annealing treatment.

Table 2. The composition of grain semi-quantitatively analyzed at 3 different points by EDS.

Analysis at	Composition of element (wt.%)				Predicted phase
	Ni	Mn	Co	Sn	
α	44.76	19.16	2.20	33.87	Ni-Mn-Co-Sn
β	41.04	36.82	2.12	20.02	Ni-Mn-Co-Sn
γ	8.60	85.97	-	5.43	Mn- Rich

In order to find out how the microstructure develop towards a single phase structure, SEM observation of the fracture surface of the bulk sample was then carried out for as cast and annealed alloys. SEM micrographs of annealed and as-cast samples are shown in Figure 2. Obviously, as-cast sample showing dendrite structure of in-complete grains which is significantly different with that for the annealed alloys. The grain size and grain morphology of annealed samples progressed with annealing time from which it was observed that the alloy phase began to form small crystallites with a diameter of about 1.5 μm-2.5 μm. The annealed sample has smaller crystallite size compared with that of as cast alloy. .

Table 2 summarizes EDS evaluation results on three different spots as indicated by α, β. γ shown in Figure 2b. The chemical composition based on microanalyses by EDS showing in some extent different with designated composition of synthesized alloy. It could be due to the change of weight for the synthesized alloys because of mass reduction during the alloy preparation process. Even though, it indicates that the chemical composition at points α and β may be of the main phase evolves during the crystal growth. Generally it is accepted that the characteristic transition temperature can be determined by the electron concentration principle [7]. In this case, it can be referred to the valence electron per atoms are respectively 10 ($3d^8 4s^2$) for Ni, 7 ($3d^5 4s^2$) for Mn, and 4 ($5s^2 5p^2$) for Sn. In connection with the volume fraction $Ni_{43}Mn_{41}Co_5Sn_{11}$ alloy, the phase of NiMnCoSn as the main phase is the largest among other phases [8,9]. In this regards, the existence of NMCS main phase in the alloy can be confirmed that the sintered alloys were formed which have a main phase of NiMnCoSn composition.

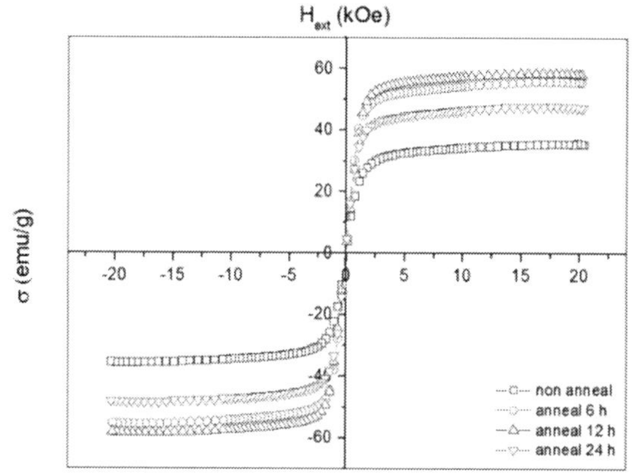

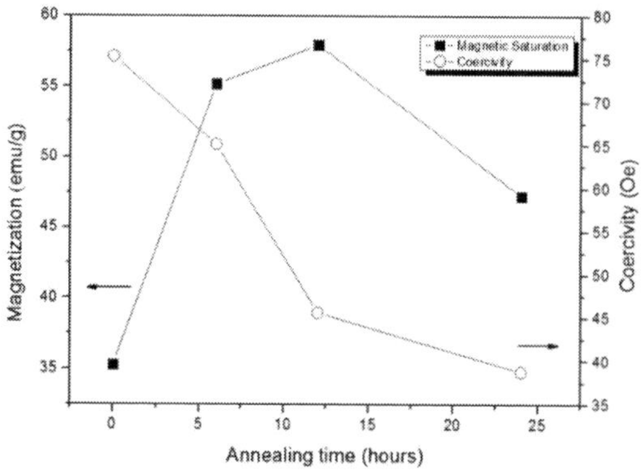

Figure 3. (a) Magnetic hysteresis(b) Magnetic saturation and coercivity of $Ni_{43}Mn_{41}Co_5Sn_{11}$ as a function of annealing time.

Table 3. Magnetic characterization of $Ni_{43}Mn_{41}Co_5Sn_{11}$dependence anneal time.

Anneal Time	M_s (emu/g)	M_r (emu/g)	Hc (Oe)	Tc
0 h	35.29	2.28	75.33	348 K
6 h	55.19	2.47	65.10	348 K
12 h	57.96	3.85	45.60	350 K
24 h	47.28	6.53	38.78	351 K

Figure 3 shows the hysterisis loop of annealed samples compared with the as-cast. All samples are showing soft magnetic characteristic which different value of total magnetization at a fixed maximum applied magnetic field in the magnetization process. The values were plotted in Figure 2b along with their respective coercivity values. It indicates that total magnetization value increases until 57.96 emu/g in the sample annealed for 12 hours and decreases again at 24 h annealing times, whereas the coercivity value shows a continuous decrease with annealing time. Values of total magnetization and coercivity for all samples are summarized in Table 3 which representing the magnetic characteristics of these samples. The curie temperature value was around at 348 K~351 K. Obviously, the T_c of NiMnCoSn phase is not affected by the microstructure. The largest value of total magnetization was obtained in a sample after 12 hours annealing time (see Table 3) in which there are two major magnetic phasein the sample respectively NiMnCoSn with volume fraction of 45.3% and Ni_2MnSn with 52.6 % in volume fractions. Whereas in the 24 hours annealed sample, there was only 1 major phase with 99.3% in volume fraction of NiMnCoSn. Although the 24 hours annealed sample showed a single phase system, the total value of magnetization is lower than that of 12 hours. In line with previous research, Ma and Dan [10,11] indicate that the total value of magnetization of Ni_2MnSn phase larger than NiMnCoSn phase. In this case, the total magnetization 'Js' of the sample may be calculated through the rule of mixture according to equation (3.1) [12].

$$Js_{\text{alloy}} = \sum_{i=1}^{n} Vi.Jsi \quad (3.1)$$

where Js_{alloy} is the saturated magnetic field of alloys, Vi is volume fraction of phases in alloys to i, and Jsi is saturated magnetic field of phases to i. This should result in total value of magnetization of the 12 hours annealed sample is higher than that of the 24 hours annealed sample.

4. Summary

The microstructure and magnetic properties of $Ni_{43}Mn_{41}Co_5Sn_{11}$ alloy have been investigated. The as-cast alloy consisted of dendrite structure which progressively transformed into a fully crystalline grain structure. Annealed temperature at 1173 K for 24 hours is necessary to homogenize the microstructure towards a single phase alloy. However, the largest total magnetization value could be obtained in a sample that annealed for 12 hours resulted in a value about 57.96 emu/g with the corresponding coercivity of 45.60 Oe.

Acknowledgement
The authors gratefully acknowledge the support of Department of Physics, Universitas Indonesia for the research facilities and Metallurgical Research Center, Lembaga Ilmu Pengetahuan Indonesia for magnetic properties measurement. This study funded by the university under research grant contract no. 1993/UN2.R12/HKP.05.00/2016.

References

[1] Zhang Y, Zhang L, Zheng Q, Zheng X, Li M, Du J, et al 2015 *Sci Rep* **5** 11010
[2] Xuan H C, Xie K X, Wang D H, Han Z D, Zhang C L, Gu B X, et al 2008 *Appl. Physics Lett.* **92** 242506
[3] Zheng H, Wang W, Xue S, Zhai Q, Frenzel J 2013 *Acta Mater* **61** 4648–56
[4] Ghosh A, Mandal K 2014 *Appl Phys Lett.* **104** 3–7
[5] Aksoy S, Acet M, Deen P P, Mañosa L, Planes A 2009 *Phys Rev B* **79** 1–4
[6] Zhong Z, Ma S, Wang D, Du Y A 2012 *J Mater Sci Technol* **28**193–9
[7] Paramanik T, Das I 2016 *J Alloys Compounds* **654** 399–403
[8] Yang L H, Zhang H, Hu F X, Sun J R, Pan L Q, Shen B G 2014 *J Alloys Compounds* **588** 46–8
[9] Graf T, Felser C, Parkin S S P 2011 *Prog Solid State Chem* **39** 1–50
[10] Dan N H, Duc N H, Yen N H, Thanh P T, Bau L V, An N M, et al 2015 *J Magn Magn Mater.* **374** 372–5
[11] Ma S C, Xuan H C, Zhang C L, Wang L Y, Cao Q Q, Wang D H, et al 2010 *Appl Phys Lett.* **97** 052506
[12] Callister W 2007 *Materials Science and Engineering An Introduction 7th edition* Joseph Hayton (New York: John Wiley & Sons, Inc) chapter 16 pp 580-581

IOP Conf. Series: Materials Science and Engineering **196** (2017) 012020 doi:10.1088/1757-899X/196/1/012020

Characterization of Precipitated $CaCO_3$ Synthesized from Dolomite

Zaenal Arifin, Nurul Fitria Apriliani, Mochamad Zainuri, dan Darminto

Avanced Materials Research Group, Department of Physics, Faculty of Mathematics and Natural Science, Institut teknologi Sepuluh Nopember (ITS), Kampus ITS Sukolilo, Surabaya 6011, Indonesia.

E-mail : zaenalw18@gmail.com

ABSTRACT. The precipitated $CaCO_3$ has successfully been synthesized from dolomite. The influence of various temperature and carbonation rate on the morphology, crystal size, phase and structure of the carbonation products were studied. The particles resulted from this process have the crystal size 400-800 nm. The calcite phase wasformed inthe synthesis at 30 °Cwithcarbonationrate of2 and 7 SCFH, and synthesis at 80 °C and 2 SCFH also produces calcite phase. Observations by SEM shows morphology of calcite as so-called schalenohedral.

1. Introduction

Calcium carbonate ($CaCO_3$) has widely been used in various industrial fields, such as papers, plastics, coatings, ink, paint and polymer products. Meanwhile, higher and special quality $CaCO_3$ has also been needed for ingredientsfrom cosmetics [1], drug delivery [2], bioactive materials [3], to supplements of nutrition [4]. Calcium carbonate is used in the industry is the type of calcium carbonate precipitates.The precipitated $CaCO_3$ (PCC) is generally prepared by precipitation process using CO_2, gas bubbling in the solution containing Ca ions, such as $Ca(OH)_2$. The starting materials $Ca(OH)_2$ may potentially be a limestone as natural resources with calcium content of around 98% [5].

In our previous study, the PCC has successfully been prepared from limestone employing a facile method of carbonisation, resulting in $CaCO_3$ with various phases and uniqe morphology [6]. However, the various processes which have been applied to prepare the PCConly produced powders having crystalline size > 1 μm. Several applications, in fact, need $CaCO_3$ with smaller crystal size to improve its better characteristics. A further study focusing the PCC with reduced crystal size is therefore a challenge.It was reported [7] thatone way to reduce the crystal size in the preparation of $CaCO_3$is to add other elements, such as Mg, which is close to Ca. The Mg^{2+} ion is effective in the synthesis of calcium $CaCO_3$with aragonite (needle-like) structure. It was also reported [8] that the PCC with calcite and aragonite phases have successfully been synthesized from dolomite using poly (methyl methacrylate) (PMMA).

In this paper, the synthesis of $CaCO_3$ from the starting material containing Mg, namely dolomite, will be presented. The Dolomite is a carbonate mineral containing Mg, or $CaMg(CO_3)_2$. The use of dolomite, where Ca and Mg coexist, is expected to be more effective to produce $CaCO_3$ with smaller crystal size.

2. Experimental Method

The raw material was dolomite obtained from a quarryin Tuban (East Java, Indonesia).The elemental content and initian phases of dolomite were characterized by X-ray flouresence (XRF)and diffractometry (XRD). The powderized dolomite was dissolved in a solution of 12 M HCl, and then stirred by using hot plate/magnetic stirrer with speed of 500 rpm for 30 minutes. Carbonation process was performed by flowing CO_2 gas into the solution by bubbling method. Synthesising process was performedin three different conditions, (i) CO_2 flow of 2 SCFH at 30°C, (ii) 7 SCFH at 30°C, and (iii) 2 SCFH at 80°C. The different synthesis conditions were applied to produce PCC with different morphologies. The solution that has been carbonated would form white precipitatesas carbonation products,which were then filtered and dried at 80 °C- 90 °C for 24 hours. The white precipitates were finally characterized by XRD and scanning electron microscopy (SEM).

3. Results and Discussion

The elemental composition of dolomite as a result of by XRF measurement is shown in Tab. 1. The content of calcium is of 92.40% and magnesium 6.5%, while the othersare impurity phases having content < 1%. The existence of magnesium in dolomite is expected to be the best alternative to producePCC with a smaller size compared to using limestone doped with elements Mg.The characterization results of dolomiteby XRD and diffraction patterns analysis using the"MATCH" Software are shown in Fig. 1. It is seen that all diffraction peaks are matched with dolomite phase without impurity. It means that Mg and Ca exactly form $CaMg(CO_3)_2$.

Table 1. The elemental content of dolomite by XRF.

Elemental content	Concentration (wt%)
Ca	92.40
Mg	6.50
Fe	0.45
Mn	0.055
Yb	0.54
Cu	0.073

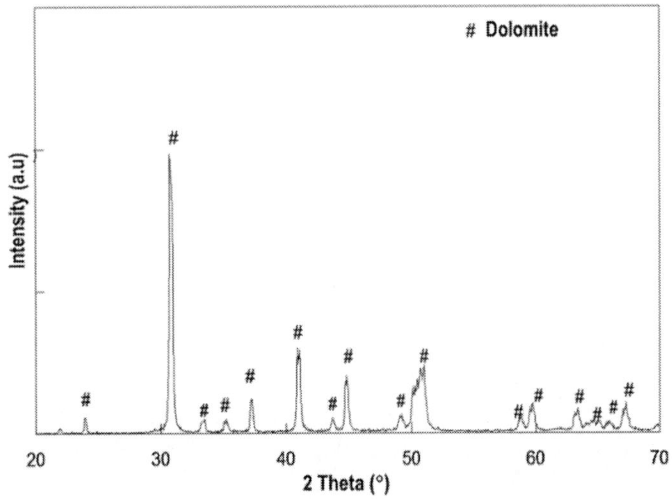

Figure 1. XRD patterns of samples of the limestone.

Fig. 2 shows the XRD patterns of the carbonation products. Three different synthesis conditions have producedthe same $CaCO_3$phase of calcite. This is certainly interesting, because in our previous study[6], these different conditions of synthesis have led to the different phases of $CaCO_3$. The Dolomite has an ideal structure [9,10] consisting of ionic layers of Ca^{2+} and Mg^{2+} with strong bond. Calcination process is not carried out in this synthesis, because it induces decomposition. Dolomite bond is still strong to still obtain calcite phase during carbonation process.The calculation results of crystal size based on the XRD data in Fig. 2 using"MAUD" Software are shown in Tab. 2.

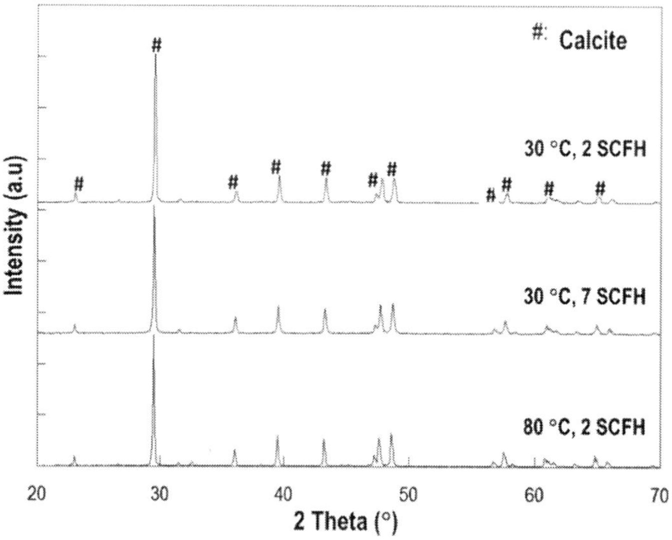

Figure 2. The XRD spectra of the carbonation products, with different processing parameters as specified.

Table 2. Crystal size of the carbonation products.

Synthesis condition	Crystal Size
30 °C, 2 SCFH	674.3±0.3nm
30 °C, 7 SCFH	417.7 ±0.4nm
80 °C, 2 SCFH	782.8 ±0.6nm

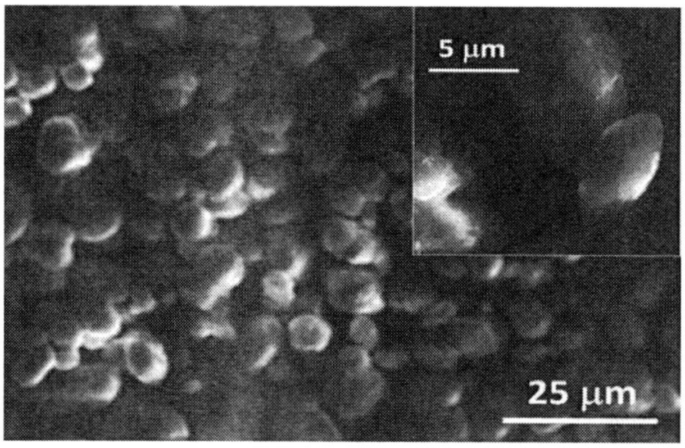

Figure 3. Morfologi of the carbonation products at 30 °C and carbonation rate of 2 SCFH with magnification of 1500 x and 7500x (inset).

Morphology of the PCC product synthesized at 30 °C and carbonation rate of 2 SCFH is given in Fig. 3. Morphology of the carbonation products in different synthesis condition also hase calcite phase with the so-called scalenohedral. Scalenohedral calcite formed if the ratio of [Ca]/[CO_3]> 1.56 [11]. When the carbonation is conducted with a smaller flow rate as in this synthesis, the ratio of [Ca]/[CO_3] will be greater, so that it will increase the formation probability of scalenohedral calcite morphology. Further, The phase of dolomite hasthe structure withalonger c-axis because ofsubstitution of Mg. The Ca-O bond length is also much longer than that in the calcite phase [12]. Pokrovsky pointed out that in the case of homogeneous precipitation from supersaturated solutions, aragonite is the ubiquitous phase at Mg^{2+} /Ca^{2+} activity ratio higher than 3, while mostly calcite is formed at lower Mg^{2+} /Ca^{2+} ratio [13].This in turn affectsthat the resulted calcite has morphology with slightly elongated shape. It is also observed in Fig. 3 that theparticles shape and size are almost uniform at all points. Furthermore, the inset to Fig. 3 shows the rhombohedral calcite particles which are formed by a collection of calcite crystals with size in hundreds nanometer or in sub-micrometer.

4. Conclusions

The precipitated calcium carbonate with calcite phase has successfully been synthesized from dolomite by carbonation methode with three different processing parameters.The product were observed with elongated shape morphology, as collection of calcite having crystalline size < 1 micrometer.

Acknowledgement
This research was partially supported by PUPT Research Grant, DRPM, Kemenristekdikti, 2015 - 2016.

References

[1] Wang C, Tong H C, Liu Zand Zeng X F 2010 *Int. J. Pharmace* **308** 160
[2] Peng C, Zhao Q and Gao C 2010 *Colloids Surf A:Physicochem Eng. Asp* **353** 132
[3] Kumar G S, 2008 *J. Colloid Interf. Sci* **349** 56
[4] Ghamgui H 2007 *Biochem. Eng. J* **37** 34
[5] Arifin Z, Pratapa S, Tririkantoro and Darminto 2013 *AIP Conf.Proc.* **1554** 90
[6] Arifin Z, Pratapa S, Tririkantoro and Darminto 2015 *Trans. Ind. Ceram. So* **74** 1
[7] Park W K, Ko S J, Lee S W, Cho K H, Ahn J, Wand Han C 2008 *J Cryst Growth* **310** 2593
[8] Mantilaka M M M G P G, Karunaratne D G G P, Rajapakse R M G, Pitawala H M T G A 2013*Powder Tech* **235** 628
[9] Land L S 1985 *J Geologedu* **33** 112
[10] Warren J K 1985 *Am. Assoc.Petrol.Geol.Bull.* **69** 1013
[11] Carmona J G, Morales J G, Clemente R R 2003 *J Colloid Interf Sci.* **261** 434
[12] Titiloye J O, de Leeuw N H, Parker S C 1998 *GeochimCosmochiActa.* **62** 2637
[13] Pokrovsky O S 1998 *J Cryst Growth* **186** 233

IOP Conf. Series: Materials Science and Engineering **196** (2017) 012021 doi:10.1088/1757-899X/196/1/012021

Defect and Magnetic Properties of Reduced Graphene Oxide Prepared from Old Coconut Shell

Kurniasari[1], A Maulana[1], A Y Nugraheni[1], D N Jayanti[1], S Mustofa[2], M A Baqiya[1] and Darminto[1,3]

[1]Advanced Materials Research Group, department of Physics, Institut Teknologi Sepuluh Nopember (ITS), Kampus ITS Sukolilo, Surabaya 60111, Indonesia
[2]Center for Science and Technology of Advanced Materials, National Energy Agency of Indonesia, Kawasan Puspiptek Serpong, Tangerang 15134, Indonesia
[3]Center of Excellence in Automotive Control and System, Institut Teknologi Sepuluh Nopember (ITS), Kampus ITS Sukolilo, Surabaya 60111, Indonesia

E-mail: kurnia.physics@yahoo.com

Abstract. Defect state may be created on the grafitic bonding of carbon atoms as the dangling bonds of the reduced graphene oxide (rGO) compound. In this study, the rGO sample has been prepared by burning the coconut shell in air. The resulted burnt samples were then heated at varying temperatures (400, 700, and 1000°C) for 5 hours in air. Further, a mechanical exfoliation process using ultrasonic vibrator was added to the samples for 5 hours. Characterization using Raman spectroscopy exhibited that the intensity ratio of defect (I_D) and graphene (I_G) increases, implying the increase of defect number, with respect to the increasing heating temperature. According to the magnetic property measurement using vibrating sample magnetometer (VSM), it is shown that the magnetization enhances with incerasing number of defects, featuring an occurrence of defect-induced magnetism in rGO.

1. Introduction

Indonesia is one of the countries with the biggest coconut producers in the world. Most the parts of coconut can be used. But there is underutilized waste from coconut product, it is coconut shell [1].Coconut shell is a natural material with a carbon content of 49.86 % (especially old coconut type) composing of cellulose (($(C_6H_{10}O_5)n$) and hemicellulose and containing the bonds related to the hexagonal crystal structure [2]. Coconut shell, which was gived a particular treatment were found graphene with high purity carbon bond [3] . Charcoal contains impure carbon produced with removing water and volatile matter. In contrast to activated carbon, this product needs activator to produce a large surface area so it can be used as adsorbent. Activated carbon and graphite has same structure. Graphite is one type of carbon material formed by carbon atoms that are in sp^2 orbital. Generally, graphite consists of him sheets with molecular bonding between carbon called graphene [4]. Graphene, a one- atom – thick planar sheet of sp^2 bonded carbon atoms that densely packed in honeycomb crystal lattice, is well known recently because of its special electronic and optical properties [5].

Reduced Graphene oxide (rGO) has properties similar to graphene, this is because the rGO is the reduction of the hygrogen and oxygen atoms from graphene which undergoes oxidation.It has been reported by Fandi et al (2014) if the reduced graphene oxide (RGO) has successfully been synthesized from coconut shell by heating at 1000°C in air, followed by rinsing treatment in distilled water [6.] Samples prepared have the peaks of Raman spectra at 1300 and 1590 cm^{-1} with the ratio of I_D/I_G is around 2.5, confirming that the resulted samples are the RGO phase. Therefore, the purpose of this work is to analyze efect of heat treatment to the defect from the coconut shell charcoal using Raman spectroscopy. After that,it has been measure a magnetic properties using Vibrating Sample Magnetometer (VSM) to identify what's the corresponding beetween defect with magnetism in rGO.

2. Experiment

The main material used in this research is coconut shell (Cocus nicifera) charcoal and distilled water. The equipment used in this study are mortar, furnace, glass beaker,hot plate, ultrasonic cleaner, digital balance, spatula, aluminium foil,filter paper, and oven. Sample preparation was done by burning on fire the coconut shell until it turns into charcoal. Then, choose charcoal which fully converted and oven overnight at temperature of 150°C to remove water content. After that the samples were crushed and drilled by mortar to obtain powder form and sifted using 500 mesh of sieve. It was heated using furnace at temperatute 400° C,700°C ,and 1000°C in atmosphere with holding 5 hours.Then, it was exfoliated by ultrasonification process in aquadest for 5 hours.Furthemore,it was drying in hotplate at 100°C to remove aquadest content. The samples were characterization by Raman Spectroscopy which is a powerful instrument that is often used to identify defect on samples. For this characterization, the samples were converted to pellet using hydrolic pressure with strength 350 bar. To identify the magnetic properties, it also characterized using vibrating sample magnetometer (VSM) in powder form.

3. Results and Discussion

Based on previous research X-Ray Diffraction (XRD) pattern for coconut shell charcoal heat at some various temperature shown in Figure 1. Furthermore analysis by comparing with reference [6] confirmed that the sample has the similar wide peaks with rGO phase at 2θ ~ 23° and 43°.

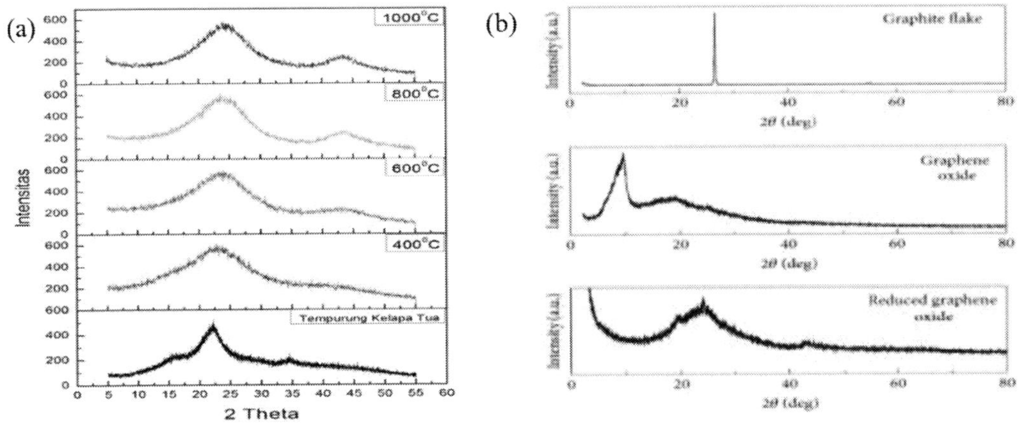

Figure 1. XRD pattern of (a) coconut shell charcoal sample in various temperatures, (b) Graphite flake, GO, and rGO [6].

Raman scattering carried out on this reduce graphene oxide (rGO) was dominated by graphene vibrational modes for the excitation wavelength of 785 nm. Raman data obtained shows the relationship between the Raman intensity and wave number, presented in the graphical form in the range 400 cm^{-1} to 2000 cm^{-1}. This is done because when the entire sample was measured above 2000 cm^{-1}, the resulting graph looks noisy, representing no phase. 2D peaks didn't appear in this work. Furthemore, raman spectroscopy was performed to indicate the structures of graphite, graphite oxide and rGO by the resulting characteristic G and D bands sensitive to defects and disorder, respectively. From the result, it clearly seen at the Figure 2 (a) that all of samples consists of two dominant peaks at 1350 cm^{-1} and 1590 cm^{-1}. The peak at around 1350 cm^{-1} is the D band. The D band is attributed to the present defects, like disruption in the sp^2 bonding because of heptagon and pentagon rings, vacancies, edge effect, and etc.The defect also caused by wrinkle formation and the presence of functional groups [7].Figure 2 (b)

IOP Conf. Series: Materials Science and Engineering **196** (2017) 012021 doi:10.1088/1757-899X/196/1/012021

was raman spectra from GO and rGO prepared from chemically reduction[8]. Based on this spectra was confirmed that the value of I_D/I_G in this research at the range from rGO.

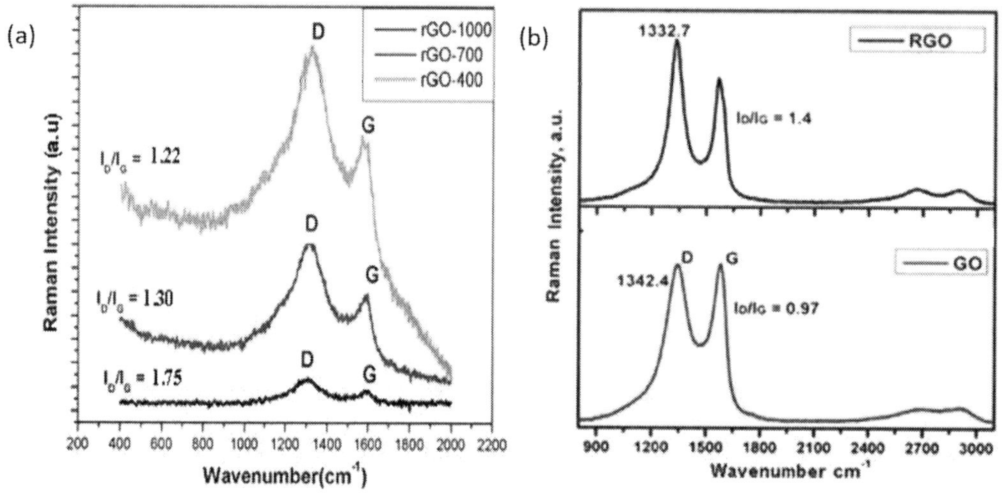

Figure 2. (a) Raman Spectra of the samples at different heating temperature prepared from old coconut shell ; (b) Raman Spectra of the GO and rGO prepared with chemically reduction [8].

The peak at around 1590 cm^{-1} is the G band. The G band is a characteristic of all graphitic structures, arising due to the inplane bond-stretching motion of sp^2 hybridized carbon atoms. The value of D and G peaks at 400°C, 700°C, and 1000°C were respectively at 1348.5 cm^{-1} and 1585 cm^{-1};1332.5 cm^{-1} and 1582.5 cm^{-1}; and 1310 cm^{-1} and 1580.5 cm^{-1}. There was a little shift of D and G peaks at different heat temperature. Raman intensity generated at different temperatures also did not same. If temperature increases, the value of raman intensity will be decreasing. That's mean if temperature increas, samples will be damaged and can be created defect. This was evidenced by increasing the value of I_D/I_G in the temperature rised. The I_D/I_G ratio increased from 1.22 at 400°C to 1.75 at 1000°C, which suggests that more sp^2 domains are formed, indicating that the defect on the rGO increasing. The increasing of the defect influence to magnetic properties from rGO.

Figure 3(a) showed a superparamagnetism of rGO having enhanced magnetization with increasing temperature or defect concentration. Higest value of magnetization given by sample rGO were prepared at temperature 700^{0}C for 5 hours with Ms = 0.26199 emu/gr, and followed by sample were prepared at temperature 1000^{0}C and 400^{0}C with Ms = 0.2579 emu/gr and Ms = 0.15726 emu/gr, respectively. The magnetization appear caused by electron from atom carbon not coupled, when induced by external magnetic field will create magnetic induction and conducted by VSM [9]. Value of magnetization will be equivalent with the number of unpaired electrons from atom. Fig.3(b) showed the correlation of I_D/I_G ratio and Magnetic Saturation at different temperature. There was thge same trend of I_D/I_G ratio and magnetic saturation, implying that the increasing defect will induce the increasing magnetism.

IOP Conf. Series: Materials Science and Engineering **196** (2017) 012021 doi:10.1088/1757-899X/196/1/012021

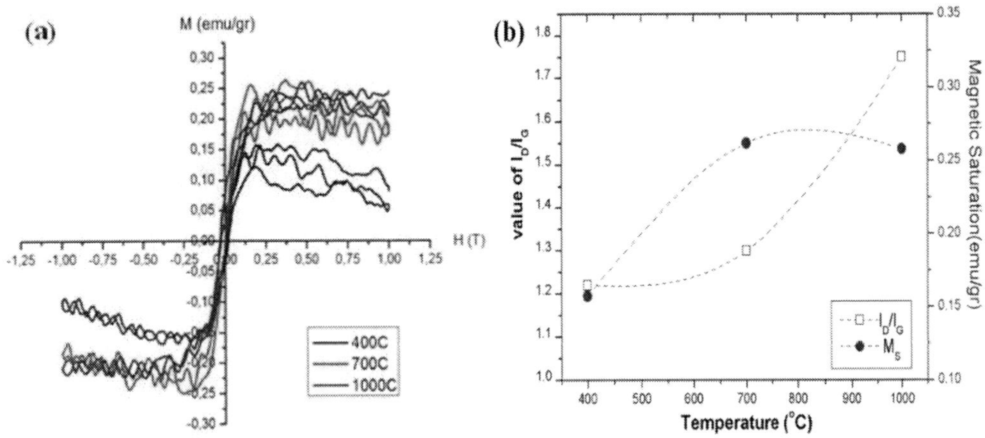

Figure 3. (a) Hysteresis curve from rGO at different heating temperature (b) The of I_D/I_G ratio and magnetic saturation at different heating temperature.

4. Conclusion

The rGO has been successfully formed from coconut shell charcoal which give heat treatment at different temperature and ultrasonic process. Characterization using Raman spectroscopy exhibited that the intensity ratio of defect (I_D) and graphene (I_G) increases, implying the increase of defect number, with respect to the increasing heating temperature. According to the magnetic property measurement using vibrating sample magnetometer (VSM), it is shown that the magnetization enhances with incerasing number of defects, featuring an occurrence of defect-induced magnetism in rGO.

Acknowledgement

This research was partially supported by "Hibah Kompetensi" Research Grant, DRPM, Kemenristekdikti, 2016.

References

[1] Mas'udah K W, Astuti F, and Darminto D 2016 *IPTEK J. Proc. Ser.* **2** 1
[2] Geim A K and Novoselov K S 2007 *Nat. Mater.* **6** 183
[3] Wachid F M, Perkasa A Y, Prasetya F A, Rosyidah N and Darminto 2014 *AIP Con. Proc.* **1586** 202
[4] Warner J, Schaffel F, Rummeli M, Bachmatiuk A *Graphene, 1st Edition* ISBN 9780123945938
[5] Soldano C, Mahmood A and Dujardin E 2010 Production, properties and potential of graphene *Carbon* **48** 2127
[6] Prasetya F A, Nasrullah M, Nugraheni A Y and Darminto 2015 *Mater. Sci. Forum* **827** 290
[7] Kudin K N, Ozbas B, Schniepp H C, Prud'homme R K, Aksay I A, and Car R 2008 *Nano Lett.* **8** 36
[8] Rajagopalan B and Chung J S 2014 *Nanoscale Res. Lett.* **9** 535
[9] Sarkar S K, Raul K K, Pradhan S S, Basu S, and Nayak A 2014 *Phys. E Low-Dimens. Syst. Nanostructures* **64** 78

IOP Conf. Series: Materials Science and Engineering **196** (2017) 012022 doi:10.1088/1757-899X/196/1/012022

Effect of Activators on Strength of Hybrid Alkaline Cement

Wiwik Dwi Pratiwi[1,2], Hamzah Fansuri[3], Januarti Jaya Ekaputri [1] and Triwulan[1]

[1]Department of Civil Engineering, Institut Teknologi Sepuluh Nopember, Surabaya, Indonesia
[2]Shipbuilding Institute of Polytechnic Surabaya, Indonesia
[3]Department of Chemistry, Institut Teknologi Sepuluh Nopember, Surabaya, Indonesia

E-mail: pratiwinoval@gmail.com

Abstract. Hybrid alkaline cement is a class of alkaline cement resulted from alkali activation of the medium calcium content of aluminosilicate materials. This paper presents an experimental analysis of alkali activators effect on strength of hybrid alkaline cement produced from 80% fly ash and 20% ordinary Portland cement. Two alkali activators were observed i.e. 5% sodium sulfate and a combination of 5% of sodium sulfate-1.1 M SiO_2 of sodium silicate solution. Compressive strength tests were performed on 20mmx 40mm cylinder paste specimens while setting time tests were conducted by Vicat needle. Scanning electron microscopy analysis and measurement of fly ash reaction degree were performed to explain the compressive strength of paste. It is concluded that addition of soluble silicate on the dry mix of hybrid cement-sodium sulfate activator reduce compressive strength and shorten the setting time. Both of activators give relative low fly ash reaction degree.

1. Introduction

Portland cement shows high performance as a binder in the concrete mixture. Nevertheless, its manufacture is energy-intensive for clinker processing at temperatures of 1450–1550°C [1]. In addition, production of one tonne of Portland cement release 0,55 tonnes of CO_2 from lime stone calcination and 0.40 tonnes of carbon-dioxide from fuel combustion [2]. Alkaline cement, which is a group of binders, have great potential as Portland cement alternatives to overcome those problems [3-6].

The system of CaO-SiO_2 -Al_2O_3 alkaline cement are divided into three main categories namely moderately calcium rich cement, low calcium cement, and hybrid alkaline cement. Moderately calcium rich cement (for example blast furnace slag) activated by relatively moderate alkaline conditions and gives C-S-H (calcium silicate hydrate) gel as the main reaction product. Low CaO contents such as metakaolin or class F fly ash requires highly alkaline media at curing temperatures of 60–200°C. The main reaction product, in this case, is an N-A-S-H (natrium aluminum silicate hydrate) or geopolymer gel. Hybrid alkaline cement or hybrid cement is the result of the alkaline activation of materials with more than 20% of CaO, SiO_2 and Al_2O_3 contents. The cement is subclassified into two sub groups i.e. with and without cement clinker. The reaction products are mixtures of C-A-S-H (containing sodium) and (N,C)-A-S-H (high calcium content N-A-S-H) gels [7].

Hybrid cement with low clinker cement content is an interesting area of study by many researchers [8-13]. Moreover, Shi et al. [3] revealed that hybrid alkaline cement system had higher possibility to encounter the aforementioned environmental issue than the other type of alkaline cement without Portland cement. One of the root cause is the quality of a prime material of alkaline cement has not been standardized yet.

Two essential factors influence the strength of alkaline cement are raw materials and alkali activators. Different anions or cations result in a distinctive effect on activation process. Common types of anion are hydroxyl (OH-), silicate, carbonate, and sulfate, while usual cations are Na^+ and K^+ [14]. Mechanical strength developed by the same raw material may significantly different depending on activators being used.

Sodium sulfate (Na_2SO_4) activator could increase the strength of hybrid cement significantly [15]. Addition of Na_2SO_4 promotes NaOH formation through reaction with $Ca(OH)_2$, expressed in equation (1) [16]:

$$Na_2SO_4(s) + Ca(OH)_2\,(s) + 2H_2O(l) \rightarrow CaSO_4{\cdot}2H_2O(s)\downarrow + 2NaOH(aq) \qquad (1)$$

In this case ($Ca(OH)_2$ is produced by Portland cement hydration. Sodium hydroxide increases the pH of the solution, accelerates the dissolution of fly ash and promotes the pozzolanic reaction.

Palomo et al. [12] studied hybrid alkaline cement activated by NaOH solution, sodium silicate + NaOH solution and pure water as a control specimen. System hydrated with sodium silicate + NaOH solution reached the highest strength. Surprisingly, NaOH activation showed the lowest strength, which was even lower than water hydration. The authors concluded that Portland cement hydration strongly influenced by OH^- concentration and the presence of soluble silicate in the medium. The finding was also supported by another research team [17]. A combination of soluble silicate and sodium hydroxide showed positive role in strength development of hybrid alkaline cement.

Silicate species in the sodium silicate solution exist in the form of the monomer, dimer, membered rings, and polymer. Minimum silicate monomer concentration for geopolymer gel formation is approximately 0.6 M SiO_2 which corresponds to 0.8M SiO_2 for Na/Al about 0.5 and 0.4 SiO_2 for Na/Al =0.75 [18].

Since NaOH demonstrates corrosive and hygroscopic behavior, it can not be mixed with prime materials in the solid state. In contrast, sodium sulfate is readily dry mixed with Portland cement-fly and gives processing advantage similar to ordinary cement. However, the combination of this activator with sodium silicate solution has not been focused by researchers yet. This study investigates comparison of compressive strength between hybrid alkaline cement activated by solid sodium sulfate versus combination of solid sodium sulfate- sodium silicate solution activator.

2. Materials and method

2.1 Materials

This study was conducted at hybrid alkaline cement system made of 80% class F fly ash and 20% ordinary Portland cement (OPC) mixture. SEM image of fly ash particles prepared with Nebula™ particles dispersion and particles size counted by Phenom illustrated in Figure 1. The chemical composition of these materials listed in Table 1 was determined by X-ray fluorescence technique. Sodium sulfate monohydrate ($Na_2SO_4.H_2O$) was used as solid activator while sodium silicate liquid 52°Baume having the composition of 15% Na_2O, 30% SiO_2, 55% H_2O was used as a liquid activator.

Table 1. Chemical compositions of fly ash and OPC

	SiO_2	reactive SiO_2 *	Al_2O_3	CaO	Fe_2O_3	TiO_2	K_2O	MgO	Na_2O	SO_3
Fly ash	55.52	26.8	14.74	6.77	15.79	1.02	0.66	3.02	<0.001	0.75
OPC	20.92		5.49	65.21	3.78			0.97		2.22

* reactive SiO_2 of fly ash was determined by gravimetric method

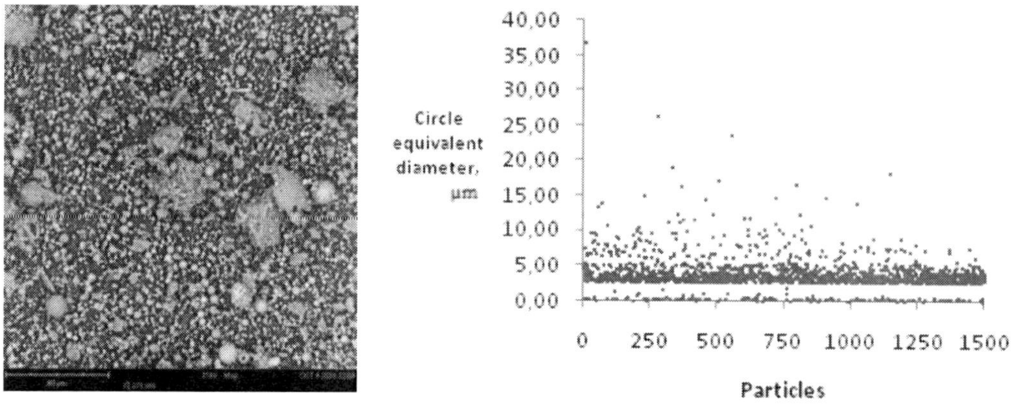

Figure 1. SEM image of fly ash (left side) and particles size pattern counted by Phenom's particle analyzer software (right side)

2.2 Methods

Fly ash, OPC, and sodium sulfate monohydrate activator were dry mixed. This blended cement-solid activator was reacted with two kinds of liquids: deionised water and solution of 1.1 M of SiO_2. The later was prepared by diluting sodium silicate liquid 52° Baume with deionized water. The SiO_2 concentration was adopted from [18] for lower Na/Al mol ratio. Total water to total solid ratio (W/S) was arranged to get similar pastes flowability. Total water is the sum of deionized water and water in sodium silicate solution, while total solid is the sum of fly ash, OPC, sodium sulfate and the solid content of sodium silicate solution. Table 2 presents the detail mixtures.

Table 2. Proportion of hybrid alkaline cement pastes

Paste	Mass proportion			W/S	$[SiO_2]$, M*	Na/Al mol ratio*
	Fly ash	OPC	$Na_2SO_4.H_2O$			
AS05	80	20	5	0.29	0	0.24
AS05-SS11	80	20	5	0.45	1.1	0.32

*in reacting solution

The two resulting pastes were casted in 20mm x 40mm acrylic cylinders molding. These specimens were de-molded after 24 hours and stored under wet-sealed plastic films for the duration of the curing process. All pastes were tested for compressive strength at 3, 7, and 28 days. Setting time of pastes was tested by Vicat needle at 28°C in accordance to ASTM C191-08 [19]. Scanning electron microscopy (SEM) was used to analyze microstructure of the pastes. The degree of fly ash reactions of paste was gravimetrically measured by Takashima attack [12] followed by chloride acid (HCl) attack [20]. Takashima attack was conducted by reacting paste with a salicylic acid-methanol solution. Portland cement hydration product and unreacted Portland cement were disrupted by this treatment, while fly ash was not influenced. The residue of Takashima attack was mixed with water : HCl 37% solution (1:20). This solution dissolve reacted fly ash and left unreacted one. The ratio of soluble fly ash to total fly ash attacked by HCl was identified as fly ash reaction degree.

3. Result and Discussion

Table 3 shows setting time, the compressive strength of paste at 3, 7, 28 days and fly ash reaction degree at 7 days. As a usual behavior of cement materials, the compressive strength was increasing with curing time for all specimens. Surprisingly, specimen activated by a sodium sulfate - sodium silicate solution (AS05-SS11 paste) had shorter setting time than that with an only solid activator (AS05 paste) while the former composed of the mixture with higher water to solid ratio. Generally, increasing water to solid ratio correlated to longer setting time[21].

Table 3. Setting time and the compressive stength of hybrid alkaline cement paste

Paste	Setting time [h:m]		Compressive strength [MPa]			Fly ash reaction degree, 7days
	Initial	Final	3 days	7 days	28 days	
AS05	4:00	5:55	17	20	28	0.31
AS05-SS11	0:35	1:05	5	9	12	0.30

Observation by simple experiment on fly ash and OPC separately mixed with two kinds of activator was also has been carried out. It was found that pure OPC set quickly when it was reacted with sodium sulfate-soluble silicate solution activator. It is highlighted that soluble silicate is an aggressive species readily reacted with Portland cement. Furthermore, it is supposed that the reacted part of OPC has hardened immediately which in turn hinder further reaction. In other word, it is hypothesized that Portland cement was not completely reacted in AS05-SS11 system.

It was found also that activation of fly ash with solid-liquid activator decrease paste flowability than another one activated by a single activator, solid sodium sulfate. This condition leads to an increasing of water demand of AS0511-SS11 to get the same paste flowability of AS05. Usually, the strength of cement paste is decrease as water to solid ratio increase [21]. It is concluded that the lower strength of AS05-SS11 system than AS05 supposed resulted by uncomplete of OPC reaction and high water to solid ratio. The exact reason might be found by deep studying the effect of activators to each component of hybrid cement raw material separately.

Figure 2 presents backscatter scanning electron microscope (BS-SEM) image of paste specimens before and after Takashima attack. Before Takashima attack, specimen AS05 (a) is denser than AS05-SS11 (c) which conform to the compressive strength test results presented in Table 3, where AS05 paste has higher strength than AS05-SS11. Denser paste is produced by the higher portion of reaction product phase or lower water to solid ratio.

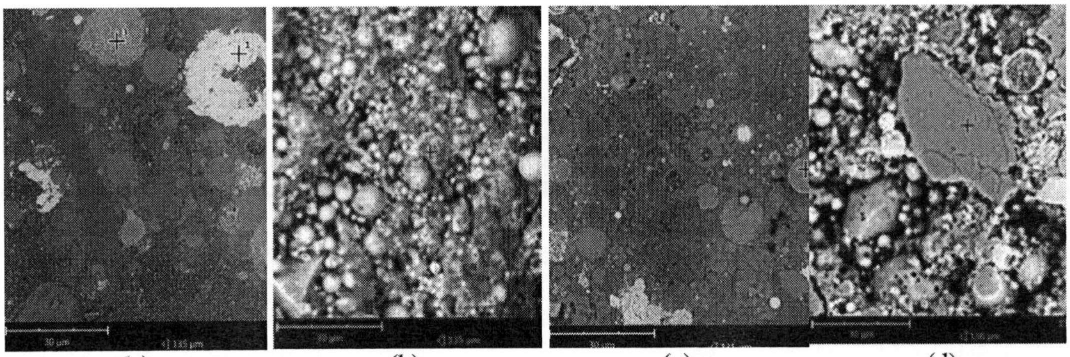
(a) (b) (c) (d)

Figure 2. BS-SEM Image, (a) AS05 paste before Takashima attack, (b) AS05 paste after Takashima attack, (c) AS05-SS11 paste before Takashima attack, and (d) AS05-SS11 paste after Takashima attack

The strength of Portland cement paste is very likely contributed by C-S-H phase, while N-A-S-H phase is responsible for strength development of alkali activated fly ash. Both phases coexist in hybrid cement but interfere each other [17].

Figure 2 also shown that both of Portland cement hydration product and unreacted cement were dissolved by Takashima attack, while activated and unreacted fly ash were unaffected by this treatment. Figure 2 (b) and (d) illustrated bond of unreacted fly ash by activated one of AS05 and AS05-SS11 paste, respectively, which the former seem tighter than the later. This fact gives good confirmation of the higher compressive strength of AS05 than AS05-SS11. Fly ash reaction degree by

HCl attack at 7 days is presented in Table 2. Only about 30% of fly ash of the two systems are converted to reaction product at 7 days. It means that the strength contributed by fly ash reaction product is quite low. Criado [20] was found reaction degree of 55% at 7 days resulted by activation of fly ash having 50.44% reactive silica content. The possible reasons of low part of reacted fly ash in the hybrid cement mixture are inappropriate activator or low reactive silica content in fly ash, i.e 26.8%. This assumption might be proven by detail alkali activation experiments for numbers of fly ash with different reactive silica content.

4. Conclusion

The following conclusions can be drawn from this study:

- The lower strength of AS05-SS11 system than AS05 supposed resulted by uncomplete of OPC reaction and high water to solid ratio. The exact reason might be found by deep studying the effect of activators to each component of hybrid cement raw material separately.
- Both systems have relative low fly ash reaction degree. The possible reasons are inappropriate activator or low reactive silica content in fly ash.

References

[1] Taylor H F W 1997 *Cement Chemistry* 2nd ed (London: Thomas Telford Publishing) p 1
[2] Davidovits, J 1994 *World Resour. Rev.* **6** 263–78
[3] Shi C, Jiménez AF, Palomo A 2011 *Cem. Concr. Res.* **41** 750–63
[4] Duxson, P, Fernandez-Jimenez, A, Provis, J L, Lukey, G C, Palomo, A, & Van Deventer, J S 2007 *J. Mater. Sci* 2007 **42** 2917–33
[5] Keyte L M 2009 *Geopolymers: Structures,Processing, Properties, and Industrial Application*ed John L Provisand Jannie S J van Deventer(Cambridge: Woodhead Publishing) p 15
[6] Provis J L 2014 *Mater. Struct.* **47** 11–25
[7] Palomo A, Krivenko P, Garcia-Lodeiro I, Kavalerova E, Maltseva O, Fernandez-Jimenez A 2014 *Mater Construcc* **64** **1**–23
[8] Donatello S, Maltseva O, Fernandez-Jimenez A, Palomo A 2014 *J.Am. Ceram. Soc.* **97** 929–37
[9] Garcia-Lodeiro I, Carcelen-Taboada V, Fernandez-Jimenez A, Palomo, A 2016 *Constr Build Mater.* **105** 218–26
[10] Garcia-Lodeiro I, Palomo A, Fernandez-Jimenez A, Macphee DE 2011 *Cem. Concr Res* **41** 923–31
[11] García-Lodeiro I, Fernandez-Jimenez A, Palomo A. 2013 *Cem. Concr. Compos.* **39** 82–92.
[12] Palomo A, Fernández-Jiménez A, Kovalchuk G, Ordoñez LM, Naranjo MC 2007 *J. Mater. Sci.* **42** 2958–66.
[13] Fernandez Jimenez A, Flores E, Maltseva O, García-Lodeiro 2013 *Rom. J Mater.* **43** 195–200.
[14] Garcia-Lodeiro I, Palomo A, Fernandez-Jimenez A 2015 *Handbook of Alkali- Activated Cement, Mortars and Concretes* ed Pacheco-Torgal (Cambridge: Woodhead Publishing) chapter 2 p 61-9
[15] Shi C, Krivenko P and Roy D 2006 *Alkali-Activated Cements and Concretes* (London and New York: Taylor & Francis) p 244-5
[16] Shi, C and Day R 2000 *Cem. Concr. Res.* **30** 51–8
[17] Garcia-Lodeiro I, Fernandez-Jimenez A, Palomo A. 2013. *Cem. Concr. Res.* **52** 112–22
[18] Rees CA, Jannie van Deventer, John Pravis, Grant C Lukey 2007 *Mechanisms and Kinetics of Gel Formation in Geopolymers* (PhD Thesis: Dep Chem Biomol Eng Univ of Melbourne)
[19] ASTM International 2013 *ASTM C191-08: Standard Test Methods for Time of Setting of Hydraulic Cement by Vicat Needle* (Conshohocken, PA)
[20] Criado M, Fernandez-Jimenez A, de la Torre AG, Aranda MAG, Palomo A 2007 *Cem. Concr. Res.* **37** 671–9
[21] Bediako M, Frimpong AO 2013 *Mater Sci Appl.* **4** 20–8

IOP Conf. Series: Materials Science and Engineering **196** (2017) 012023 doi:10.1088/1757-899X/196/1/012023

Tensile Properties and Deflection Temperature of Polypropylene/Sumberejo Kenaf Fiber Composites with Fiber Content Variation

S.L. Ollivia[1], A.L. Juwono[1] and Seto Roseno[2]

[1]Department of Physics, Faculty of Mathematics and Natural Sciences, Universitas Indonesia, Depok 16424, Indonesia

[2]Agency for Assessment and Application of Technology, Serpong 16340, Indonesia

E-mail: ariadne.laksmidevi@ui.ac.id

Abstract. The use of synthetic fibers as reinforcement in composites has disadvantage which are unsustainable and an adverse impact on the environment. An alternative reinforcement for composites is natural fiber. Polypropylene and Sumberejo kenaf fibers were used respectively as the matrix and reinforcement. The aim of this research was to obtain the optimum tensile properties and deflection temperature with the variation of kenaf fiber fractions. Polypropylene/kenaf fiber composites were fabricated by hot press method. The kenaf fiber was soaked in NaOH solution before being used as the reinforcement and polypropylene was extruded before being used as the matrix. The weight fractions were varied to produce composites and pristine polypropylene samples were also prepared for comparison. The optimum tensile strength, modulus and deflection temperature were found in the composites with the 40 wt% kenaf fiber fraction with an increase up to 80% and 170% compared to the pristine polypropylene with the values of (60.3 ± 4,3) MPa and (159.1 ± 1,8) °C respectively. The Scanning Electron Microscope observation results in the fracture surface of the composites with the 40 wt% fiber fraction showed a relatively good bonding interface between fibers and the matrix and the failure modes were fiber breakage and matrix failures.

1. Introduction

Synthetic fibers have been used in industry but these materials are not environmentally friendly and appear to be environmental issues such as waste disposal and pollution from combustion [1]. The use of natural fibers as reinforcement in composites have been widely researched and developed in recent years. Utilization of natural fibers has increased due to the relatively low cost of materials, high specific modulus, relatively light compared to glass fiber, lower energy requirements, the availability of abundant, biodegradable, resistance to deforestation and other benefits [2]. Natural fibers are expected to be an alternative to synthetic fibers and give good impact to the environment [3]. Automotive and aircraft industry requires a renewable and environmentally friendly materials [4]. Types of natural fibers used in the automotive industry is hemp, sisal, wool and kenaf. Kenaf fiber began to be developed in the world such as BMW, Toyota, and Ford as a package tray and door panels in the car [5]. Not only in the automotive field, kenaf fiber is also used in textile, agriculture, medicine and other fields. In an effort to save the forest wood, kenaf fibers along the stem has been developed for the manufacture of good quality pulp [6]. Indonesia is a tropical country with a variety of natural resources that have the potential to develop natural fibers such as jute, flax, cotton, abaca, agave, linum, roselle and kenaf [7]. Kenaf plant has many benefits because it can be cultivated using a simple technology, the relatively short lifespan of about 4-5 months [8].

The content of fiber determines the mechanical properties, thermal properties and morphology of the composite [3]. Lee et al. [9] have done research on long and discontinuous kenaf fiber reinforced polypropylene composites, it was concluded that the kenaf fiber reinforced PP composites have an optimum nominal fiber fraction of 30% by weight at the which the tensile and flexural modulus are the highest, while the reduction in strength is minimal. Zampaloni et al [10] reported that chopped kenaf fiber content of both 30% and 40% by weight has been proven to provide adequate reinforcement to increase the strength of the polypropylene powder. In the research by Wambua et al [11], the best results of tensile strength lies in composites with a weight fraction of 50 wt% using randomly fibers oriented. Bledzki et al [12] conducted a research on a composite with some natural fiber reinforcement, including kenaf fibers. Weight fraction of the fiber used in this research was 40 wt% with a size of 2 mm kenaf fiber.

Natural fibers require alkali treatment prior to be used as reinforcement in composites. Alkali treatment on fiber is the fiber immersion in a solution of NaOH. Alkali treatment on natural fibers can increase the stiffness, strength, modulus of dynamic flexibility in the composite, and improve the interface bond strength and adhesion between matrix and fiber. Alkali content used was 5% because provides the optimal tensile strength kenaf fiber reinforced polypropylene composite [13].

In this research kenaf fiber derived from Sumberejo, East Java, Indonesia was used as the reinforcement for polypropylene, so that it can provide added value in its use. Continuous oriented of kenaf fiber used in this research to aim obtaining the optimal mechanical properties and thermal properties.

2. Experimental Method

Preparation of Materials. The materials used in this research were kenaf fibers (KF), polypropylene (PP), and NaOH flakes. The 5% NaOH solution was prepared by mixing aquabidestilation with NaOH flakes with a ratio of 1:20. The kenaf fibers were soaked for 24 hours and then rinsed with destilated water. The fiber was dried at room temperature for 48 hours and dried at the temperature of 60 °C for 24 hours. Polypropylene pellets were extruded to make sheet forms of polypropylene so that the PP easily dispersed and wetted the KF.

Fabrication of Composite. KF And PP in sheet form were arranged according to a certain weight fraction. Weight fraction of KF in PP used were 20 wt% (PP/Kenaf20), 30 wt% (PP/Kenaf30), 40 wt% (PP/Kenaf40), 50 wt% (PP/Kenaf50) and 60 wt% (PP/Kenaf60). KF and PP were weighed then arranged into a mold in the form of lamina. KF and PP then hot press conducted at 190 °C for 5 min at a pressure of 50 Bar. The composites for tensile test specimen prepared according to the ASTM D638 type IV standard. The specimens for heat deflection temperature test were cut according to standard of ISO 75-2.

3. Result and Discussion

Single fiber tensile test results of kenaf before alkalization treatment is amount to 215.9 MPa and after alkalization treatment is amount to 111.5 MPa. Figure 1 shows a tensile strength value of each weight fraction of fiber. For pristine PP has a value of tensile strength of 32.7 MPa, values for tensile strength of the composite increased when adding fiber weight fraction of up to 40 wt%. The maximum tensile strength value obtained on a sample of PP/ Kenaf40. The increased value of tensile strength due to the greater fiber content will strengthen the composite, but after passing through the fiber fraction of 40 wt%, the value of tensile strength decreased due to the PP/Kenaf50 and PP/Kenaf60 because there are many fiber content so that PP difficult to wetting the fibers. Young modulus increase up to the optimum ratio of weight fraction in the sample PP/Kenaf40 and then decreased in PP/Kenaf50, as can be seen in Figure 1. Young modulus of pristine PP is 1.8 GPa, PP/Kenaf40 has a modulus of young of 6.4 GPa.

Lee *et al* have done research on long discontinuous kenaf fiber [9]. From this research, the value of the tensile strength of pristine PP amounted to (33 ± 2) MPa with a maximum tensile strength value in the composite with a weight fraction of 40 wt%. According to Lee et al, the value of tensile strength increasing due to interaction of the interface adhesion to the matrix and fibers. The interaction is a physical, because the composite fabrication process does not use the coupling agent. The use of coupling agent indicate a chemical changes. During a tensile test, interface bonding between matrix and fiber is not strong enough to resist the load and it causes the lower load transfer and obtained a fracture. Composite parts that are not filled with fibers may be another cause of low tensile strength because the void is a source of stress.

Figure 2 shows the deflection temperature of the weight fraction of fibers in composite samples of pristine PP. The pristine PP has a deflection temperature of 59 °C, the addition of fiber to 40 wt% in the samples of PP/Kenaf40 lead to higher deflection temperature composite became 159.1 °C, the sample of PP/Kenaf50 obtained a deflection temperature amonted to and the PP/Kenaf60 has a deflection temperature amounted to 94.2 °C. The values obtained with optimum deflection temperature is the weight fraction of fiber composites with 40 wt%. Bledzki et al have done research on the effect of heat on a composite polypropylene/kenaf fiber [14]. A significant increase in temperature with increasing weight fraction of fiber. Bledzki assume that the parameters Heat deflection Temperature is affected by two things: the load on the fiber and matrix interface strength. The difference value Heat Deflection Temperature test results are also affected by the weight fraction in the composite.

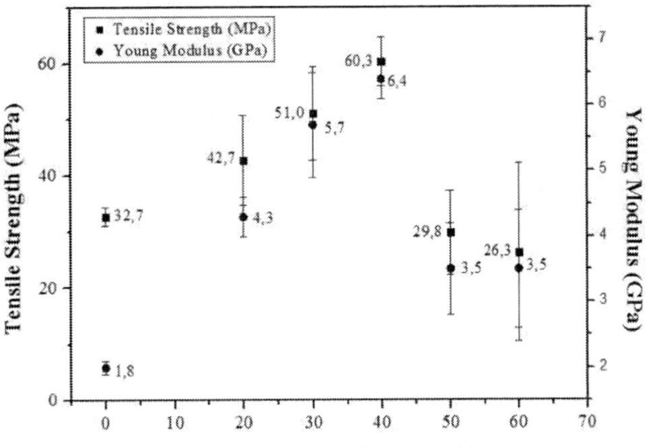

Figure 1. Tensile strength and young modulus of PP/KF composites versus fiber fractions.

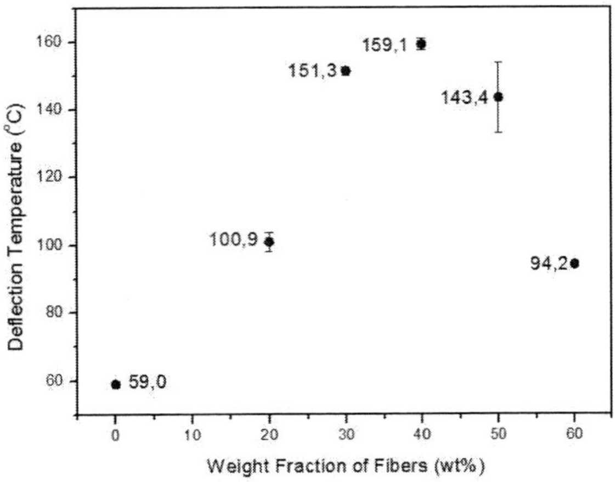

Figure 2. Deflection temperature of PP/KF composites versus fiber weight fractions.

Figure 3 shows the Scanning Electron Microscope (SEM) images of sample PP/Kenaf20, PP/Kenaf30, PP/Kenaf40 and PP/Kenaf50 before and after the tensile test. In the sample of PP/Kenaf20 (Figure 3 (a)) before the tensile test shows voids and not homogeneous layer between matrix and fibers, it can affect the mechanical and thermal properties of composites. Figure 3 (b) shows the SEM images of PP/Kenaf20 after the tensile test, it shows there are fibers pull-out from matrix. In Figure 3 (c) shows the SEM images before tensile tests on composites PP/Kenaf30, it shows the layer between fiber and matrix is homogeneous. In Figure 3 (d) there are a failure modes in the form of fiber pull-out from the matrix and fiber breakage. SEM image before tensile test on a sample of PP/Kenaf40 (Figure 3 (e)) shows the number of relatively less voids and layers between the matrix and fiber are more homogeneous than samples PP/Kenaf20. Figure 3 (f) shows the SEM images a sample of PP/Kenaf40 after the tensile test which shows that less fibers pull-out than the PP/Kenaf20 and PP/Kenaf30. SEM image before tensile test on a sample of PP/Kenaf40 (Figure 3 (e)) shows the number of relatively less voids and layers between the matrix and fiber are more homogeneous than samples PP/Kenaf20. Figure 3 (f) shows the SEM images a sample of PP/Kenaf40 after the tensile test with a magnification of 50 times which shows that less fibers pull-out than the other samples. Figure 3 (g),(h) show the SEM images of PP/Kenaf50 before and after tensile test. There are voids and fiber failure such as fiber breakage and fiber is not completely equitable distribution. Figure 4 shows the fiber is not completely wetted and separated between the matrix and fibers relatively much in amount than the other samples. Samples PP / Kenaf60 value less than the tensile strength of the composite pristine PP and the other PP/kenaf fiber, this is due to the failure mode of this fiber composition is larger in amount than the other samples.

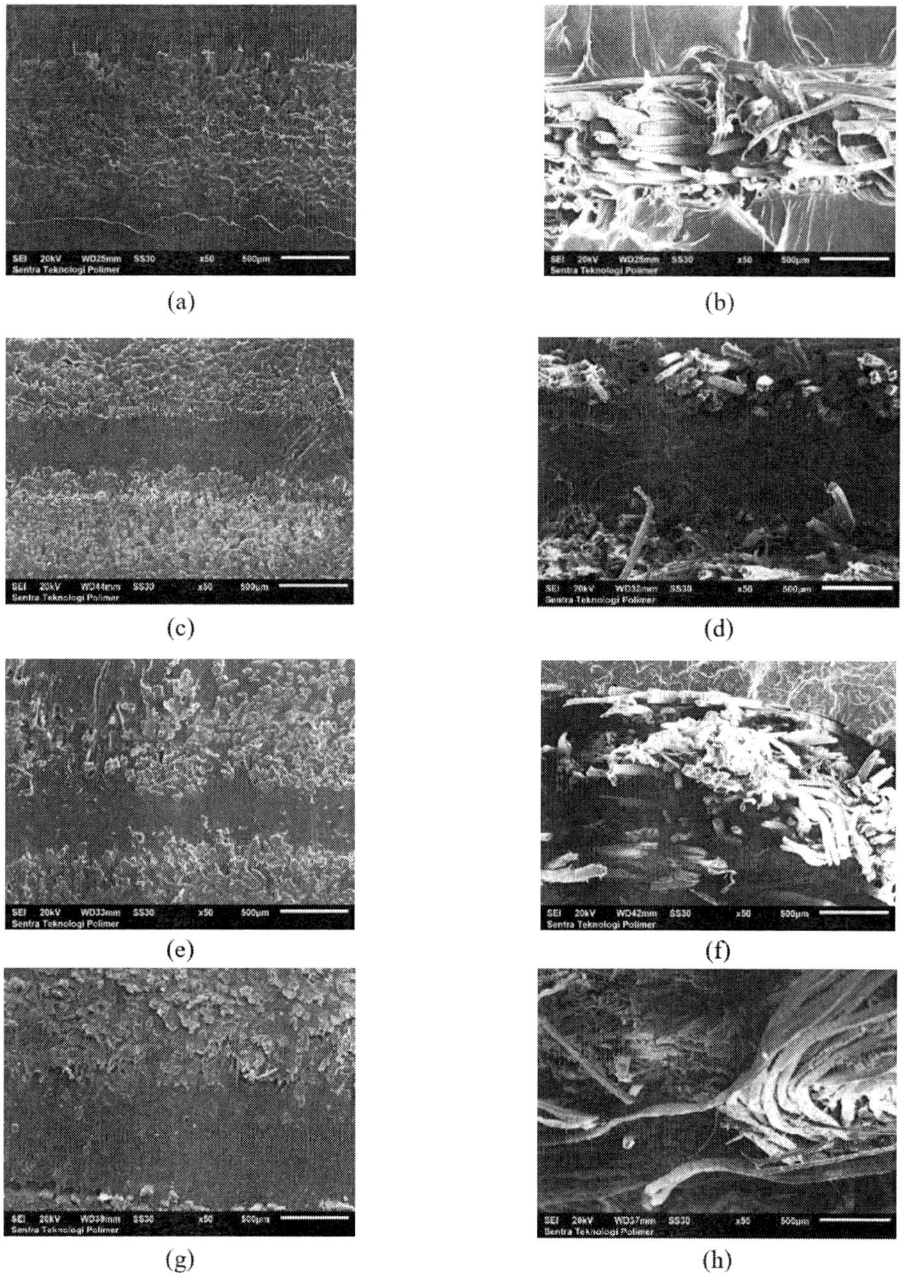

Figure 3. SEM image of PP/KF. (a) PP/Kenaf20 before tensile test (b) PP/Kenaf20 after tensile test (c) PP/Kenaf30 before tensile test (d) PP/Kenaf30 after tensile test (e) PP/Kenaf40 before tensile test (f)PP/Kenaf40 after tensile test (g)PP/Kenaf50 before tensile test (h)PP/Kenaf50 after tensile test.

Surface bond between the matrix and fibers determines the mechanical properties of the composite. The load is transferred from the matrix to the fibers through the interface, the interface bonding which is required to reach the optimal reinforcement, natural fiber has hydrophilic properties and matrix has hydrophobic properties so that cause a not good interface bonding as to restrict the mechanical strength of the composite. The bond formed between the matrix and the fiber matrix can be influenced by the ability of wetting the fibers. The fibers are not wetted properly will result in defects on the bonding interface so can obtain stress on the composite [15].

IOP Conf. Series: Materials Science and Engineering **196** (2017) 012023 doi:10.1088/1757-899X/196/1/012023

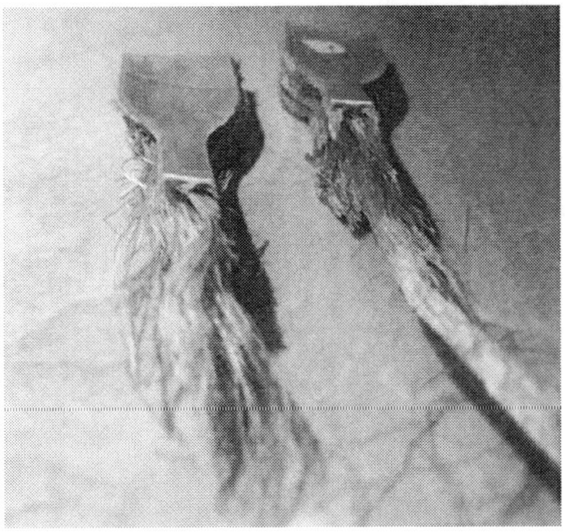

Figure 4. Sample of PP/Kenaf60 after tensile test.

4. Conclusion

From the tensile test and Heat deflection Temperature test for pristine PP and composites of PP/kenaf fiber obtained results of tensile strength and deflection temperature which increases with fiber additon in the composite up to 40 wt% fiber fraction, then the value of tensile strength and deflection temperature decreased after addition of the 50 wt% and 60 wt% fibers content. The highest value for tensile strength and deflection temperature are in composite with weight fraction of 40 wt% which is amounted to (60.3 ± 4.3) MPa and (159.1 ± 1.8) ° C respectively, these values increase each 80 % and 170% when compared to polypropylene. This is also supported by observations of Scanning Electron Microscope which showed the good interface bonding between matrixand fiber in sample PP/Kenaf40.

Acknowledgment
This work was funded by Universitas Indonesia for PITTA grant in 2016

References

[1] Haameem M J A, Abdul Majid M S, Afendi M, Marzuki H F A, Fahmi I. and Gibson A G 2016 *Compos. Struct.* **136** 1
[2] Saba N, Paridah M T and Jawaid M 2015 *Constr. Build. Mater.* **76** 87
[3] Pickering K L, Efendy M G A and Le T M 2016 *Compos. Part A Appl. Sci. Manuf.* **83** 98
[4] Szolnoki B, Bocz K, Sóti P L, Bodzay B, Zimonyi E, Toldy A, Morlin B, Bujnowicz K, Wladyka-Przybylak M and Marosi G 2015 *Polym. Degrad. Stab.* **119** 68
[5] John M J, Bellmann C and Anandjiwala R D 2010 *Carbohydr. Polym.* **82** 549
[6] Marjani, Sudjindro, Purwati R D 2009 *Jurnal Littri* vol. 15. 2 53
[7] Sudjindro 2011 *Perspektif* **10** 92
[8] Singh R K, Dubey S R and Srivastava R K 2013 *Global Journal of Biology, Agriculture & Health Sciences* **2** 72
[9] Lee B H, Kim H J and Yu W R 2009 *Fibers Polym.* **10** 83
[10] Zampaloni M, Pourboghrat F, Yankovich S A, Rodgers B N, Moore J, Drzal L T, Mohanty A K and Misra M 2007 *Compos. Part A Appl. Sci. Manuf.* **38** 1569
[11] Wambua P, Ivens J and Verpoest I 2003 *Composites Science and Technology,* **63** 1259
[12] Bledzki A K and Gassan J 1999 *Prog. Polym. Sci.* **24** 221
[13] Asumani O M L, Reid R G and Paskaramoorthy R. 2012 *Compos. Part A Appl. Sci. Manuf.* **43** 1431
[14] Bledzki A K, Franciszczak P, Osman Z, and Elbadawi M 2015 *Ind. Crops Prod.* **70** 91
[15] Liu W, Drzal L T, Mohanty A K and Misra M 2007 *Compos. Part B Eng.* **38** 352

IOP Conf. Series: Materials Science and Engineering **196** (2017) 012024 doi:10.1088/1757-899X/196/1/012024

The Use of Mushroom Growing Media Waste for Making Composite Particle Board

Tjahjanti P H[1], Sutarman[2], Widodo E[1] and Kusuma AT[1]

[1] Department of Mechanical Engineering, Universitas Muhammadiyah Sidoarjo, Indonesia
[2] Department of Agriculture, Universitas Muhammadiyah Sidoarjo, Indonesia

E-mail: prantasiharmi@umsida.ac.id

Abstract. One of the agricultural waste is mushroom growing media, where the number of this waste is huge and accumulated after mushroom harvest. This accumulation is caused by using the new planting medias. Meanwhile, the old planting medias have thrown out and been as solid matter to widen the slope of cliff area. Therefore, this research aims to use mushroom growing media waste (MGMW) as a composite particle board added by polyester resin and mekpo (methyl ethyl ketone peroxides) as catalys. Some physical and mechanic tests which are done, namely: density, moisture content, thickness swelling after immersion in water, strength of absorption water, internal bonding, modulus of elasticity, modulus of rupture and screw holding power. The composition of 75% MGMW + 24% polyester resin + 1% catalyst mekpo suitable to the physical and mechanic tests and accordance with SNI 03-2105-2006 and JIS A 5908-2003.

1. Introduction

Composite particle board is a part of particle board, until now, the basic material for making particle board is still wood, either wood fickle or wood waste, such as sawdust. Meanwhile, the stock of wood in the forest is limited. In the period of eight years later, the use of wood as raw material in the timber industry is descending. In 2005-2013, the fullfilment of raw material from natural forest decreased from 20.5 million m^3 to 5.54 million m^3. On the other side, the increasing of wood material from planting forests, including the industrial planting forest and public forest is from 11.22 million m^3 to 39.8 million m^3. Thus, it makes the condition of forest in Indonesia is getting worse [1].

One of the agricultural waste is Mushroom Growing Media Waste (MGMW) where the number of this waste is huge and accumulated after mushroom harvest. This accumulation is caused by using the new planting medias. Meanwhile, the old planting medias have thrown out and been as solid matter to widen the slope of cliff area. This research aims to use MGMW as a composite particle board must be in accordance with the Standart. The application will be used for audio equipment such as the manufacture of box speakers.

2. Experimental Studies

The basic ingredients used in producing composite particle board are MGMW mixed by polyester resin and added by catalyst mekpo (methyl ethyl ketone peroxides). A composite particle board is made from the three compositions, namely: (1) Sample A: 80% MGMW + 19% polyester resin + 1%

catalyst mekpo, (2) Sample B: 75% MGMW + 24% polyester resin + 1% catalyst mekpo, and (3) Sample C: 70% MGMW + 29% polyester resin + 1% catalyst mekpo. The first is drying the MGMW under the sunray for about two days and then it will be sieved by size 1.5 mm. The second step is the MGMW is dried again in the oven at temperature of 90°C for six hours. If the the MGMW is really dry (free from the humidity), it can be mixed with polyester resin and catalys, and poured into a mold length x width x height (40 x 40 x 1.5) cm and then is pressed by the hot press machine with pressure of 50 kg/cm^2 at 130 °C for 30 minutes. After it reversed and press again for ± 30 minutes at the same temperature. The results of hot press is composite particle board is shown in **Figure 1(a)** and further subdivided to form the sixth section (**Figure 1b**) on physical and mechanical testing on composite particle board according to the JIS standard A 5908-2003 [2] and Standar Nasional Indonesia (SNI) 03-2105-1996 [3] includes some tests, they (for physical testing) are density and moisture content (number 5), internal bonding/IB (number 3), swelling in thickness after immersion in water, and strength of absorbtion water (number 4). Mechanical testing are modulus of elasticity/MOE (number 1), modulus of rupture/MOR (number 2), and strength of screw holding (number 6).

Figure 1(a). Composite particle board.

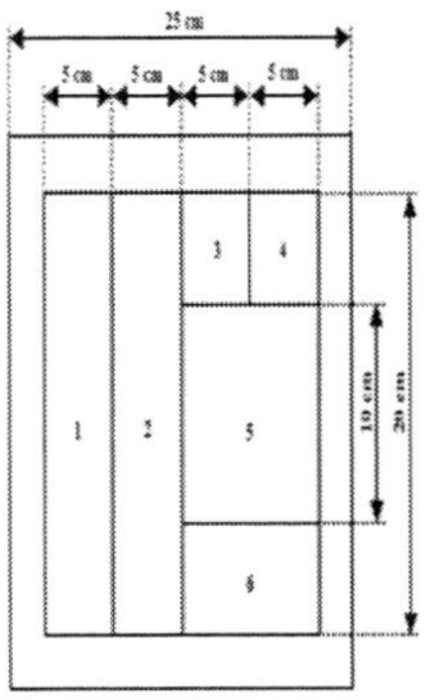

Figure 1(b). Subdivided of composite particle board (JIS A 5908-2003).

3. Result and Discussion

The result of density, moisture content, internal bonding, swelling in thickness after immersion in water, strength of absorbtion water, modulus of elasticity, modulus of rupture, and strength of screw holding for the three compositions: (1) Sample A, (2) Sample B, and (3) Sample C, which can be seen on the **Table 1**.

Table 1. Result of physical and mechanical testing on composite particle board.

Composition Composite Particle Boards	Density (gr/cm^3)	Moisture Content (%)	Swelling in Thickness after Immersion in Water (%)	Strength of Absorbtion Water (%)	IB (kgf/cm^2)	MOR (kgf/cm^2)	MOE (kgf/cm^2)	Strength of Screw Holding (kgf/cm^2)
Sample A	0.82	5.07	6.09	53.18	2.33	92.00	7740.24	41.95
Sample B	0.87	5.14	5.13	40.24	4.00	123.11	22228.52	62.91
Sample C	0.81	4.81	3.43	39.32	5.67	89.56	12425.88	80.06

Result of physical testing are density, moisture content, swelling in thickness after immersion in water, strength of absorbtion water (**Table 1**) shown that density of all sample (A,B, and C) suitable with SNI 03-2105-2006 and JIS A 5908-2003 and include in high density. Composition of 75%

MGMW + 24% polyester resin + 1% catalyst of composite particle board is the most highest of density with value 0.87 gram/cm^3. Differences value of density is affected by the number of particles MGMW and adhesive levels in the mix with the same volume. It may be occured the particles spread unevenly and also become widen in the process of hot compression. This causes the particle mass on each side of composite particle board is not the same, so the pressure and the heat is also different on the whole of sheet surface. The density of particle board is rarely uniform toward its thichness [4].

Composition of 75% MGMW + 24% polyester resin + 1% catalyst of composite particle board is the most highest of moisture content with value 5.14%, The high of moisture content is mean that adhesives used in the making of composite particle board in small quantities. It is not permitted when results of below 5% . Because will be need much more adhesives used. If the adhesive is increasingly being used then the contact between the particles will become tight so that the water would be difficult to enter between in the composite particle board.

All sample (A,B, and C) suitable with SNI 03-2105-2006 and JIS A 5908-2003 for swelling in thickness after immersion in water. Result of swelling in thickness after immersion in water will decrease when adhesive is added. However, the mechanism of thickness swelling after immersion in water of composite particle board is more complex because the particle in the board is bonded with an adhesive that can prevent the thickness swelling. The thickness swelling in the composite particle board is caused by the combination of the potential thickness recovery from the particles densification and destruction of adhesive bond web (the bond strength among particles or the pressure on adhesive bonding). Koch, 1985 [5] also states that the changes of the board dimensions are influenced by the variables of product manufacture itself, such as the density of the raw material, the thickness of the particle, the amount of compression given, adhesive level, and the number of pressure given on the board.

Another physical testing is strength of absorbtion water, explain that decrease of it because of adhesive is added. According to Maloney's statement (1993) [6], the longer the time of immersion in the water, the water absorption will be the greater. This is because the water that goes into the composite particle board is increasing as more as the adhesive substituted. Djalal (1984) [5] defines that in addition to the water absorption properties of the raw materials used and the adhesive resistance to the water, there are other factors that influence the increase of moisture content on the particle board. These factors includes the volume of empty space that can accommodate water amongthe particles, the presence of a capillary channel that connects the empty space with each other, the broad of particle surface area, and the broad of particle surface area can not be covered by adhesive. If the process of compression does not run maximally, it will make the particle broad compression is not quite good.So it will cause the water absorb easily in the particle cracks. A high absorption can also be caused by the spread of the particles unevenly and the board compression does not work optimally. It causes the particles will be loose and it will be easier to be placed by the water.

Result of physical testing are internal bonding, modulus of elasticity, modulus of rupture, and strength of screw holding. Composition of 75% MGMW + 24% polyester resin + 1% catalyst is most highest value for MOR **(Figure. 2a)**. Maloney (1993) states that the MOR values are influenced by the content and the type of adhesive used, adhesive holding capacity and particle size. Koch (1985) states that the factors that affect the value of MOR is the density of the wood particle board, particle geometry, adhesive level, moisture content of the board, and the process of compression. If the density is too high, it will cause the difficulties in processing adhessiveness. Because of the thickness of the cell wall and the small amount of cavity volume. Thus,it will make the adhesive does not penetrate easily, so the penetrating adhesive will be limitede in the a certain depth.

For MOE, only composition of 75% MGMW + 24% polyester resin + 1% catalyst suitable with SNI 03-2105-2006 and JIS A 5908-2003 with value is 22,228.52 kgf/cm^2 (SNI Minimum 15,000 kgf/cm^2 and JIS Minimum 20,000 kgf/cm^2) **(Figure. 2b)**. MOE value is influenced by the content and the type of adhesive used, adhesive holding capacity and particle size. Bowyer et al. (2003) [7] states that in addition to the density and levels of adhesive, the particle geometry is the main feature that determines the result of board properties.

All samples (A,B, and C) suitable with SNI 03-2105-2006 and JIS A 5908-2003 for internal bonding. Djalal (1984) states that the increase of density causes the particle bonding will be stronger.

The higher of density value (in accordance with the standards), the bond internal value that be resulted also getting bigger. Maloney (1993) states that the resin content gives an effect to the internal bonding. If the higher levels of resin in a certain limit, the internal bonding board will increase. This is caused by the greater the amount of the constituent molecules which reacts with wood during the adhessiveness process. However, the increase of density that exceeds the standards will make dissolved particles during compression so that it will produce a weak internal bonding.

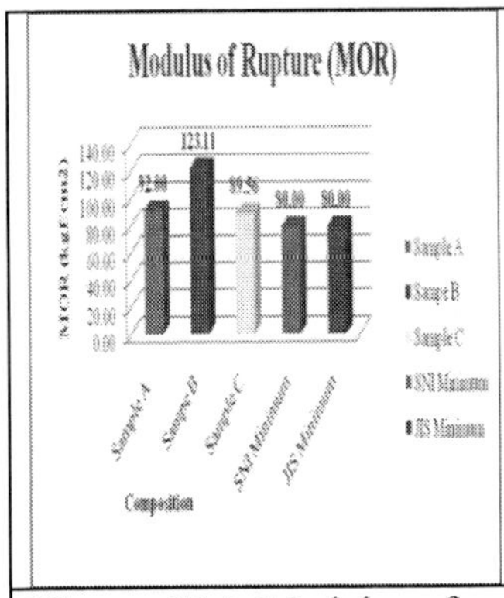

Figure. 2(a). Modulus of Rupture of sample A, B and C.

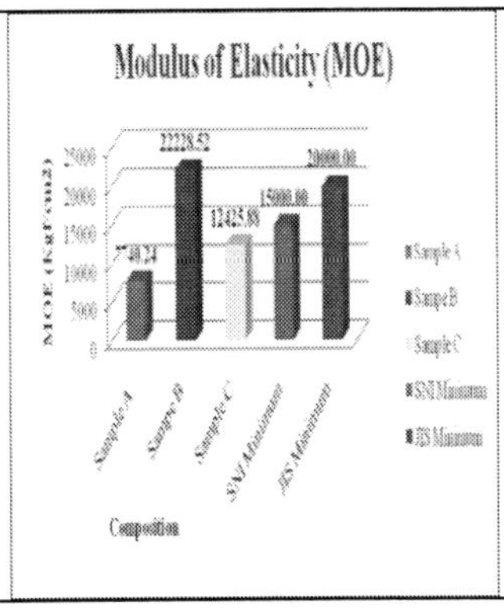

Figure. 2(b). Modulus of elasticity of sample A, B and C.

Samples A,B, and C suitable with SNI 03-2105-2006 and JIS A 5908-2003 for strength of screw holding. Bowyer et al. (2003) states that the value strength of screw holding is influenced by the density of the particle board, adhesive level, and the spread of adhesive. The high density of particle board will generate a high value of strength of screw holding.

4. Conclusion

From the discussion above, it can be concluded that optimum composition is on the composite particle board which has 75% MGMW + 24% polyester resin + 1% catalyst, suitable with SNI 03-2105-2006 and JIS A 5908-2003, can making composite particle board. The results of physical test and mechanical testing, has been appropriate with standard of JIS A 5908-2003 and Standar Nasional Indonesia (SNI) 03-2105-1996.

Acknowledgment

The authors would like to thank to the Ministry of Research and Technology National Education Indonesia (Kemenristekdikti) to grant aid to the Competence Research First Year (2015) and Second Year (2016) and University of Muhammadiyah Sidoarjo (UMSIDA) for support this research.

References

[1] Rusmantoro W 2013 *Nasib Hutan Kita yang Semakin Suram*, vol 3, Copyright Â© 1992-2014 Pelangi Indonesia
[2] Japanese Standards Association 2003 *Japanese Industrial Standard (JIS) A 5908Particleboard* (Tokyo: Japanese Standards Association)
[3] Standar Nasional Indonesia (SNI) 03-2105-1996
[4] Djalal M 1981 Tesis Bogor: Program Pasca sarjana IPB
[5] Koch P 1985 *Utilization of The Southern Pines* (Washington DC: U.S. Department of Agriculture Forest Service)
[6] Maloney T M 1993 *Modern Particleboard and Dry-Process Fiberboard Manufacturing* San Fransisco: Miller Freeman Inc. Sanfransisco)
[7] Bowyer J L, Shmulsky R, Haygreen J G 2003 *Forest Products and Wood Science: An Introduction* Ed ke-4 Ames (Iowa: Iowa State Press)

Synthesis and Properties of Iron Oxide Particles Prepared by Hidrothermal Method

T Saragi, A S Santika, B Permana, N Syakir, M Kartawidjaja and Risdiana

Department of Physics, Padjadjaran University, Jl. Raya Bandung-Sumedang km.21 Jatinangor, Sumedang 45363, Indonesia

E-mail: t.saragi@phys.unpad.ac.id

Abstract. Iron oxide of hematite (α-Fe_2O_3) has been successfully synthesized by hydrothermal method. The starting materials were $Fe(NO_3)_3.9H_2O$, 2-methoxyethanol, diethanolamine and n-hexane. The optical, morphology and crystal structure were measured by UV-VIS, TEM and XRD, respectively. From UV-VIS measurement, it was found that the band-gap of sample was 4.17 eV. The morphology of particle was plate-like form. The sample which sintered at 1100°C has high quality crystal with hexagonal structure of α-Fe_2O_3 phase.

1. Introduction

Iron oxides are the most industrially important iron ores. It is well known that there are three iron oxides in Fe-O system, namely, iron (II) oxide (FeO), iron (II,III) oxide or magnetite (Fe_3O_4) and hematite (α-Fe_2O_3) or maghemite (γ-Fe_2O_3) which is ferromagnetic mineral with the spinel group. Hematite is most stable iron oxide under ambient conditions with rhombohedral structure and antiferromagnetic below its Morin transition (T_M) of about 260 K. Magnetic properties of hematite depend on pressure, particle size, and magnetic intensity. Hematite is also *n*-type semiconductor with E_g = 2.1 eV. Maghemite can be change to be hematite phase at high temperature. Maghemite with spherical nanoparticles (NPs) and diameters less than 20 nm has advantages properties, such as chemical stability, biocompatibility and can be used for biomedical applications due to their ability to affect the relaxation rate of water proton. The relaxation rate can be decreased signal intensity resulting darkening effect in magnetic resonance image [1,2].

Various of chemical methods have been done to synthesize magnetic nanostructure including the precipitation method [3], sol–gel process [4], hydrothermal synthesis [5], high temperature reaction [6], microwave irradiation synthesis [7], polyol methods [8] and seed mediated growth method [9,10]. To control the size and morphology of nanostructure, all the methods manipulate various experimental parameters such as the pH of the solution, ionic strength, capping agent, reaction temperature, composition and pressure. However, research to obtain nanoparticles with controlled size and morphology is still open wide challenges. Recently, Kazeminezhad and Mosivand [1] have been synthesized iron oxide by electro-oxidation method and explored the sintering effect to produce magnetite, maghemit and hematite. However the sintering temperatures are limited in the range of 100°C – 1000°C. The sintering temperature above 1000°C is still attracting much interest to find other structure properties of iron oxide. In this paper, we reported the synthesis of iron oxide prepared by hydrothermal method and explored the crystal structure of iron oxide on the sintering temperature at 1100°C.

2. Experimental Method

Iron oxide prepared by hidrothermal method. This method is a simple, reliable and cheap process for small and large scale laboratory. The starting materials were $Fe(NO_3)_3.9H_2O$, 2-methaoxyethanol and diethanolamine. The raw materials were dissolved in 2-methaoxyethanol and diethanolamine (DEA) was droped wisely to control the value of pH at 10. The solutions were reacted in hydrothermal chamber at 100°C for 72 hours, following by washing with *n*-hexane. The NH_4OH added to the solution until reaching a pH value of 5, and resulting precipitate particles. Particles were then re-dispersed in 2-methaoxyethanol solution. The sample was characterized by UV-VIS and TEM to observe the optical and morphological particle, respectively. The value of band-gap was calculated from UV-VIS data by using simple equation of $E_g = hc/\lambda_{cut\text{-}off}$. To study structure and magnetic properties, sample was dried at 200°C for 6 hours, pre-sintered at 400°C for 4 hours, and sintered at 1100°C for 6 hours, then characterized by XRD and VSM measurement.

3. Result and Discussion

Figure 1 shows the wavelength dependent of absorbtion from UV-Vis measurements. From the graphics, it was found that $\lambda_{cut\text{-}off}$ is 297.36 nm. Using the equation of E_g, the band-gap of the sample was 4.17 eV. The large value of band-gap in this sample indicated the magnetite characteristic. This indication is also supported by the black colour of particles [1,11].

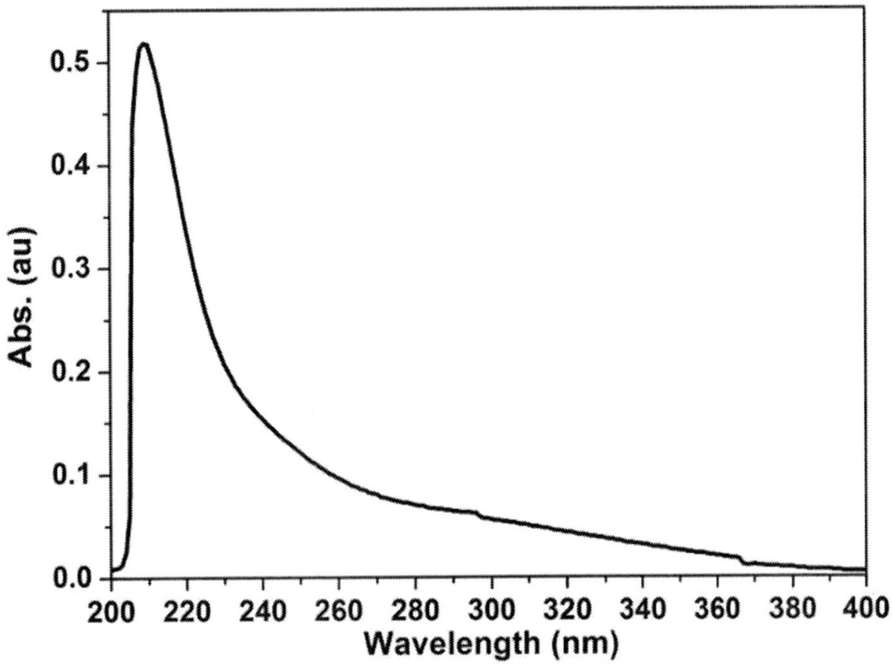

Figure 1. UV-Vis spectra of magnetite (Fe_3O_4) prepared by hydrothermal.

Figure 2 shows TEM image of iron oxides with the bar scale of 100 nm. The morphology of sample is mainly plate-like particles. The average size of particle has been calculated at 115 nm × 47 nm. The particles of iron oxide were still in submicron size and need to be optimized to obtain magnetite nanoparticles.

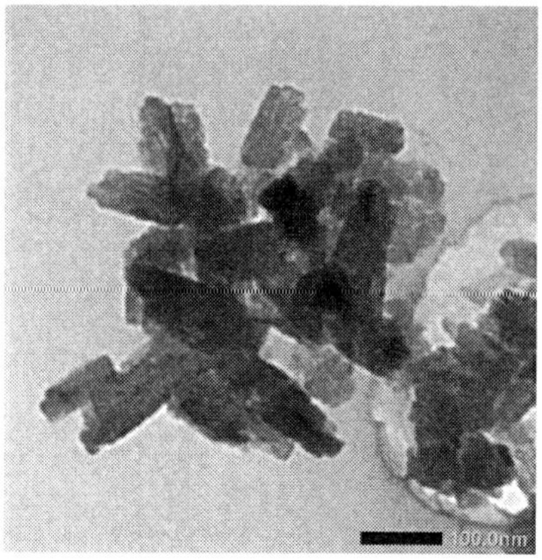

Figure 2. TEM image of sample iron oxides prepared by hydrothermal.

Figure 3 shows the XRD pattern of iron oxide sample sintered at 1100°C for 6 hours. For comparison, the XRD patterns of Fe_3O_4 (JCPDS#75-0033) and α-Fe_2O_3 (JCPDS#86-0550) are also displayed. It was found that the sample has high quality crystal with hexagonal structure of hematite α-Fe_2O_3 phase without impurity peaks. Compared with the experimental result reported by Kazeminezhad and Mosivand [1], the crystal structure sintered at temperature in the range of 800°C – 1000°C was rhombohedral. However, in our experiment, we observed the hexagonal phase structure when it was sintered at 1100°C. The changes of structure from rhombohedral to hexagonal phase probably due to the arrangement of oxygen position during sintering process. The hexagonal phase of α-Fe_2O_3 has more advantages and potential in drug delivery system. It can also provide a slow drug release matrix and protect protein from chemical degradation [12]. The lattice parameter *a* and *c* are 5.0356 Å and 13.7489 Å, respectively, which is similar with the lattice parameter observed in JCPDS#86-0550.

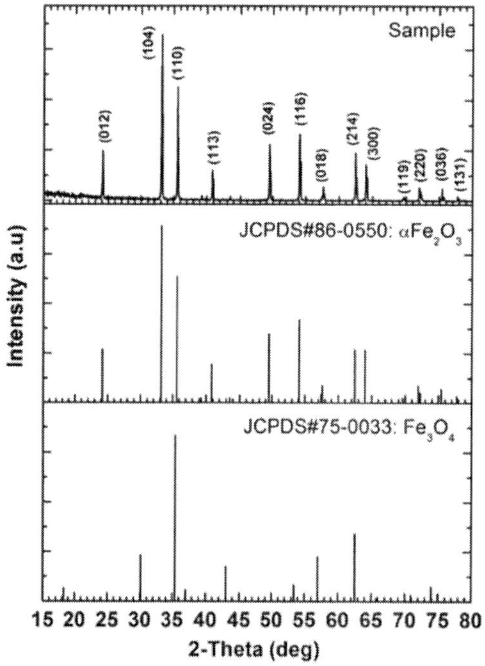

Figure 3. XRD pattern of powder α-Fe_2O_3 sintered at 1100°C for 6 hours.

From XRD data, the average crystallite size can also be calculated using the Scherrer equation below [13]:

$$D = \frac{K\lambda}{\beta \cos\theta}$$

where D is the average size of the crystallites, K is the Scherrer constant (0.9) [14], λ is the wavelength of radiation (1.54186A°), β is full width at half maximum (FWHM) of diffraction lines and θ corresponds to the peak position. The average size of crystallite is about 15.79 nm.

4. Conclusion

Iron oxide of hematite (α-Fe_2O_3) has been successfully synthesized by hydrothermal method. It was found that the band-gap of sample was 4.17 eV with the plate-like form of 115 nm × 47 nm, in dimension. The sample sintered at 1100°C for 6 hours has high quality crystal with hexagonal structure of α-Fe_2O_3 phase with lattice parameter of 5.0356 Å and 13.7489 Å, for *a*-axis and *c*-axis, respectively. The average size of crystallite is about 15.79 nm. The present result gave a good information to develop a research in drug delivery system.

Acknowledgment
We would like to thanks to Universitas Padjadjaran for financial supporting by Hibah Internal Program.

References

[1] Kazeminezhad I and Mosivand S 2014 *Acta Physica Polonica A*. **125** 1210
[2] Revia R A and Zhang M 2016 *Materials Today*. **19** 157
[3] Ghandoor H E, Zidan H M, Khalil M M H and Ismail M I M 2012 *Int. J. Electrochem. Sci.* **7** 5734
[4] Shirsath, S E, Mane M L, Yasukawa Y, Liu X, Akimitsu M 2013 *J Nanopart Res*. **15** 1976
[5] Hayashi H and Hakuta Y 2010 *Materials* **3** 3794
[6] Roca A G, Morales M P, Grady K O, Serna C J 2006 *Nanotechnology* **17** 2783
[7] Khollam Y B, Dhage S R, Potdar H S, Dhespande S B, Kulkarni S D, Date S K 2002 *Mater. Lett.* **56** 571
[8] Wei C and Wan J 2007 *J Colloid Interface Sci **305** 366*
[9] Park J, Kang B, Kim B, Suh J S, Huh Y M, and Haam S 2015 *Nanoconvergence*. **3** 1
[10] Joshi H M, Lin Y P, Aslam M, Prasad P V, Schultz-Sikma E A, Edelman R, Meade T, and Dravid V P 2009 *J. Phys. Chem. C.* **113** 17761
[11] Cornell R M and Schwertmann U 1996 *The iron oxides* VCH Press
[12] Clogston J and Caffrey M 2005 *Journal of Controlled Relase* **107** 97
[13] Trif L, Tolnai G, Sajo I and Kalman E 2008 *Progr. Colloid Polym Sci.* **135** 38
[14] Xiao-Liang N, Wen L Z, Zhong Y, Sun K, Zhong L L 2007 *Trans. Nonferrous Met. Sci. China* **17** S854

IOP Conf. Series: Materials Science and Engineering **196** (2017) 012026 doi:10.1088/1757-899X/196/1/012026

Improving Photocatalytic Performance in Degradation of Methylene Blue using Magnetite Fe-doped ZnO/Montmorillonite Nanocomposite

M I Pratiwi[1,2], N Afifah[1,2] and R Saleh[1,2]

[1]Departemen Fisika, Fakultas MIPA, Universitas Indonesia, Kampus UI Depok, Depok16424, Indonesia
[2]Integrated Laboratory of Energy and Environment, Fakultas MIPA, Universitas Indonesia, Kampus UI Depok, Depok 16424, Indonesia

E-mail: rosari.saleh@ui.ac.id

Abstract. In this work, we have successfully synthesized Fe:ZnO/Montmorillonite (MMT) nanocomposite with various loading of magnetite (Fe_3O_4) using co-precipitation method. The as-prepared samples were characterized by some measurements such as X-ray Diffraction (XRD), Burneur-Emment-Teller (BET) surface area analysis, and Fourier Transform Infrared Spectroscopy (FTIR). The XRD result shows that the diffraction pattern of nanocomposites exhibit characteristic from hexagonal wurtzite structure of ZnO and cubical spinel structure of Fe_3O_4. However, the diffraction pattern from MMT could not be detected using XRD measurement. The presence of MMT in the nanocomposite was further confirmed in the FTIR spectra. Furthermore, the photocatalytic performance of the samples was also checked for degrading methylene blue (MB) under UV light irradiation. The result shows that incorporation of Fe_3O_4 in Fe:ZnO/MMT has better photocatalytic efficiency than Fe:ZnO/MMT alone. Stability of the nanocomposites was also monitored after several cycle processes.

1. Introduction

Photocatalysis has been a popular technique to degrade organic dyes for the past few years. Moreover, photocatalysis is low cost and environmental friendly [1]. One of many tools that is used in the photodegradation of dyes is semiconductor catalyst. We had studied photocatalytic performance of Fe-doped ZnO (Fe:ZnO) and Fe:ZnO modified with montmorillonite in our previous works [2,3,4]. Doping method is advantageous for inhibiting the recombination rate of electron-hole pairs because the dopant acts as the electron trap [5]. In addition, modifying the catalyst with MMT could increase the specific surface area of the catalyst. Therefore, the contact probability between irradiated light and the catalyst increases, resulting in enhanced photocatalytic activity.

Stability of catalyst has attracted so much interest among scientists. It is known that incorporation of magnetite (Fe_3O_4) nanoparticle could stabilize and increase the magnetic saturation of the non-magnetic catalyst [6]. So an external magnetic field could be used to separate the catalyst that has been used in the first cycle of ohotocatalytic activity from the aqueous solution. The separated samples was then reused in the next runnings. Therefore, in this work, we synthesized Fe:ZnO/MMT/Fe_3O_4 nanocomposite by using co-precipitation method with five weight ratio of the magnetite in Fe:ZnO/MMT by 0.05, 0.1, 0.3, 0.5, and 1. Some characterizations were conducted in this work such as Diffuse Reflectance Spectroscopy, X-Ray Diffraction, Burneur-Emment-Teller and Fourier Transform Infrared Spectroscopy.

2. Experimental Details

2.1. Materials

Zinc (II) sulfate hepta hydrate ($ZnSO_4.7H_2O$), iron (II) sulfate hepta hydrate ($FeSO_4.7H_2O$), ethylene glycol, acetic acid, and sodium hydroxide (NaOH) were purchased from Merck while montmorillonite from Nanocor. They were analytical grade and used without further purification. Distilled water was used throughout the experiment.

2.2. Synthesize of the Fe:ZnO/MMT/Fe_3O_4 nanocomposites

The Fe:ZnO/MMT nanocomposites were prepared using co-precipitation method as described in our previous work [4]. The Fe:ZnO/MMT/Fe_3O_4 nanocomposites were synthesized five times with weight ratio of Fe:ZnO/MMT to Fe_3O_4 by 1:0.05, 1:0.1, 1:0.3, 1:0.5, and 1:1. These are the following steps: first, the Fe:ZnO/MMT and Fe_3O_4 were dissolved into ethanol. Then, the solution was mixed with distilled water and ultrasonicated for two hours. Next, a centrifugation process was conducted to separate the solutions and precipitations. After that, the precipitations were allowed to stand under room temperature for 18 hours before being dried in the oven at 100°C in vacuum conditon for one hour.

2.3. Characterization

The optical, structure, magnetic, and surface area of the samples were characterized using X-ray diffraction (Rigaku Miniflex 600) operated at 30 kV and 15 mA, vibrating sample magnetometer (Oxford type 1.2 T), infrared spectroscopy (FTIR) and Burneur-Emment-Teller (BET).

2.4. Catalytic activity

The photodegradation of methylene blue (MB) was investigated using Fe:ZnO/MMT/Fe_3O_4 for two hours at the interval of 15 minutes. The dye's degradation rate was monitored by using UV-Vis Hitachi UH5300 spectrophotometer. A 40 W UV lamp was used as light irradiation source. The effect of pH was also examined in this work. The catalyst was run for four cycle processes to check its stability.

3. Result and Discussion

The XRD was used to characterize the crystal structure of the nanocomposites. The diffraction patterns of the nanocomposites are shown in Figure 1. The magnetite, pure ZnO, and Fe:ZnO/30 wt% MMT diffraction patterns are also pointed out as comparisons. As can be observed in the figure, diffraction peaks of Fe:ZnO/MMT/Fe_3O_4 at 2θ = 30.25° (220), 35.55° (311), 43.13° (400), 53.64° (442), 57.20° (511), and 62.81° (440) represent the cubic spinel structure of the magnetite [7], while at 2θ = 31.82°(100), 34.55° (002), 36.29° (101), 47.57° (102), 56.65° (110), 62.85° (103), 66.49° (200), 67.99° (112), and 69.13° (201) represent the hexagonal wurtzite structure of ZnO [8]. It means that the samples were successfully produced. As can be shown in the figure, the magnetite intensity increases as its load increases in Fe:ZnO/MMT. The lattice parameter obtained by the Rietveld refinement technique using the program MAUD and crystal size of the nanocomposites calculated using Debye-Scherer formula [9] are tabulated in Table 1. pecific surface area of nanocomposites were calculated using BET, as shown in Table 2. It indicates that the addition of magnetite nanoparticle into Fe:ZnO/MMT could decrease the surface area value of the nanocomposites.

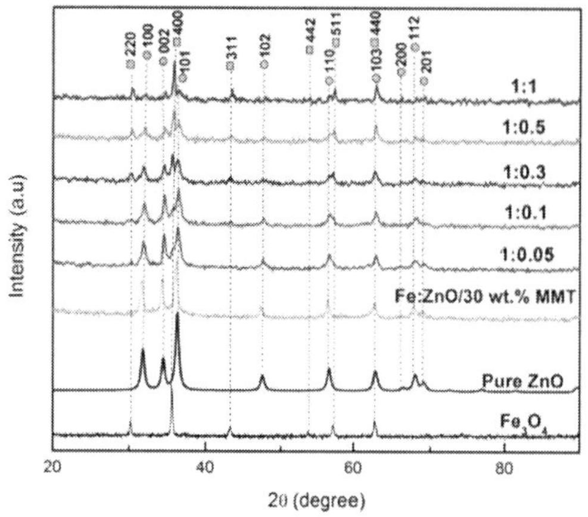

Figure 1. XRD Patterns of Fe_3O_4 nanoparticles, ZnO nanoparticles, Fe:ZnO/MMT, and Fe:ZnO/MMT with Various Weight Ratio of Fe_3O_4.

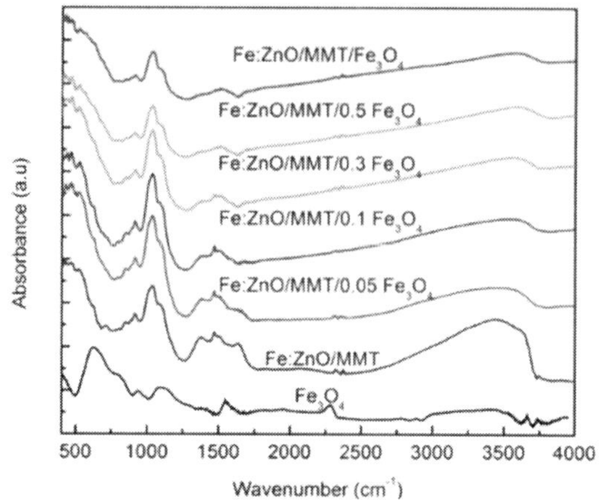

Figure 2. FTIR Spectra of MMT, Fe_3O_4, Fe:ZnO/MMT, and Fe:ZnO/MMT with Various Weight Ratio of Fe_3O_4.

Table 1. Lattice Parameters, Crystallite Size, and Unit Cell Volume of The Samples

Sample	Weight Ratio	Lattice Parameter			Crystallite Size (nm)		Unit Cell Volume (Å³)	
		ZnO		Fe_3O_4	ZnO	Fe_3O_4	ZnO	Fe_3O_4
		a=b (Å)	c (Å)	a=b=c (Å)				
Fe_3O_4	-	-	-	8.365	-	45	-	585.326
Fe:ZnO/30 wt.% MMT	-	3.261	5.225	-	16	-	48.126	-
Fe:ZnO/MMT/Fe_3O_4	1:0.05	3.244	5.196	8.459	14	18	47.361	605.281
	1:0.1	3.237	5.196	8.461	11	22	47.163	605.710
	1:0.3	3.244	5.197	8.536	10	29	47.076	621.961
	1:0.5	3.213	5.207	8.586	9	37	46.586	632.954
	1:1	3.192	5.217	8.61	7	40	46.05	638.277

Absorption spectra of Fe:ZnO/MMT/Fe_3O_4 nanocomposites with various weight of the magnetite are displayed in Figure 2. The existence of MMT in the samples can be confirmed by the Si-O-Al bending vibration and Si-O-Si stretching vibration found at wave number about 970 cm^{-1} and 1090 cm^{-1}, respectively [10,11]. Moreover, there is also absorption peak assigned to stretching vibration of Fe-O at wave number 594 cm^{-1} [12], which derive from the dopant ion Fe and Fe_3O_4. Furthermore, the absorption peaks at wave numbers 1049, 1384, and 1723 cm^{-1} derive from the stretching vibrations of C-O, O-H, and C=O [13]. Then, the broad absorpiton peak at wave numbers range 3000-3500 cm^{-1} is attributed to vibration mode of the O-H bond [14].

The photodegradation of methylene blue has been examined using Fe:ZnO/MMT/Fe_3O_4 nanocomoposites as photocatalysts. Figure 3 exhibits the rapid degradation rate of MB using the photocatalyst with various weight of the magnetite. The MB degradation rate using Fe:ZnO/MMT without Fe_3O_4 is also shown. As can be seen in the figure, the photocatalyst with ratio of 1:0.05 performs the best photocatalytic activity. This result is corresponding to their surface area, where the nanocomposite with higher surface area could interact with MB more effectively when irradiated.

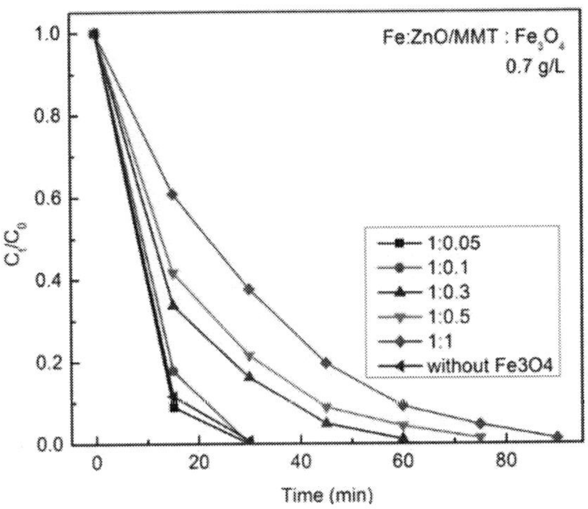

Figure 3. MB Degradation Curve Using Fe:ZnO/MMT and Fe:ZnO/MMT with Various Weight Ratio of Fe_3O_4.

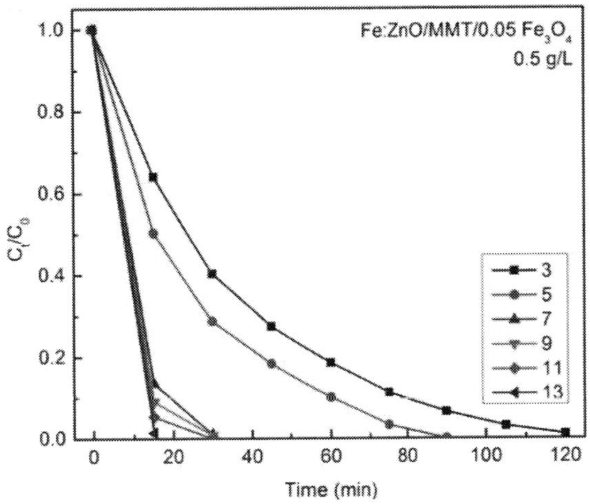

Figure 4. MB Degradation under different pH Value.

Table 2. Surface Area of the Samples.

Sample	Weight Ratio	Surface Area (m^2/g)
	1:0.05	39.326
	1:0.1	34.242
$Fe:ZnO/MMT/Fe_3O_4$	1:0.3	28.502
	1:0.5	25.196
	1:1	21.702

The reduced degradation may also due to increased magnetite loading which can cover the active sites of ZnO nanoparticles.

It was [15] reported that ZnO has zero point charge of 9±0.3. It causes the surface of ZnO positively charged when the pH is below this value and negatively charged above this value. Moreover, MB is cationic dye and positively charged when dissolved into water. When the pH value of the aqueous solution of MB increases, its degradation is further increased, as presented in Figure 4. The explanation is as follows. In acidic and neutral condition, both catalyst surface and MB are positively charged, resulting repulsive force and reducing photocatalytic activity because MB cannot provide hydroxyl group to form hydroxyl radical [16]. In contrast, opposite charge on the surface of the catalyst and MB in alkaline condition are attracted to each other so that the MB degradation increases.

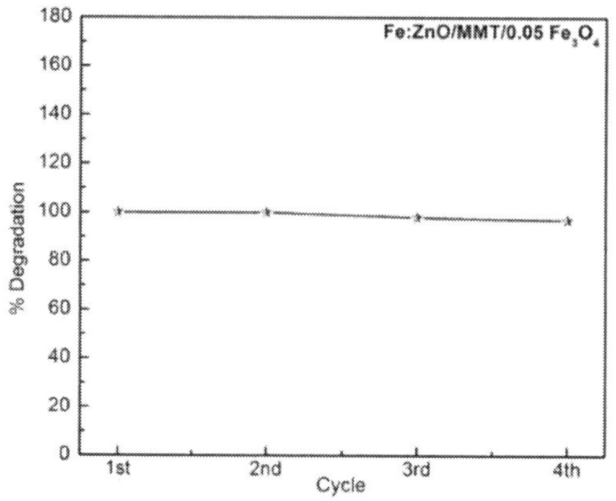

Figure 5. Degradation Percentage of The Nanocomposite For Four Cycles of Photocatalytic Activity.

Reusing the photocatalyst for four times processes results in great MB purification in aqueous solution, as shown in Figure 5. It can be seen that the photocatalytic activity exhibits 100% MB degradation for the first and second cycle, 98% for the third and 97% for the fourth. It indicates that Fe:ZnO/MMT/Fe_3O_4 nanocomposite is a stable photocatalyst.

4. Conclusion

The Fe:ZnO/MMT/Fe_3O_4 nanocomposites had been synthesized successfully by using co-precipitation method. The XRD patterns showed that the incorporation of magnetite nanoparticle could change the structure of ZnO. Maximum discoloration of methylene blue was performed by Fe:ZnO/MMT/0.05Fe_3O_4 sample. After four cycle processes of photocatalytic activity, the sample exhibited good stability in degrading methylene blue.

References

[1] Li C, Sun Z, Huang W and Zheng S 2016 *J. Taiwan Inst. Chem. Eng.* **66** 363
[2] Saleh R and Djaja N F 2014 *Superlatt. Micros.* **74** 217
[3] Afifah N, Andriani S, Djaja N F and Saleh R 2015 *Adv. Mat. Res.* **1123** 295
[4] Afifah N, Djaja N F and Saleh R 2015 *Mat. Sci. For.* **827** 19
[5] Djaja N F and Saleh R. 2014 *Spect. Acta A: Mol. Biomol. Spec.* **130** 581
[6] Fauzian M, Jalaludin S, Taufik A and Saleh R 2016 *J. Phys: Conf. Ser.* **710** 012006
[7] Siva C, Baraneedharan P, Nehru K and Sivakumar M 2016 *J. Alloy. and Compd.* **665** 404
[8] Estévez-Hernández O, Santiago-Jacinto P and Reguera E 2013 *Mater. Focus* **2** 438
[9] Khan E M, Alam F, Parveen A and Naqvi AH 2013 *J Adv Micros Res* **8** 45
[10] Du E, Yu S, Zuo L, Zhang J, Huang X and Wang Y 2011 *Appl. Clay Sci.* **51** 94
[11] Slavutsky A M, Bertuzzi M A and Armada M 2012 *Brazil J Food Tech* **15** 208
[12] Chong S, Zhang G, Tian H and Zhao H 2016 *J. of Enviromental Sciences* **44** 148
[13] Stankovich S, Piner R D, Nguyen S T and Ruoff R S 2006 *Carbon* **44** 3342
[14] Xing S, Zhou Z, Ma Z and Wu Y 2011 *Appl. Catal. B Environ.* **107** 386
[15] Zhang Y, Wan J and Ke Y 2010 *J. Hazard. Mater.* **177** 750
[16] Pouretedala H R, Norozi A, Mohammad H K and Semnani A, 2009 *J. Hazard. Mater.* **162** 674

Thermal Reduction Study of Graphene Oxide Paper

Fitrilawati[1], M B Perkasa[1], N Syakir[1], A Aprilia[1], L Safriani[1], T Saragi[1], Risdiana[1], S Hidayat[1], A Bahtiar[1], R Siregar[1], R R Sihombing[2], A Nugroho[2]

[1] Department of Physics University of Padjadjaran
Jalan Raya Jatinangor Km 21, 45363 Sumedang, West Jawa, Indonesia
[2] Faculty of Mathematics and Natural Sciences, Institut Teknologi Bandung
Jalan Ganesha no. 10, 40132 Bandung, Indonesia

E-mail: fitrilawati@phys.unpad.ac.id

Abstract. We report the preparation, reduction and characterization of GO paper that obtained by assembly of individual GO sheets. The free standing GO paper was prepared from 4 mg/ml GO dispersed in water, and following by evaporation of water and then detach from the petri dish. In order to obtained graphene-like film, the free-standing GO film then thermally reduced by heating the samples at 250°C at varied heating time. Characteristic and properties of the thermally reduced GO paper was measured by means of infrared spectroscopy, X-ray diffraction and EDS.

1. Introductions

The unique electronic property of graphene provides potential applications such as electromechanical resonator, protective layers, chemical filters, components of electrical batteries or supercapacitors [1]. Some those applications require a free-standing paper-like or foil-like materials. It is reported that graphene oxide (GO) and reduced graphene oxide (RGO) papers have been investigated as an electrode for actuator [2], sensor [3], and flexible supercapasitor [4]. Also, it has previously been reported a preparation of graphene oxide paper [5] and reduced graphene oxide paper [6] using vacuum filtration technique. It is our main concern to justify if the GO and RGO thin film properties will remain if the films become thicker and forming a paper.

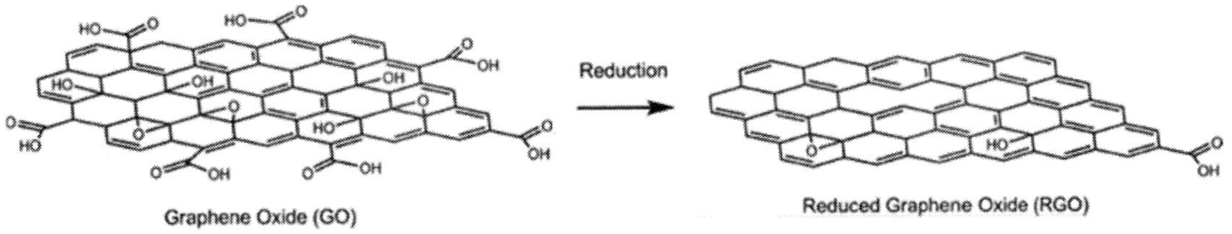

Figure 1. Schematic illustration of the reduction of graphene oxide (GO) to reduced graphene oxide (RGO) [7]

In this paper, we report a preparation of GO paper and its reduction process by means of thermal heating at 250 °C, as well as its characterization results. Reducing GO to produce RGO is a vital process since it will be effect a quality of the produced RGO. The reduction process will remove oxygen functional group of GO as described in Figure 1 [7]. We have previously studied a thermal reduction of GO thin film using optic characterization [8]. We found that in the form of thin film, the absorption of RGO is increased significantly after thermal reduction process. It is related to a recovery of conjugation bonds after thermal reduced process that has removed oxygen functional groups. It is interesting to find out if the RGO properties will remain if the thin films become a paper.

2. Experiment

We synthesize graphite oxide from natural graphite using modified Hummers by added graphite powder into a solution mixture of concentrated sulfuric acid (H_2SO_4, 95 – 97 %, MERCK) and sodium nitrate ((NaNO3, 30 %, AnalaR) that placed in the ice bath. Then the potassium permanganate (KMnO4, MERCK) was slowly added to the reaction mixture and then stirred at room temperature for several days. The peroxide acid solution (1 % H_2O_2 in water) was added to the dark brown paste, then the mixture was filtered and washed with chloric acid and water. The resulting black solid material was then further washed with excessive water to remove residual salts and acids. The brown suspension was dispersed in water. In order to obtained graphene oxide (GO), exfoliation was carried out by sonicating of 0.4 mg/mL graphite oxide dispersion under ambient condition for 30 minutes.

The free-standing GO paper was prepared in petri dish from 4 mg/ml GO dispersed in water, and following by evaporation of water in atmospheric and room until it is almost dried. It was further dried at 50°C the vacuum oven overnight and then detached from the petri dish. In order to obtain graphene-like film, the free-standing GO film then thermally reduced by heating the samples at 250 °C inside the pyrex tube that flow with Argon gas with various heating time.

We performed Energy Dispersive Spectroscopy (EDS) measurement using Hitachi Tabletop Microscope TM3000 to probe ratio of C/O of the prepared samples. Fourier transform infrared (FTIR) spectroscopy was carried out in absorption mode to measure structure of prepared samples using Prestige 21 Shimadzu. X-ray diffraction experiments were performed using SmartLab (Rigaku) diffractometer (Cu Kα radiation, X-ray wavelength $K\alpha 1 = 1.54056$ Å, $K\alpha 2 = 1.5444$ Å, ratio = 0.5 operating at 40 keV, cathode current of 40 mA) under normal laboratory conditions in the range of $2\theta = 5^{o}$ to $2\theta = 40^{o}$ and with step size of 0.02.

3. Results

The prepared GO paper can be easily detached from the petri dish and its shape was a circle with a diameter of 10 cm, that the same as the petri dish. The GO samples were cut in size of 1.5 cm x 3 cm for further characterizations. The EDS measurements show that before reduction the amount of percentage atomic of carbon and oxygen of GO paper is 62.3 % and 36.7 %, respectively. After reduction for 1 hour, the atomic percentages of carbon and oxygen of RGO are changed to 78 % and 21 %. The ratio of C/O is increased from 1.69 to 3.73 after thermal reduction of GO paper that was done by heating at 250 °C for 1 hour. Notably, the longer reduction time can significantly increase the C/O ratio. This result shows that the thermal reduction had reduced its oxygen functional groups from the GO paper.

Figure 2 shows FT-IR spectra of GO and RGO that thermally reduced for 30 minutes and 1 hour. These spectra of GO paper show a present of the broad and intense peak of O–H groups that centered at 3400 cm^{-1}, strong C=O peak at 1735 cm^{-1}, the O–H deformation peak at 1400 cm^{-1}, the C–OH stretching peak at 1200 cm^{-1}, and the C–O stretching peak at 1000 cm^{-1}. When it was thermally reduced and becomes RGO, the intensity of hydroxyl O-H group at 3400 cm^{-1} are almost completely removed after heating for 1 hour. Figure 2 shows a magnification spectra of samples between wavenumbers of 2000 and 900 cm^{-1}. It is obvious that the intensity at 1400 cm^{-1} that related to carboxyl are significantly reduced and intensity of C=C (aromatics) at 1600 cm^{-1} are increased. The increasing of C=C vibration intensity indicates a recovery of conjugated bonds in the RGO film.

However, several carboxyl and carbonyl of C=O groups at 1735 cm^{-1}, epoxy C-O at 1200 cm^{-1}, alkoxy C-O at 1000 cm^{-1} are still remained on the RGO films. This results confirm a formation of RGO by thermal reduction of GO paper, although the reduction process need to be improved in order to further reduce the oxygen contain.

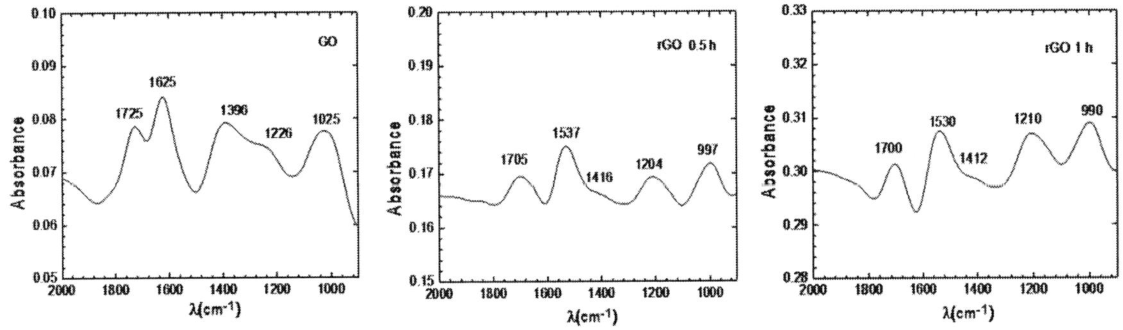

(a) (b) (c)

Figure 2 FTIR spectra of GO paper (a), RGO reduced for 0.5 hour (b), RGO reduced for 1 hour (c)

Figure 3 shows XRD patterns of GO paper and RGO papers that thermally reduced for 30 minutes and 1 hour. The X-ray pattern of GO has a peak centered at $2\theta = 11.4°$, corresponding to the (002) inter-planar spacing. After reducing for 30 minutes the peak value that centered at $2\theta = 11.4°$ reduced significantly and then appeared two new peaks at $2\theta=19°$ and 23.8°. When the reduction time was increased to 1 hour, the two peaks emerge and becomes one broad peak with the centre 2θ at 23.5°. Since the peak is corresponding to the layer-to-layer distance (*d*-spacing) [9], shifting of X-ray peaks of RGO to higher angle result in a decreasing *d*-spacing, it shows that reduction process decreases the inter-planar spacing. Sharp peak on the right-hand side at $2\theta = 28°$ is artefact and should be disregarded. This results imply a removing of oxygen were taken part during reduction process, that are in-line with characterization results using EDS and FTIR spectroscopy.

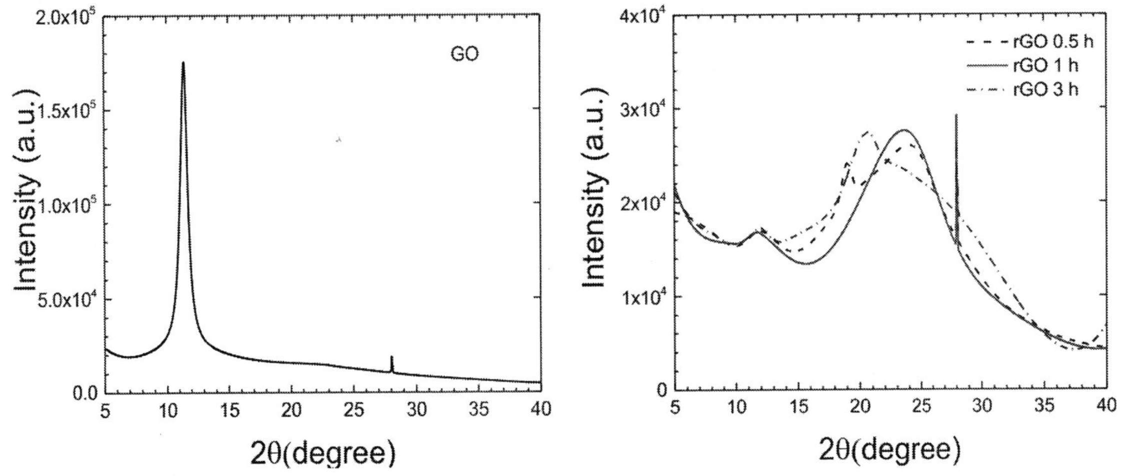

(a) (b)

Figure 3 XRD patterns of GO paper (a) and RGO reduced at varied heating time (b)

4. Conclusions

We have successfully prepared GO paper by using solution casting technique. The GO paper can be converted into RGO paper by means of reduction thermal at 250 °C in argon atmosphere. The EDS measurement confirmed removing of oxygen functional groups by increasing of C/O ratio after thermal reduction process. The FTIR spectra show a recovery of C-O and O-H bond to C=C bond, and XRD measurement show a decreasing of interlayer spacing of RGO layer that also confirmed removing of oxygen group. This show that we still can use thermal reduction process to obtain RGO paper from GO.

Acknowledgment

This work was funded by Academic Leadership Grant (ALG) Universitas Padjadjaran contract number 480/UN6.RKT/Kep/HK/2016

References

[1] Novoselov K S, Geim A K, Morozov S V, Jiang D, Zhang Y, Dubonos S V, Grigorieva I V and Firsov A A 2004 *Science* **306** 666
[2] Selvakumar D, Alsalme A, Alghamdi A and Jayavel R 2017 *Materials Letters* **191** 182
[3] Zhang M, Halder A, Hou C, Ulstrup J and Chi Q 2016 *Bioelectrochemistry* **109** 87
[4] Boruah B D and Misra A 2016 *Energy StorageMaterials* **5** 103
[5] Dikin D A, Stankovich S, Zimney E J, Piner R D, G. Dommett H B, Evmenenko G, Nguyen S T, Ruoff R S, 2007 *Nature* **448** 457
[6] Xu Y, Bai H., Lu G, Li C and Shi G 2008 *J. Am. Chem. Soc.* **130** 5856
[7] Li H and Bubeck C 2013 *Macromol. Research* **21** 290
[8] Fitrilawati, Syakir N, Aprilia A, Liu Z, Feng X and Bubeck C 2015 *Materials Science Forum* **827** 317
[9] Tokarczyk M, Kowalski G, Witowski A M, Kozinski R, Librant K, Aksienionek M, Lipinska L and Ciepielewski P 2014 *Acta Physics Polonica A* **126** 1190

Effect of Spreading Time on Contact Angle of Nanofluid on the Surface of Stainless Steel AISI 316 and Zircalloy 4

D H Prajitno, V Trisnawan and D G Syarif

PSTNT-BATAN Jl. Tamansari 71 Bandung 40132, Indonesia

E-mail: djokohp71@yahoo.co.id

Abstract. The solid surface tension plays an important role in the heat and mass transfer system for heat exchanger equipment. In the nuclear power plant industry, the stainless steel AISI 316 and Zircalloy 4 have been used for long time as structure materials. The purpose of the experimental is to study solid state surface tension behavior by measure contact angle Nano fluid contain nano particle alumina on metal surface of stainless steel AISI 316 and Zircalloy 4 by sessile drop method. The experiment is to measure the static contact angle and drop nano fluid contains nano particle alumina on stainless steel 316 and zircalloy 4 with different spreading time from 1 to 30 minute. It was observed that stainless steel 316 and zircalloy 4 lose their hydrophobic properties with increasing elapsed time during drop of nano fluid on the surface of alloy. As a result the contact angle of nano fluid on surface of metal is decrease with increasing elapsed time. While the magnitude diameter of drop nano fluid and wetting surface is increase with increasing elapsed time on the surface of the stainless steel SS 316 and Zircalloy 4.

1. Introduction

Nanofluids are colloidal mixture of nano particles in a base fluid. The properties of nanofluids are improve the heat transfer coeficient and conductivity attributes from the original fluid. These nano particles usually metal or metal oxide, increase the conduction and convection coefficients of base fluid (1). Experimental studies on nanofluid single-phase heat transfer have been reported in the number of literature such as alumina-water and titania-water are studies in turbulent convective heat transfer in tubes. Laminar convective heat transfer and viscous pressure loss of alumina–water and zirconia–water nanofluids have are been studied. From this study nanofluid improved heat transfer compared with based fluid (2,3). The studies of nanofluids for nuclear power plants application such as primary and secondary cooling system and emergency core cooling systems(ECCSs) were carried out by a number of researcher. This studies were characterized by the improvement of critical heat flux (CHF) of nanofluid after injection in experiment of loss of coolant accidents (LOCA). The benefits of nanofluid improved CHF during simulation accident conditions (4). It is clear that the nanofluid engineered ECCSs should be compatible with conventional systems during normal operations to make nanofluid technologies practical in Nuclear Power Plants.The interaction of a fluid with a solid surface of metal is referred to as wettability. When the fluid coverages spontaneously out as a thin film over the surface of metal , it is said as wetting the surface of metal. When the interactions are weak, the fluid drops on the solid surface of metal and the liquid locally wets the surface. The wetting phenomena plays an important role in the mass and heat transfer for engineering structure such as heat exchanger, steam generation, waste heat recovery and emergency cooling system. An index of wetting properties is defined by the magnitude of contact angle between the liquid and solid surface of metal (5,6). Solid metal provides planar surface that allow the geometric measurement of a contact angle on the metal. If the contact angle is lower than 90o the material is hydrophilic otherwise it is hydrophobic. Contact angle value can be assessment and calculated by Young equation (7) and this equation is valid for the ideal surface such as chemically homogenous, non-reactive and insoluble. In the experiment the values of the contact angle were determined by use of a goniometer or from captured image of a droplet of liquid. By assembling on the digital images of the droplets of liquid, geometrical approach and developed tangent method were applied to calculate the value of contact angle.

The purpose of the present study is to examine the contact angle of demineralized water contain nanoparticle alumina on the metal surface of stainless steel 316 and zircalloy 4 with different spreading time in nanofluid solution will be studied. The stainless steel 316 and zircalloy 4 as an engineering material is used as nuclear material structure in the secondary cooling system and primary cooling system respectively (8,9). The contact angle is highly influenced by surface finish composition of the metal and nanofluids. Contact angle fluid and solid of stainless steel 316 and zircalloy 4 in nanofluids media contain nanoparticle alumina is rarely studied. The present work is to investigate the contact angle behavior of nanofluid contain 0,01 gpl nanoparticle alumina on stainless steel 316 and zircalloy 4 surface with different spreading time using a sessile drop methods for evaluation.

2. Experimental

2.1 Material and nanofluid

The specimens used in this experiments were circular discs of stainless steel 304 having 1.6 Cm in diameter and 2 mm in thickness. The specimens are stainless steel 316 and zircalloy 4 with the major chemical composition (in wt%) 18.21 % Cr, 8.40% Ni, 0.39% Si, 1.56% Mn, 0.45%Mo, 0.025% C balanced Fe and 98%Zr, 1,4Sn, 0.24Fe 0.15Cr, respectively. The specimens were roughed by wet grinding with different emery paper 230,500, 800 and 1000 grit, degreased with detergent after that cleaned with ultrasonic cleaner in alcohol solution. Finally, the specimens dried in air freely. In the present study Al2O3 was used as nanoparticles (10) while demineralized water was applied as a base fluid. Nanofluid is produced by mixing aqua demineralized water (adm) and 0.01 gpl Al2O3 nanoparticles. The nanofluids were prepared by incorporating 0.01 gram of Al2O3 nanoparticles with an average particle size of 30-70 nm mixed into 1 L aqua demineralized water by magnetit stirrer for 30 minute. Ultrasonic vibrator operating at a frequency of 44 kHz was used to stabilise nanofluid for 20 min in order to the Al2O3 particles dispersed and stabilized in aqua demineralized water. Figure 1 show micrograph of nanoparticle Al2O3 examination by TEM with magnificiant 150000. The particle size of nanopartivle Al2O3 relatively homogen.

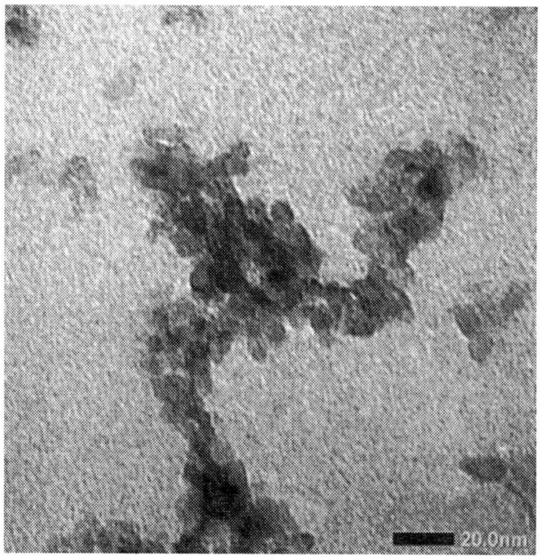

Figure 1. Micrograph nanoparticle Al2O3 examination by TEM with magnificiant 150000

2.2 Sessile drop measurement

Contact angle measurement was carried out to evaluate the wettability characteristics of the stainless steel. The contact angle can be defined as the angle between the liquid phases formed on the surface of a stainless steel and the line tangent to the droplet radius from the point of contact with the surface of stainless steel. The Contact angle perform have been conducted by sessile drop method (11) equipment system from Kruss GmbH goniometer with a resolution of 0.1° in the measuring range of 1-180°. This equipment has been provided by software drop shape analysis (DSA 4) for controlling the experiment and analyzing the drop shape to calculate the contact angle. Droplets of known volume of 2 μL were dropped on a substrate of SS 316 and zircalloy 4 using an automatic dispenser and injected slowly onto the solid surface by a syringe. The experiments were conducted under similar air conditioned laboratory environments. The measured values of the contact angle, the images of the droplets were taken directly after 20 second deposition in order to eliminate the impact of droplet evaporation. The contact angle tests were conducted at an ambient temperature of 27 °C .

2.3 Sample characterization

Surface morphology of the stainless steel 316 and zircalloy 4 after wet grinding was examined by using a SEM microscope to obtained different image.

3. Results and Discussions

3.1 Optical image of stainless steel 316 and zircalloy 4

Figure 2(a) and (b) shows surface morphology of the stainless steel 316 and zircalloy 4 that was examined by using a SEM microscope after wet abraded by emery paper. Surfaces morphology of the sample relative the same roughness were produced by wet grinding stainless steel 304 zircalloy 4.

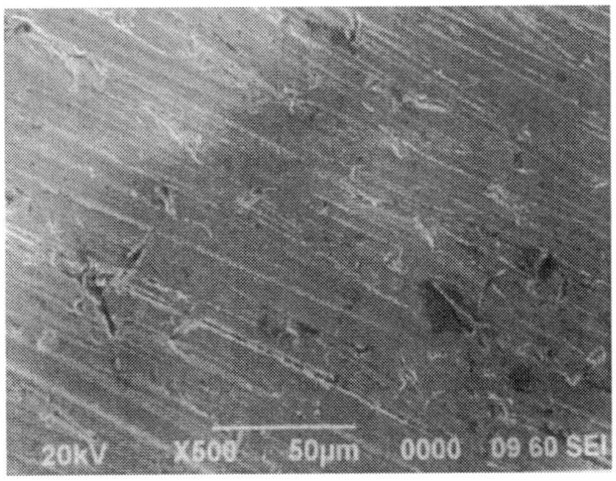

a

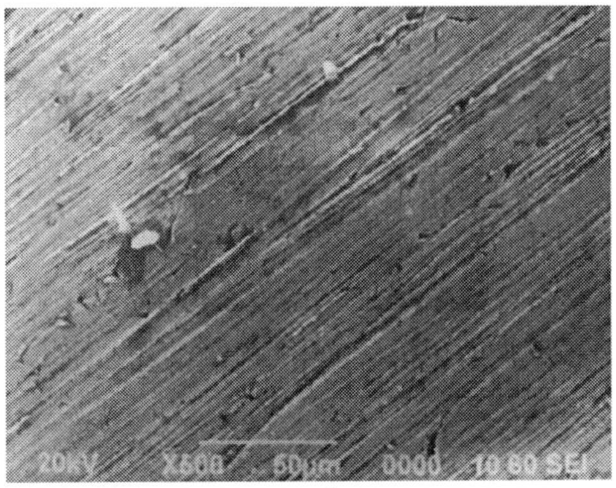

b

Figure 2. Surface morphology of a)stainless steels 316 b) zircalloy 4 after wet abraded with emery paper

3.2 Effect of Al2O3 in nanofluid on the wettability of Stainless Steel 316 and Zircalloy 4

In Figure 3(a) and (b) showed the contact angle of droplet aqua demineralized and nanofluid on Stainless Steel 316 and Zircalloy 4 as a function of spreading time multiple measurement. From Figure 3(a) and (b) it can be seen that the contact angle value of demineralized water higher compare with nanofluid at different spreading time on the surface of two of materials stainless steel SS 316 and zircalloy 4. From figure 3(a), it is shown that early-stage spreading the contact angle of low-viscosity drops of demineralized water different significantly with nanofluid on a partially wetting substrate of stainless steel 316. For long spreading time, the wetted area is found to grow and the contact angle decrease for all considered wettability's of demineralized water and nanofluid but different of contact angle significantly between demineralized water and nanofluids. From figure 3(b), it is shown that early-stage spreading the liquid drop of low-viscosity demineralized water different significantly with nanofluid on a partially wetting substrate of zircalloy 4. For long spreading time, the wetted area is developing and the contact angle decrease for all considered wettability's of demineralized water and nanofluid. The contact angle of demineralized water and nanofluid on the surface of zircalloy is not significantly different for long spreading time. It can be concluded that wetabilities of nanofluid show good performance compare with demineralized water on the surface of stainless steel SS 316 and zircalloy 4.

Figure 4 depicted the different of the contact angle of nanofluid on the surface of Stainless Steel 316 and Zircalloy 4. From figure 4, it is shown that early-stage spreading the liquid drop of nanofluid different significantly on a partially wetting substrate of stainless steel 316 compared with zircalloy 4. For long spreading time, the wetted area is developing and the contact angle decrease for all considered wettability's of nanofluid. Is not significantly different for stainless steel 316 compared with zircalloy 4. It can be concluded that wetabilities of nanofluid show good performance on stainless steel SS 316 compare with zircalloy 4.

(a)

(b)

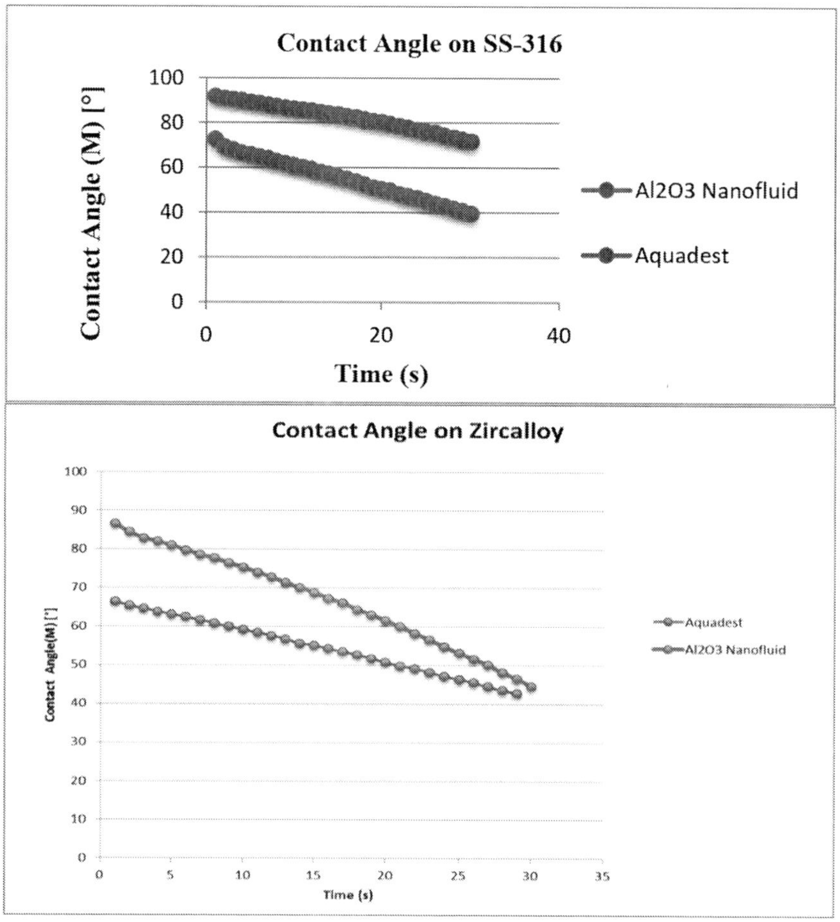

Figure 3. Multiple measurements at ambient temperature of the spreading of aqua demineralisez and nanofluid on a) Stainless steels 316 b) Zircalloy.

The droplet image of nanofluid on Stainless Steel 316 and Zircalloy 4 has been measured by using the sessile drop method is represented in Fig. 5. This figure showed that the diameter of droplet nanofluid for Stainless Steel 316 and Zircalloy 4 decrease with increasing spreading time. The wettability properties of nanofluid on the surface of Stainless Steel 316 and Zircalloy 4 are not different significantly due to the function of two materials Stainless Steel 316 and Zircalloy 4 serve as heat exchanger equipment (9). The diameter of droplet decrease with increasing spreading time due to gravitational force at the surface of nanofluid droplet.

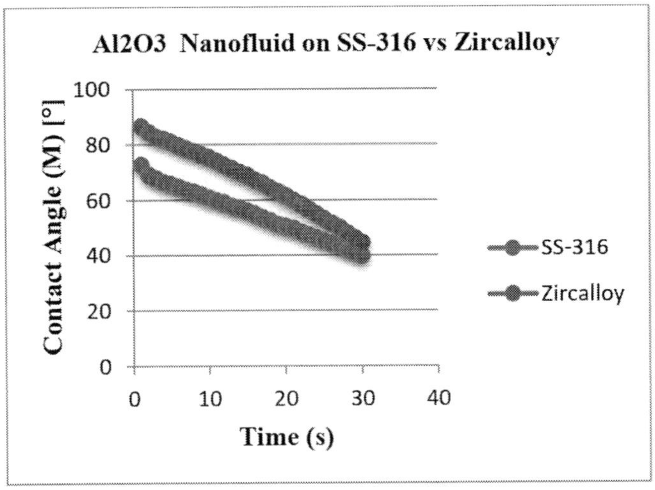

Figure 4. Measurements of the spreading of nanofluid on Stainlees Steel 316 and Zircalloy 4.

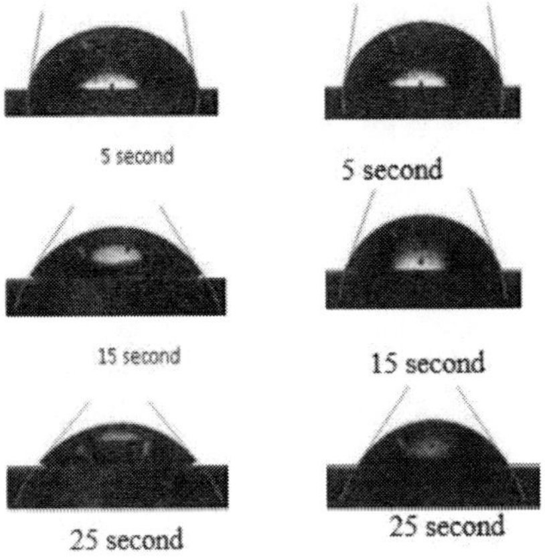

Figure 5. A droplet of nanofluid on Stainless Steel 316 and Zircalloy 4 for 5,15, 25 second.

4. Conclusion

The wettability behavior of Nano fluids contains 0.01 gpl nanoparticle alumina on the surface of stainless steel SS 316 and zircalloy 4 samples was studied using sessile drop methods. The results obtained can be summarized as follows:
(i) The wettability of nanofluid with 0.01 gpl nanoparticle alumina on the surface of stainless steel SS 316 and zircalloy 4 much better compare with demineralized water
(ii) The values of contact angle decrease gradually with increasing spreading time
(iii) The wettability of the nanofluid and demineralized water on surface of the stainless steel 316 is better compared with zircalloy 4
(iv) The diameter of droplet decrease with increasing spreading time

Acknowledgments
The authors would like to extend thanks to National Nuclear Energy Agency for financial support, and the director of Research Center for Nuclear Science and Technology.

References
[1] Prajitno D H and Syarif D G 2016 *Sessile Drop Method, Journal of Physics*: Conference Series 739 012029
[2] Saidur R, Leong K Y L and Mohammad H A 2011 *Renewable and Sustainable Energy Reviews* **15** 1646
[3] Albadr J, Tayal S, Alasadi M 2013 *Case Studies in Thermal Engineering* **1** 38
[4] Rea U, Mc K T, Hu L W, Buongiorno J 2009 International Journal of Heat and Mass Transfer **52** 2042
[5] Gajewski A 2008 *International Journal of Heat and Mass Transfer* **51** 4628.
[6]. Belaud V *et al* 2015 *Tribological International* **82** 343
[7] Koen G W, Joost H W, Antonin E and Jacco H S 2012 *Physical Review E* **85** 055301
[8] Martins C M B, Moreira J L and Martins J I 2014 *Engineering Failure Analysis* **39** 65
[9] Zinkle S J *et al* 2009 *Structural Materials for fusion and fission energy Material Today* **12** 12
[10] Prajitno D H and Syarif D G 2016 *2ndMaterials Research Society of Indonesia (MRS-Id) Bandung 2016*

Effect of Weight Fractions of Jute Fiber on Tensile Strength and Deflection Temperature of Jute Fiber/Polypropylene Composites

S Nabila[1], A L Juwono[1] and S Roseno[2]

[1] Department of Physics, Faculty of Mathematics and Natural Sciences, University of Indonesia, Depok-16424, Indonesia
[2]Agency for the Assessment and Application of Technology, Puspitek, Serpong-15314, Indonesia

E-mail: ariadne.laksmidevi@ui.ac.id

Abstract. Jute is one of eco-friendly natural fiber with relatively low cost and high volume production. This study aimed to determine the effect of weight fractions of jute fiber as a reinforcement in polypropylene (PP) to obtain an optimum properties of PP/jute fiber composites. Jute fiber was pre-treated through alkalization. The PP was initially produced by extrusion process, followed by fabricated the composites by compiling the PP matrix and jute fibers into lamina using a hot-press method. The results of tensile test and heat deflection temperature test showed that the addition of 40wt% jute fiber to the PP increased the tensile strength about 19.7 % up to (38.2±4.9)MPa, the Young modulus about 79.8 % up to (3.20±0.26)GPa, and the heat deflection temperature about 143% up to (143.3±1.14)°C compared to pristine PP. Based on Scanning Electron Microscopy observation on the fracture surfaces, it was shown that the mode of failure on the composites failure surfaces was "fiber pull-out", which due to the poor interface bond between the fiber and the matrix.

1. Introduction

Over the last few years the application of natural fibers in composites is increasingly used in various fields like the automotive, furniture and packaging [1]. Natural fibers have several advantages compared to synthetic fibers, such as relatively low cost, relatively low density,renewable and biodegradable. In addition, natural fibers showed a tendency to replace glass fiber reinforced polymer composites. One of the reasons is that natural fibers have a comparable strength and a similar specific modulus with fiber glass [1].The strength of natural fibers depends on the fiber structure. The properties of natural fibers are influenced by the location, climatic conditions, age of the plants and extraction methods that are used. More over, natural fibers generally have a composition of cellulose, hemicellulose, lignin, pectin and waxy substance that allows absorption of water from the environment, and these substances will form a poor interface bonding with the matrix [2]. One of the plant fibers that is often used as reinforcement in composites is jute. As one of the cheapest natural fiber with the highest volume production, jute shows the advantages to produce material with relatively low production costs and high availability [3].

Compatibility between the hydrophilic nature of natural fibers and hydrophobic nature of polymer matrix is a major problem. Incompatibility between natural fiber and polimer caused poor interface bonding, that decrease composite strength. Therefore, natural fiber modifications such as chemical treatment, coupling agent and acetylation are considered in modifying the fiber surface properties to improve their adhesion with different matrices. Natural fiber modification improves the interface and wettability between the matrix and fiber; and at the end to improve mechanical properties of composites [4]. In this study, alkalization of fiber with a 5% solution of sodium hydroxide (NaOH) was performed before the fiber used as reinforcement in composites. Furthermore, the tensile strength and the influence of heat on the mechanical properties of PP/Jute composites were investigated by adding various fractions of jute fiber.

2. Experimental

2.1 Materials
Materials used for this study were jute fiber, 5% NaOH and homopolymer PP types cosmoplene FY3012E from The Polyolefin Company (Singapore) Pte. Ltd.

2.2 Alkali treatment
Jute fiber was prepared in ± 20 cm long for alkalization. Jute fiber bundle soaked in 5 % NaOH solutions for 60 min with a weight ratio of fibers in a solution 1:15 at a temperature of 30 ± 2°C. Then, the jute fiber in the bundle were rinsed with distilled water until the fiber was rough. The drying process was conducted in 48 hours at room temperature and then at 100°C for 6 hours until the sample was completely dry. The dried fibers were stored in airtight containers to keep the humidity of the sample. Single fiber test was conducted before and after alkali treatment.

2.3 Sampel Preparation
Continuous directionally aligned PP reinforced composites with a variation of jute fibers compositions were manufactured by hot-press method. The compositions of the jute fiber were 30, 40 and 50 wt%. The composites were made by preparing the jute fiber in parallel direction between PP sheets. Jute and PP were prepared and placed in a steel die. The dies, that was pre-cleaned by a mold released agent, were closed by applying force of 40 barr by hot-press machine at 190 °C for 5 minutes to produce a single mat of 3-4 mm. Thickness the composites were then cooled and the specimens were cut to the required size based on ASTM D638 Type 4 for tensile testing and standard ISO 75 type 2 for Heat deflection Temperature (HDT) testing.

3. Results and discussion

3.1 Tensile Test Results
Jute fiber tensile test results showed that the tensile strengths before and after alkalization were about 24.3 MPa and 132.2 MPa. It was clear that alkalization increased the tensile strength value of the jute fiber. This occured because the alkalization removed the lignin and waxy substances but the cellulose which is the main content of the jute physically was not affected by the NaOH solution [2,9]. Tensile test results of pristine PP and PP/Jute composites are shown in Figures 1 and 2. As shown in Figure 1, the tensile strength and Young's modulus of the PP/40wt% jute composites increased up to 19.7 % and 79.8 % compared to the pristine PP. The tensile strength and Young's modulus were then decreased when the jute content was 50 wt%. This indicates the poor interface bonding between the fiber and the matrix.This lead to the formation of poor interface, which resulted in poor mechanical properties of the composite [10]. The strain at break of the composites, as exhibited in Figure 2, decreased along with increasing weight fraction of jute. This showed that the composite was more brittle than pristine PP which was highly elastic.

Tensile strength on composites increases when fibers and matrix are well bonded, where the load transfer from matrix to fibers spread out well on entire surface of composite fibers and causes resistance of composite is increased against a given load. In composites with low weight fraction, the composite strength will be lower than composites with high weight fraction [5]. Therefore, the addition of weight fraction in composite will increase tensile strength until the optimum value, so the load transfer from matrix to fibers in the composite will be more evenly distributed. Despite the addition of fiber can increase the tensile strength of composite, the tensile strength will be decreasing when it reach optimum value. Weight fraction of fibers that exceed a certain value will make it does not bond perfectly with the matrix, it because the fibers are not fully wetted by the matrix as a filler in the composite which ends up causing a lot of voids that will decrease the tensile strength of the composite [5].

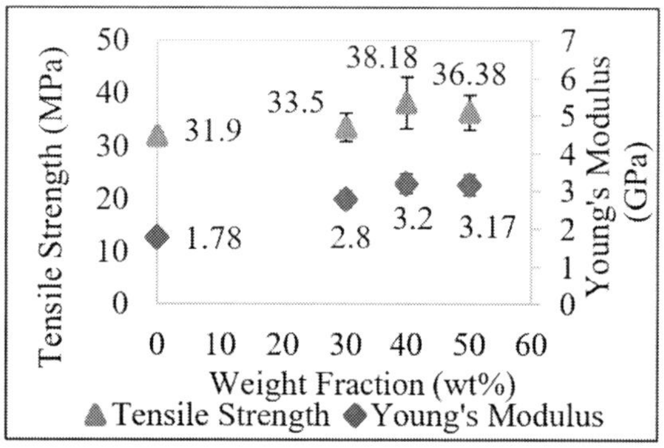

Figure 1. Tensile strength and Young modulus vs weight fraction of jute fiber.

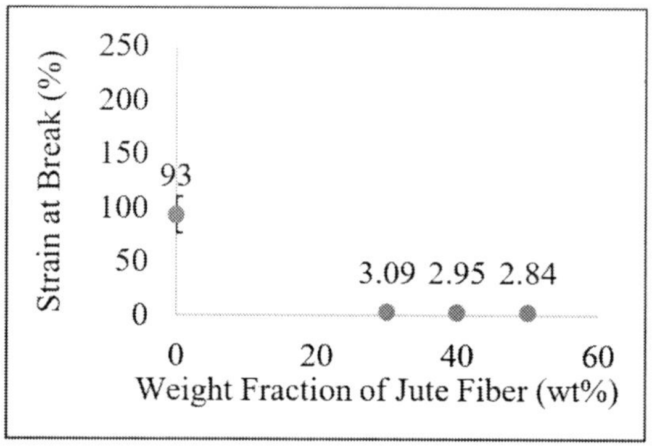

Figure 2. Strain at break vs weight fraction of jute fiber.

By using a modification of the matrix and different methods from this study, the tensile strength of PP/Jute is 34-73 MPa [6]. The same result on this study is also showed by using 2 wt% MAHgPP as a matrix modification of the PP/jute yarn composite. By using injection molding method, the tensile strength of PP/30wt% jute yarn composite were about 48 MPa [7].

a. *Heat Deflection Temperature Test Result*

Figure 3 shows the heat deflection temperature of pristine PP and PP/Jute composites. The highest heat deflection temperature value was obtained in the PP/40 wt% jute composites, which was (143.3 ± 1.14)°C. This value increased significantly of about 143% compared to pristine PP. Based on previous study [10], the best result of deflection temperature was 134°C belong to a PP/jute composite. This difference is influenced by dissimilar PP used in the two studies.

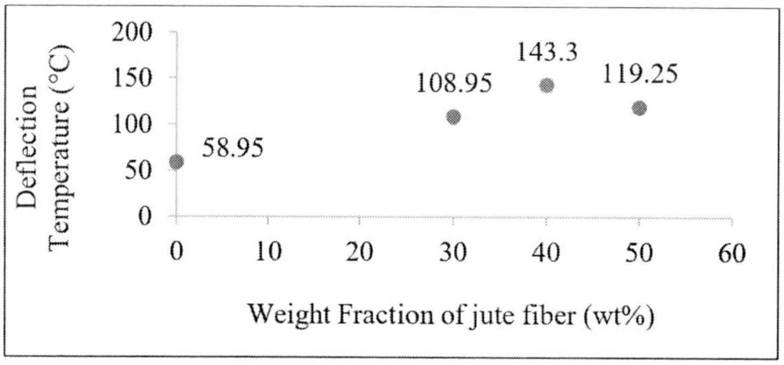

Figure 3. Deflection temperature vs weight fraction of jute fiber on PP/jute composite.

b. *SEM Observation Result*

Figures 4 (a) and (b) show the SEM images before and after tensile test of PP/40wt% jute composite surfaces. The SEM images before tensile test of PP/40wt% jute composite surface shows a clear difference layer between the matrix and fiber. This indicated imperfect wettability between the fibers and the matrix that causing poor bonding interface. Moreover, the poor interface bonding between the fiber and the matrix also caused by the existence of a void on the composite surface. Figure 4(a) shows the existance of void on PP/40wt% jute composite. Large concentration of void on the composite surface caused poor interface bonding between the fiber and the matrix and resulted low mechanical properties. Figure 4(b) shows the matrix cracks and the fiber pull-out of PP/40wt% jute composite. High damage due to the load of tensile test showed by fiber pull-out on the composite surface breakage occurs when the fiber on composite can not longer hold the applied load.

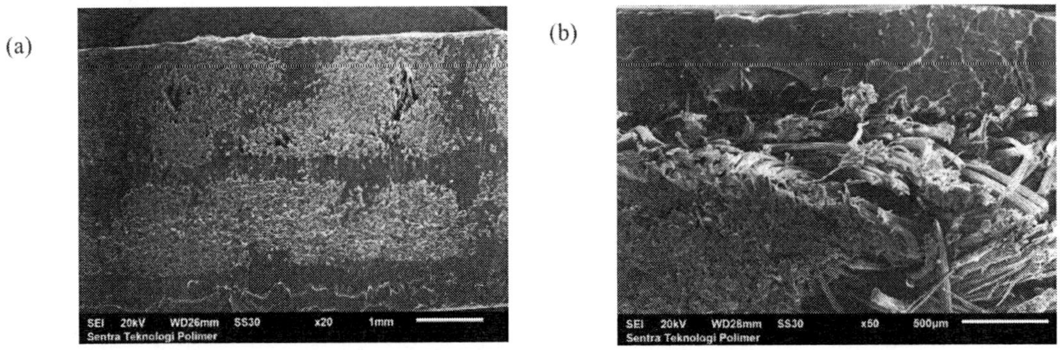

Figure 4. SEM images (a) PP/40 wt% jute before tensile test, (b) PP/40 wt% jute after tensile test.

4. Conclusion

The best mechanical and thermal properties of PP composite reinforced jute was obtained in the PP/40 wt% jute composites. Tensile strength, Young's modulus and heat deflection temperature of PP/40 wt% jute composite were (38.2 ± 4.9) MPa, (3.20 ± 0.26) GPa, and (143.3 ± 1,14)°C respectively. These values increased up to 6%, 50.2% and 143.0% respectively compared the pristine PP. Fiber pull-out occurs on all composite samples after the tensile test. This mode of failure showed a poor interface bonding between jute and PP.

Acknowledgment
This work was funded by University of Indonesia for PITTA grant in 2016.

References

[1] Hojo T, Xu Z, Yang Y, Hamada H 2014 *Energy Procedia* **56** 72
[2] Faruk O, Bledzki A K, Fink H-P Fink, Sain M 2012 *Progress in Polymer Science* **37** 1552
[3] KaradumanY, Sayeed M M A, Onal L, Rawal A 2014 *Viscoelastic properties of surface modified jute fiber/polypropylene nonwoven composites* Composites : Part B (2014) : 111
[4] Masruchin, Nanang 2012 *Pengembangan Komposit Polipropilena Berpenguat Serat Sisal Dan Serat Sabut Kelapa Untuk Material Komponen Otomotif* Universitas Indonesia
[5] Raiardi, Anhar 2013 Universitas Indonesia
[6] Rana A K, Mandal A, Mitra B C, Obson R J A C , Rowell R, Banerjee A N 1997 *Short Jute Fiber-Reinforced Polypropylene Composites:Effect of Compatibilizer* University of Calcutta
[7] Doan, Loan T, Gao S L, Ma¨der E 2006 *Composites Science and Technology* **66** 952
[8] Bledzki A K, Franciszczak P, Osman Z, Elbadawi M 2015 *Industrial Crops and Products* **70** 91
[9] Khan J A and Khan M A 2015 *The use of jute fiber as reinforcements in composite* Bangladesh
[10] Das B S and Goyal G K 2008 *effect of hygrothermal treatment on the tensile properties of hotpressed jute fibre composites* Department of Metallurgical and Materials Engineering National Institute of Technology Rourkela

IOP Conf. Series: Materials Science and Engineering **196** (2017) 012030 doi:10.1088/1757-899X/196/1/012030

Cross-Sectional Investigations of Oxide Scale Nanocrystalline FeCr Alloys after High-Temperature Oxidation Test at 900°C

H Saryanto[1], D Sebayang[2], P Untoro[3]

[1]Gajah Tunggal Polytechnic, Kel. Pasir Jaya, Tangerang-Banten, Indonesia
[2]Mercu Buana University, Jln. Meruya Selatan, Jakarta Barat, Indonesia
[3]Surya University, Jln. Boulevard Gading Serpong, Banten-Indonesia

E-mail: hsaryanto@gmail.com

Abstract. The cross-sectional examinations of oxide films formed by oxidation on the surface of FeCr alloys with various crystallite sizes were observed and investigated. X-ray diffraction (XRD) analysis was used to characterize the oxide scale morphology and to identify the phases and oxidation products. Furthermore, Scanning electron microscopy (SEM) coupled with energy dispersive X-ray spectroscopy (EDS) have been used to study the cross-sectional oxides produced by specimens after oxidation process.The cross-sectional investigation shows that the oxide scale formed on the surface of FeCr alloys consisted roughly of Cr_2O_3 with a small amount of FeO mixture. The outward diffusivity of Chromium to form Cr_2O_3 protective layers vary significantly occurrences on the surface of FeCr alloy with smallest crystallite size (38.51 nm), the scale had an enriched Cr content which improves the adherence of the oxide scale to the substrate, in another word, it increases the oxidation resistance. While the oxide scale formed on the surface of FeCr alloy with largest crystallite sizes (76.60 nm) had an enriched Fe content which reduces the resistance to oxidation, and adherence to the substrate. The thickness of oxide scale formed on nanocrystalline FeCr alloy with smallest crystallite sizes was found around 8 μm thick, which three-time thinner than FeCr alloy with largest crystallite sizes.

1. Introduction

The interconnect is a critical component in the stack of solid oxide fuel cells (SOFC), which act as a physical separation between anode and cathode gasses and provide the electronic connection between the single cells of the stack. Technical requirements to be used as interconnect in SOFC should fulfill a number of specific requirements, i.e. a high oxidation resistance, a high electrical conductivity of the surface oxide scales, gas tightness, and a coefficient of thermal expansion matched to the electrolyte and the electrodes [1,2]. Cr based alloys and high-Cr ferritic steels seem to be the most promising metallic interconnect materials [1].

The oxidation performance of FeCr used in high-temperature applications (such as Interconnector in SOFC) depends on their potential to form and maintain a Cr_2O_3 oxide scale under SOFC atmospheres [2, 7]. The main phenomena taking place during reactions of metals with oxygen. Initially, the gas absorbs on the metal surface as atomic oxygen; then oxide nucleation and growth to form a complete thin film [3]. Scale growth results from the concomitant moving of electrons, oxide ions, metal ions and oxygen diffusion through the scale [4]. As the scale grows up, cavities and micro cracks may be created from generated stresses, leading to failure of the scale [5, 6, 12]. When Cr based alloys and high-Cr ferritic steels are exposed to an oxidizing environment at elevated temperatures, an oxide layer is formed on the surface, acting as a barrier between the metal and the gas. Chromium increases the oxidation resistance of stainless steels by the formation of a chromia (Cr_2O_3) scale on the surface. Other oxides which are often encountered in the layers are iron oxides (wüstite FeO, magnetite Fe_3O_4, or hematite Fe_2O_3). Refer to our previous study [7, 9], the effect of the smallest crystallite sizes can significantly improve the high-temperature oxidation and corrosion resistance of FeCr alloy that oxidized at 900°C. However, it focused only on investigating of oxidation kinetic, oxide phase and surface morphology characterization of specimens, but did not investigate the characterization of oxide scale on a cross-sectional analysis and the growth mechanism of oxide scale. Therefore, the objective of this study is to investigate the cross-sectional characteristics of oxide scale that formed on the surface FeCr alloys after being oxidized at 900°C.

2. Specimens and Experimental Procedure

2.1. Specimens

The studied specimens of developed nanocrystalline FeCr alloys were prepared by the following process. As starting element materials, high purity (99.99%) powders of Fe and Cr with mean particle size of about 2 μm were prepared by powder metallurgy technique, the powders of Fe and Cr with nominal composition of 80wt%Fe and 20wt%Cr were mechanically alloyed into RETCH PM400 planetary ball mill with variations of milling time for 40 h, 60 h, and 80 h. The ball milled alloy powder specimens were then hot compacted at 1000°C into samples of the dimension of 33 mm diameter and 3 mm thickness, under a uniaxial pressure of 25 Mpa pressure in a vacuum atmosphere for 45 min.

2.2. Oxidation Test

The samples of oxidation test were cut from the specimens to the size of 4x5x3 mm. The samples were subjected to oxidation tests in laboratory air at atmospheric pressure in a PROTHERM box furnace at 900°C for 100 h in laboratory air at the atmospheric pressure. The linear heating and cooling rates were controlled at 5°C/min. The electronic analytical microbalance with an accuracy of 0.01 mg was used to determine the weight of the specimens after each interrupted cycle.

2.3. Preparation of cross-section specimens

Cross-section samples are mounted using "cold-mounting" resins. Samples were placed on a commercial sample holder (cylindrical), fixed through a drop of wax. Resin (plus hardener and accelerant) is then poured in the sample. Next, after 12 hours, the sample and resin becoming a hard block. Furthermore, the samples were polished on the cross-sectional face up to 1500 grade SiC paper and final polishing using 0.05 μm diamond paste.

2.4. Microstructure Characterization

The oxidized alloys were investigated by a BRUKER D8 Advance X-ray diffraction (XRD) equipped with CuKα radiation and phase identification as referred to the ICDD PDF-2 database. The surface and cross-sectional morphology analyses of oxidized specimens were carried out with a JEOL JSM-6380 scanning electronic microscope (SEM) equipped with an energy-dispersive X-ray spectroscopy (EDS) analyzer. The EDS analyses were performed as point analysis and line/mapping analysis.

3. Results and Discussion

3.1. Previous Results

The previous results on developed of FeCr Alloys, oxidation test, and parabolic rate were collected in Table 1 and Fig.1 [7, 8, 9, 10 & 11], where the parabolic rate constant (k_p) is determined by the mass gain (ΔW) per unit surface area of a specimen (A) and exposure time (t) [9-10].

$$\left({}^{\Delta W}/_{A}\right)^2 = k_p t \quad (1)$$

Table 1. Mass gain change and parabolic rate constant (k_p) of FeCr alloys after oxidized at 900°C.

Specimens	Crystallite size (nm)	Mass Gain (mg/cm^2)	Parabolic Rate, k_p (mg^2/cm^4)
FeCr "A"	53.33	21.81	6.191
FeCr "B"	38.51	7.72	0.794
FeCr "C"	76.60	35.18	13.58

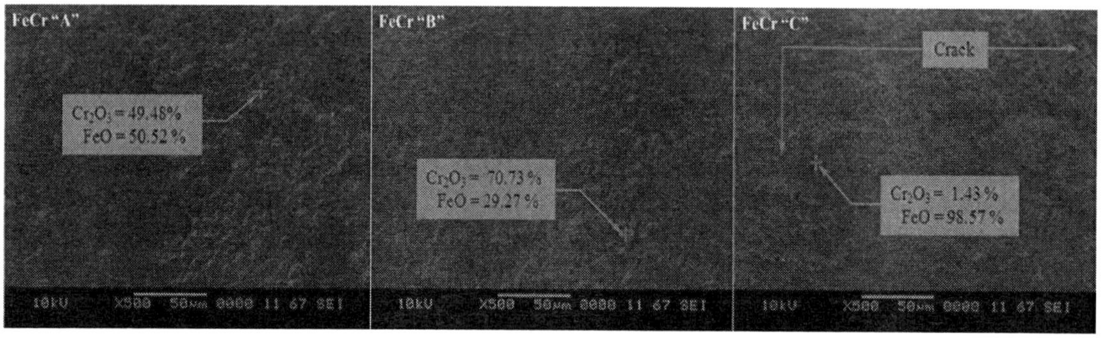

Figure 1. Surface morphology characterization of oxide scale FeCr alloys oxidized at 900°C [10].

Fig. 1 shows the surface morphology of oxide scale on the specimens that oxidized at 900°C. The surfaces of FeCr "A" and FeCr "B" were covered with compact oxide scale consisting of fine grain scales. The oxide scales showed that the mainly consisted of Cr_2O_3 which indicates the scale had an enriched Cr content which improves the adherence of the oxide scale to the substrate. This occurrence can be explained due to the chromia growth mechanism, where chromium preference diffused outward to react with oxygen. However, the oxide scale that formed on the surface FeCr "C" indicated covered and distributed evenly with the main composition of FeO approximately 98.57 at%. This indicates that Cr_2O_3 as the protective layer becomes non-protective, due to the formation of cracks and volatile Cr_2O_3 upon exposure to oxidizing atmospheres for extended periods [10].

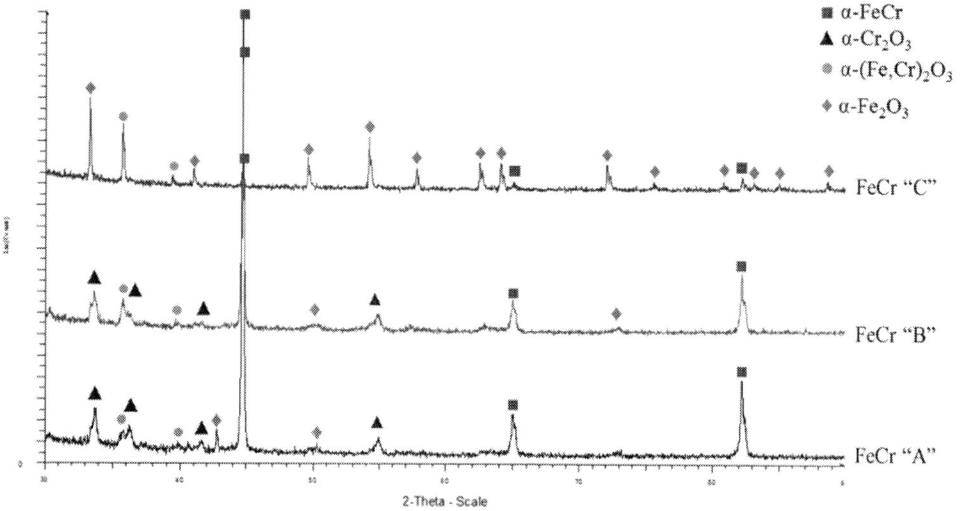

Figure 2. X-rays diffractograms of FeCr alloy oxidized at 900°C

Fig. 2 above shows the diffraction patterns of FeCr "A" and FeCr "B' that were oxidized at 900°C. The x-rays diffractograms show some peaks of the original alloys and the corundum oxides of Cr_2O_3. The presence of the peak of the original alloy indicates that oxide scale was formed as a thin layer which dominated by the presence of Cr-rich oxide and small peaks of solid solution chromium iron oxide $(Fe,Cr)_2O_3$ [10-11]. On the contrary, FeCr "C" after oxidized 900°C, all strong peaks were dominated by the presence of hematite Fe_2O_3. The results have also identified the presence of mixed oxides and a solid solution of chromium iron oxide $(Fe,Cr)_2O_3$ are present. While the presence of the peaks of the original alloys still found with a small intensity which indicates that oxide scale was growing rapidly with the dominating presence of Fe-rich oxide due to largest crystallite size effect.

3.2. Results of the current study

The cross-sectional morphologies of the oxide scale formed on FeCr alloys that oxidized at 900°C for 100 h in air atmosphere are presented in Fig 3.

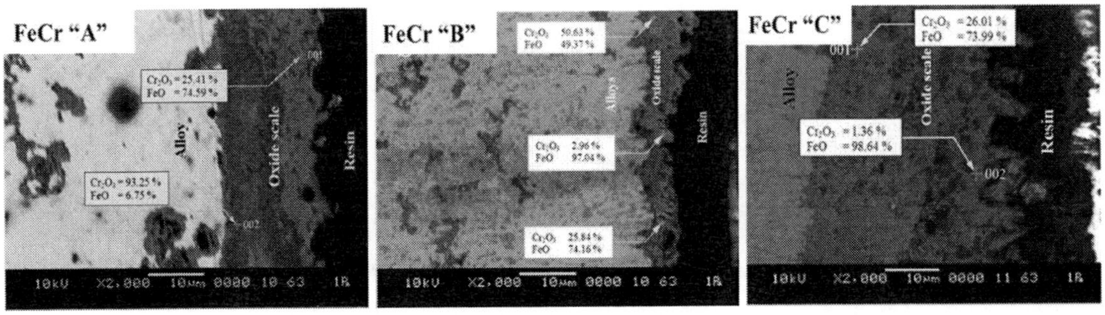

Figure 3. Morphological characterization of oxide scale FeCr "A" oxidized at 900°C.

Fig. 3 shows the SEM/EDS analyses on the cross-sectional view for oxidation of FeCr alloys, Fig.3 clearly shows that the internal layer of oxide scale was dominated by the presence of Cr_2O_3. While on the outer layer of the oxide scale was dominated by the presence of FeO. SEM micrograph of cross-sectional investigation of the specimens have also shown the thickness of oxide scale formed on FeCr "A" and FeCr "B" was found around 20 and 8 µm thick respectively, while the thickness of oxide scale formed on FeCr "C" as the specimen with the largest crystallite sizes was found around 30 µm thick. This indicates that the crystallite sizes have a significant effect to improve the oxidation resistance of the alloys. In order to understand more detail about the mechanism of oxide growth on the surface of specimens, EDX line/mapping analysis was used to verify the results in the formation of the oxide scale as shown in Fig.4, 5 and 6.

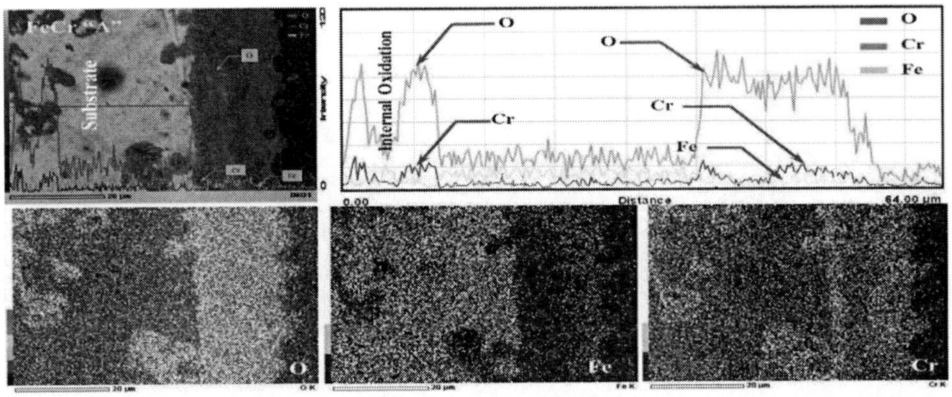

Figure 4. EDS line mapping analysis across the interface of the FeCr "A" oxidized at 900°C.

Fig.4 above shows that the scale formed on FeCr "A" consisted of outer iron oxide layers, while the inner layer of oxide scale is richer in Cr. Based on the image of EDX mapping analyses, the results showed the occurrence of internal oxidation in the substrate of the alloy. This phenomenon indicates that the diffusion is predominated by oxygen ion diffusion into the substrate alloy and reacting with Cr^{3+} ions then form a local oxidation of Cr_2O_3.

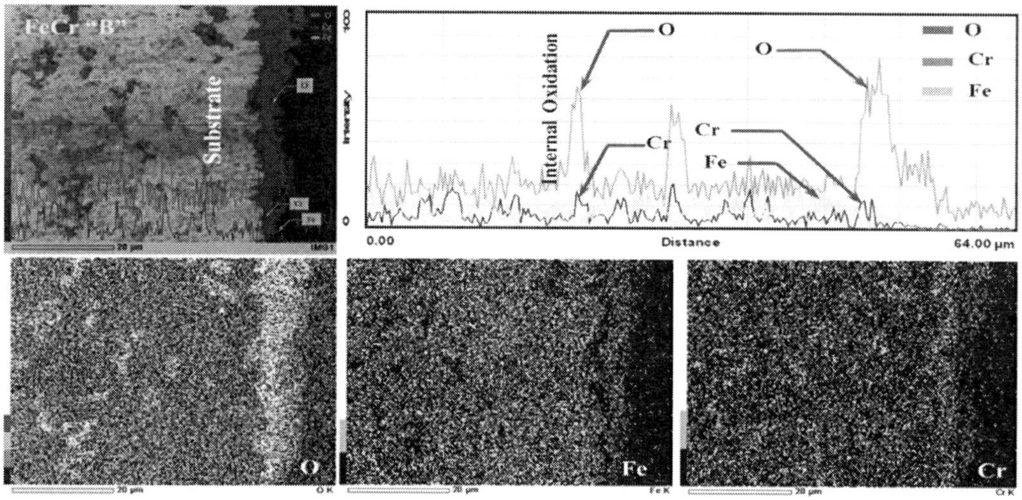

Figure 5. EDS line mapping analysis across the interface of the FeCr "B" oxidized at 900°C.

Fig. 5 above shows that the inner layer of oxide scale is richer in Cr, which can be clearly seen from the cross-sectional scale morphology of the specimens. The Cr, Fe, and oxygen as measured from the substrate/oxide interface to the outer surface of the oxide scales are similar in both of the specimens. The oxide scale that formed on the surface of FeCr "A" and FeCr "B" where the enhancement of Cr diffusion guarantee the stable growth of the Cr-rich in the inner of the oxide scales, which significantly improve oxidation resistance, even a Cr_2O_3 scale may be formed at higher temperatures owing to the higher diffusion rate of Cr. However, EDS line/mapping analysis indicates that the presence of internal oxidation in FeCr "A" and FeCr "B". It may be caused by oxygen that through diffusing into the alloy and reacting with Cr^{3+} ions that higher oxygen affinity before Cr^{3+} ions diffuses to the surface of the specimens.

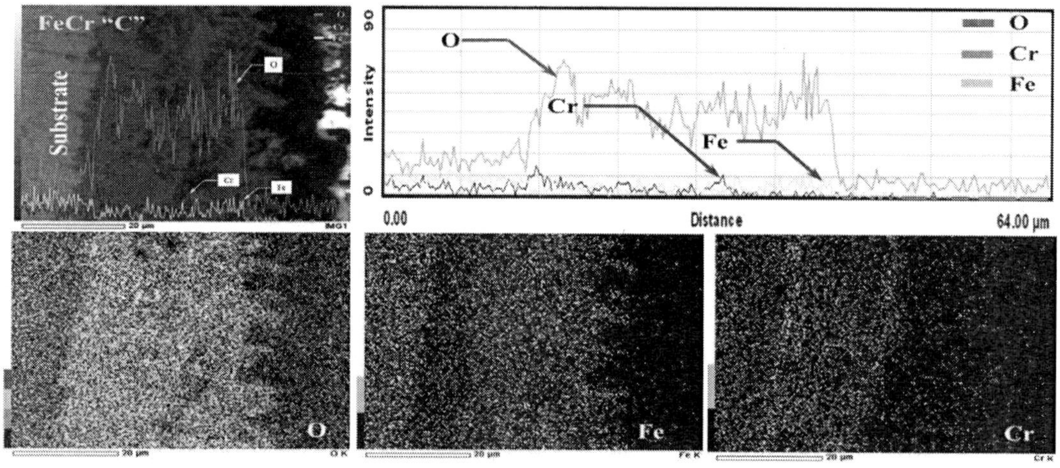

Figure 6. EDS line mapping analysis across the interface of the FeCr "C" after oxidized at 900°C.

Fig.6 above shows that the Cr concentration in FeCr "C" is still not sufficiently high to allow the formation of a more protective Cr_2O_3 layer, while the iron ions dissolve in and diffuse rapidly through the Cr_2O_3 scale and eventually an outer layer of fairly pure iron oxides will result. It may be caused by Fe and Cr ion transport is predominant then make the scale growth will proceed outwards, the oxygen transports through the FeO layer to the Cr_2O_3 layer and the Cr ions diffuse to the same layer; both of them react with each other, which make the oxide scale become thicker. Furthermore, due to the loose and incompact scale, thus causing the crack in the surface of the specimen (see Fig.1) which increase the rate of Cr evaporation. The oxygen may diffuse to the oxidation/alloy interface through the cracks and pores in the oxide scale. This phenomenon was confirmed by XRD results as shown in Fig. 2, where the result shows that the oxide phase that formed on the FeCr "C" is dominated by the presence of Fe_2O_3. It's also reinforcement by mass gain change and parabolic rate constant (k_p) as presented in Table. 1, where FeCr "C" was found as the specimens with the largest mass gain change and parabolic rate (k_p) which indicates that the oxide scale has grown rapidly on the surface of the specimen. It can be noticed that a discontinuous Cr_2O_3 scale was formed, but could not protect the alloy from internal oxidation. At the same time, Fe became enriched and diffused outwards, which caused the formation and rapid growth of FeO.

4. Conclusion

Based on the results obtained in the oxidation of nanocrystalline FeCr alloys at 900°C for 100 h the following conclusions can be formulated:

1. The specimens developed with smaller crystallite size have a significant effect to improve the oxidation resistance and the adherence of the oxide scale to the substrate of the alloys and also relatively less mass gain is found in the developed alloy with the finer crystallite size of FeCr "B".
2. The FeCr "B" show better promoting the formation of Cr_2O_3 in large amount compositions which indicated that chromium preference diffused outward to form a Cr_2O_3 layer. The thickness of oxide scale formed on nanocrystalline FeCr "B" was found around 8 μm thick, three times thinner than FeCr "C" which indicates that developed specimen with smallest crystallite size can significantly reduce the growth of the oxide scale and reduce the rate of Cr evaporation.
3. The oxide scale that formed on the surface of FeCr "C" as the larger crystallite size was formed as thicker oxide scale which significant increase in oxidation weight gain. It can be noticed that a discontinuous Cr_2O_3 scale was formed, but could not protect the alloy from internal oxidation. At the same time, Fe became enriched and diffused outwards, which caused the formation and rapid growth of FeO. Blocked by Cr_2O_3, FeO nucleated and grew at the discontinuous areas such as the incompact Cr_2O_3 scale or its micro-defective areas; leading to some bulges and cracks.

References

[1] Quadakkers W J *et al* 2003 *Materials at high temperatures* **20** 115
[2] Froitzheim J *et al* 2008 *Journal of Power Sources* **178** 163
[3] Birks N M G H and Pettit F S 2006 *Introduction to the high temperature oxidation of metals* (Cambridge University Press)
[4] Srisrual A 2013 *Photoelectrochemical characterization of thermal oxide developed on metal and model alloys* (Université Grenoble Alpes)
[5] Young D J *High temperature oxidation and corrosion of metals*. (Amsterdam: Elsevier) p 71.
[6] Norby, T R U L S 1987 *Advances in Ceramics 23* 107
[7] Khaerudini D S *et al* 2012 *Corrosion Engineering, Science and Technology* **47** 536
[8] Saryanto H *et al* 2010 *Advanced materials research* **129** 999
[9] Sebayang D *et al* 2016 *AIP Conference Proceedings* **1394** 90
[10] Sebayang D *et al* 2011 *Key Engineering Materials* **474** 2134
[11] Sebayang D *et al* 2011 *Applied Mechanics and Materials* **52** 2197
[12] Anžel I 2007 *Metalurgija- Journal of Metallurgy*, **13** 325

IOP Conf. Series: Materials Science and Engineering **196** (2017) 012031 doi:10.1088/1757-899X/196/1/012031

Study of Tensile Properties and Deflection Temperature of Polypropylene/Subang Pineapple Leaf Fiber Composites

R Hafizhah[1], A L Juwono[1] and S Roseno[2]

[1]Department of Physics, Faculty of Mathematics and Natural Sciences, Universitas Indonesia, Depok, 16424, Indonesia
[2]Agency of Assesment and Application of Technology, Serpong, Tangerang, 15314, Indonesia

E-mail: rana.hafizhah@alumni.ui.ac.id

Abstract. The development of eco-friendly composites has been increasing in the past four decades because the requirement of eco-friendly materials has been increasing. Indonesia has a lot of natural fiber resources and, pineapple leaf fiber is one of those fibers. This study aimed to determine the influence of weight fraction of pineapple leaf fibers, that were grown at Subang, to the tensile properties and the deflection temperature of polypropylene/Subang pineapple leaf fiber composites. Pineapple leaf fibers were pretreated by alkalization, while polypropylene pellets, as the matrix, were extruded into sheets. Hot press method was used to fabricate the composites. The results of the tensile test and Heat Deflection Temperature (HDT) test showed that the composites that contained of 30 wt.% pineapple leaf fiber was the best composite. The values of tensile strength, modulus of elasticity and deflection temperature were (64.04 ± 3.91) MPa; (3.98 ± 0.55) GPa and (156.05 ± 1.77) °C respectively, in which increased 187.36%, 198.60%, 264.72% respectively from the pristine polypropylene. The results of the observation on the fracture surfaces showed that the failure modes were fiber breakage and matrix failure.

1. Introduction

The applications of natural fibers as reinforcement in composites are used as the automotive, home furnishings, construction and clothing [1]. Natural fibers are widely used because of its advantages, such as lighter, less damage during processing, a relatively good mechanical properties, renewability and biodegradability [2]. Natural fibers can produce lightweight composites, showed a tendency to replace glass fibers as the reinforcement of polymer-based composites [3]. The mechanical properties of polymers have many similatities in fulfilling structural functions. However, these limitations can be overcome by using natural fibers as the reinforcement of polymer composites [4]. The properties of fiber-reinforced composites depend on many factors such as the bonding between the fibers and the matrix, the fiber volume fraction, the fiber aspect ratio, fiber orientation and efficiency of load transfer at the interface [5].

Pineapple leaf fibers (PALF) derived from the leaves of the pineapple plant (Ananas comosus). The major manufacturers of pineapple plants are Thailand, Philippines, China, Brazil, Indonesia and India, a total of approximately 2.1 million acres of the world [6]. Pineapple leaves nowadays have been wasted in most regions where pineapple plants grow, as only pineapple utilized. Indonesia is one of the major world producers of PALF [7].

In this study, the objective was to find the best weight fraction of PALF, that was grown in Subang, West Java, Indonesia, as the reinforcement of Polypropylene (PP) composite. PALF were alkalized by 5% Sodium hydroxide (NaOH) solution. PP/PALF composite was fabricated by a hot press method and the weight fractions of PALF were varied. Pristine PP was also used as a comparison. PP/PALF composite was tested with tensile test and Heat Deflection Temperature (HDT) test, then the fracture surfaces were observed by Scanning Electron Microscope (SEM).

2. Experimental

2.1 Materials

This experiment used PALF derived from a one year old pineapple plant as the reinforcement and polypropylene FY3012E from The Polyolefin Company (Singapore) Pte. Ltd as the matrix.

2.2 Alkaline Treatment

PALF were immersed in 5% NaOH solution at 25 ° C for 5 hours. The PALF was then washed with distilled water until the pH became neutral and kept at room temperature for 24 hours, then dried at a temperature of 75 ° C for 24 hours. Single fiber tensile test was conducted to know about the difference between tensile strength before and after alkalization.

2.3 Sample Fabrication

Polypropylene granules were modified into sheet form using a machine called an extruder at the temperature of 250 ° C for 30 minutes. The fabrication process of the composites used a Collin 300P Hot Pressed machine by heating to a temperature of 190 ° C for 15 minutes and then 35 barr pressure for 5 minutes, followed by cooling for 15 minutes. After the fabrication, the composite was cut to the required size of ASTM D638 type IV for tensile test and ISO 75 type II for Heat Deflection Temperature (HDT) test. The fracture surfaces of tensile tested specimens were observed using a Scanning Electron Microscope (SEM).

3. Results and Discussion

The results of the single fiber tensile test before and after alkalization were 359.96 MPa and (284.47 ± 40.66) MPa respectively. The tensile strength of PALF after alkalization was lower than that before alkalization because the lignin in PALF was dissolved. Lignin is soluble in an alkaline solution [8]. The poor fiber-matrix interface lowers tensile strength so that alkalization was an important treatment before the PALF was used as a reinforcement. The tensile strength and modulus elasticity versus the weight fractions of PALF in the composite are shown in Figure 1. It is clear that the tensile strength and modulus elasticity increase when using PALF as the reinforcement. The highest tensile strength and modulus elasticity belong to PP/PALF30% in the values of (64.04 ± 3.91) MPa and (3.98 ± 0.55) GPa respectively, which are 187.36% and 198.60% greater than pristine PP. The tensile strength and modulus elasticity of PP/PALF40% decreased by 10.99% and 16.08% respectively than those the PP/PALF30%. The excessive PALF can make the fibre acted as flaws and crazing occured, thus creating stress concentration area which lowering the stiffness of the composites and fibres were more easily expose to environmental degradation [9].

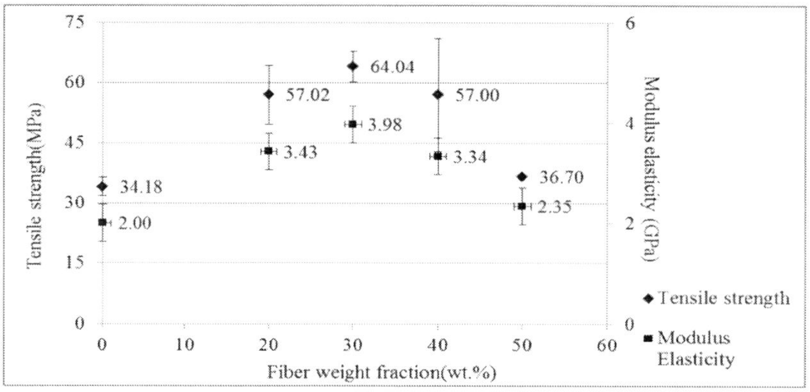

Figure 1. Tensile strength and modulus elasticity versus PALF weight fraction.

Figure 2 displays the strain at break versus fiber weight fraction. From the figure, it can be seen that PALF lowers the strain at break of PP/PALF composite compared to pristine PP. It showed that the PP/PALF composite was more brittle than pristine PP. Strain at break of pristine PP is very high because PP is elastic.

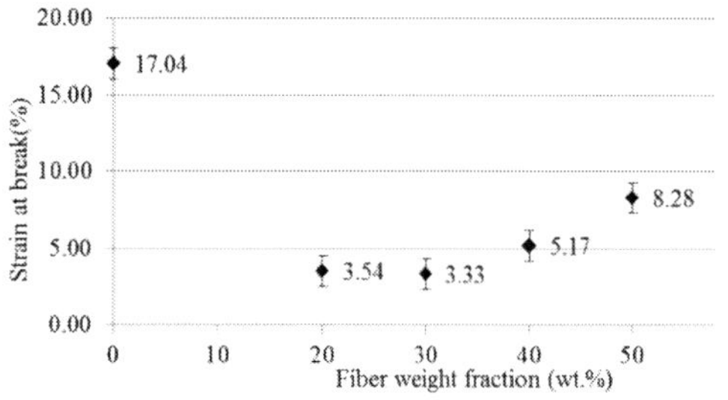

Figure 2. Strain at break versus PALF weight fraction

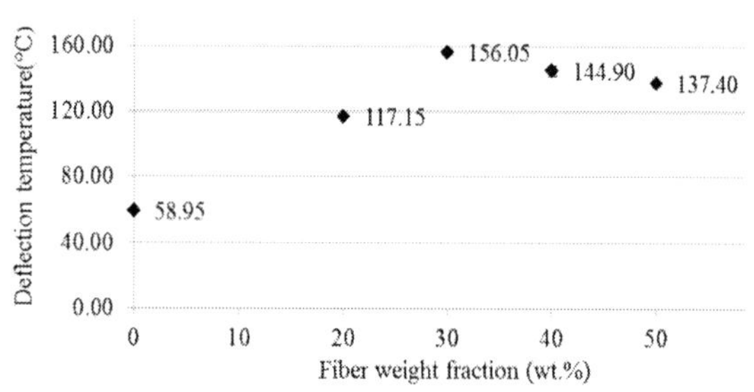

Figure 3. Deflection temperature versus PALF weight fraction

Figure 3 presents the deflection temperature versus the weight fraction of PALF of the composites. It indicates that the addition of PALF as the reinforcement increased deflection temperature of the composite. The highest deflection temperature sample belongs to PP/PALF30% composite with the value of (156.05 ± 1.77) ° C which was 264.72% greater than pristine PP. PALF had a bridging ability that could retard deformation of the PP matrix at high temperature [10].

Figures 4 (a) to (f) show the SEM images before (a, b, c) and after (d, e, f) tensile test of PP/PALF20%, PP/PALF30%, and PP/PALF40% composites. Figure 4 (a) and (c) show a few voids in the PP/PALF20% and PP/PALF40% composites. Meanwhile, in the PP/PALF30%, PALF were fully wetted by the PP and both had a strong interface bonding which can be seen in Figure 4 (b). Fiber pull-out, fiber breakage and matrix failure are observed in Figure 4 (d) to (f). In more detail, fiber pull-out can be seen in PP/PALF20%. Fiber breakage is exhibited in PP/PALF20%, PP/PALF30% and PP/PALF40% and matrix failure is observed in PP/PALF30%.

From SEM images, it is an evidence that the PP/PALF30% composites are the best sampel and this result agrees well with the tensile and HDT test. PP/PALF20% composites had a poor interfacial bonding because the PP did not completely wett the PALF, so that the tensile strength was lower than PP/PALF30%. It could be proven that the interfacial bonding between the PP and PALF, voids and fiber distribution could affect the strength [11].

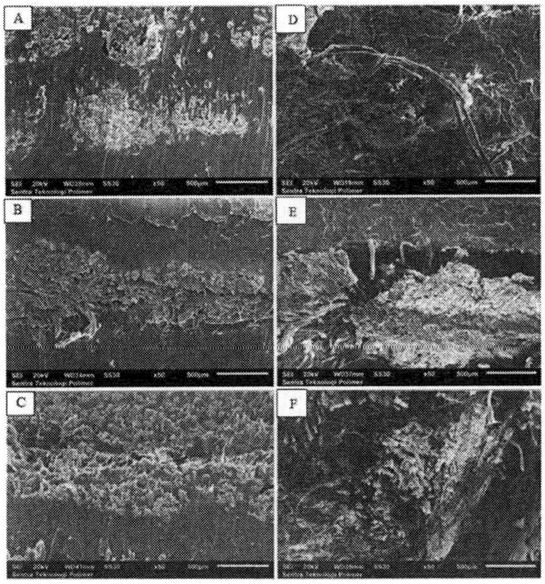

Figure 4. SEM images of (a)PP/PALF20%, (b)PP/PALF30%, (c)PP/PALF40 and SEM images after tensile test of (d)PP/PALF20%, (e)PP/PALF30%, (f)PP/PALF40%.

4. Conclusion

The optimum weight fraction of Subang PALF after alkalization treatment in PP/PALF composites was 30 wt.%. The results of tensile and HDT tests showed that PP/PALF30% composites had the greatest tensile strength, modulus elasticity and deflection temperature with the values of (64.04 ± 3.91) MPa, (3.98 ± 0.55) GPa, and (156.05 ± 1.77) °C respectively which are 187.36%, 198.60% and 264.72% greater than pristine PP. In the PP./PALF30%, Subang PALF were fully wetted by PP and there was no void in the composites. The PP/PALF30% composites experienced matrix failure and Subang PALF breakage which was confirmed that bond between PP and Subang PALF was very strong.

Acknowledgement
This work was funded by Universitas Indonesia for PITTA grant in 2016.

References

[1] Hojo T, Xu Z, Yang Y, Hamada H 2014 *Energy Procedia* **56** 72
[2] Wambua P, Ivens J and Verpoest I 2003 *Composites Science and Technology,* **63** 1259
[3] Akil H M, Omar M F, Mazuki A A M, Safiee S, Ishak Z A M, Abu A B 2011 *A review* Materials and Design **32** 4107
[4] Abilash N, Sivapragash M 2016 *Journal of King Saud University - Engineering Sciences* **28** 92
[5] George J, Bhagawan S S, Prabhakaran N, Thomas S 1995 *J. Appl Polym Sci* **57** 843
[6] Mishra S, Misra M, Tripathy S S, Nayak S K, Mohanty A K 2001 *J Reinf Plast Compos* **20** 321
[7] Leão A, Cherian B, Narine S, Thomas M SS 2015 *Biofiber Reinforcements in Compos. Mat.* 211
[8] Phuong N T, Sollogoub C, Guinault A 2010 *Journal of Reinforced Platics and Compo.* 3244
[9] Mokhtar M, Rahmat A R, Hassan A 2007 *Research VOT* **75147** Universitas Teknologi Malaysia
[10] Kengkhetkit N, Amornsakchai T 2013 *Materials & Design* **55** 292
[11] Arib N, Sapuan M, Ahmad M, Paridah M T, Zaman K 2006 *Materials & Design* **27** 391

IOP Conf. Series: Materials Science and Engineering **196** (2017) 012032 doi:10.1088/1757-899X/196/1/012032

The Role of Annealing Temperature in Photocatalytic Performance of Fe_3O_4/SnO_2 nanocomposites

Valentinus Paramarta[1,2], Ardiansyah Taufik[1,2], Rosari Saleh[1,2]

[1]Departemen Fisika, Fakultas MIPA-Universitas Indonesia, 16424 Depok, Indonesia

[2]Integrated Laboratory of Energy and Environment, Fakultas MIPA-Universitas Indonesia, 16424 Depok, Indonesia

E-mail: rosari.saleh@gmail.com

Abstract. In this work, SnO_2 nanoparticles with the variation of annealing temperature (400°C-800°C) were used as photocatalyst for remove organic dye from the aqueous solution. SnO_2 nanoparticles were synthesized using sol-gel method. For enhancing the stability and the photocatalytic activity of the samples, magnetite materials (Fe_3O_4) were combined with SnO_2 nanoparticles. The prepared samples characterized by X-ray Diffraction (XRD). The X-ray diffraction shows tetragonal structure of SnO_2 and cubic spinel of Fe_3O_4 as the components of the prepared nanocomposites. The photocatalytic activity of the samples was studied using Methylene Blue (MB) as a model organic pollutant. The photocatalytic degradation of MB by SnO_2 and Fe_3O_4 nanocomposites under UV light irradiation was examined by varying the operational parameters such as catalyst dosage and scavengers. Among the variation annealing temperature of SnO_2 nanoparticles, the 700°C annealing temperature of SnO_2 showed the highest photocatalytic activity. The repeatability of photocatalytic activity was also tested.

1. Introduction

Dyes widely used in many industries often create severe environmental pollutions in the form of colored wastewater discharged into environmental water bodies [1]. Therefore, in recent years, the treatment of various organic dye effluents has become an important topic in the research field on environmental pollution and control [2]. Of all various method, advance oxidation processes (AOPs) such as photocatalysis has attracted remarkable interest because it offers a sustainable pathway to drive chemical reactions such as water splitting carbon fixation, and degradation of organic pollutants [3-5].

Among various semiconductors, SnO_2 is an n-type semiconductor that has been applied in many applications and potentially to be ideal catalyst due to their several advantages [6-8]. But, high rate recombination electron holes pairs and reusability limitation of those semiconductor could limit the efficiency as a catalyst. Therefore, by adding with other material such as Fe_3O_4 could resolve the problem. So the material is possible to do the magnetic separation from the solutions for reusing and could inhibit the recombination of electron holes pairs [9,10]. Besides of recombination rate of the sample, still, have many things that could be affected by the catalytic activity of the sample. Many studies have reported a correlation between catalytic activity and annealing temperature, crystallinity and charge recombination rate, which significantly affects the properties of the sample [11].

In this study, Fe_3O_4/SnO_2 nanocomposite with the various of annealing temperature (400°C, 500°C, 600°C, 700°C, 800°C) were prepared using ultrasonic assisted method. The sample was characterized

IOP Conf. Series: Materials Science and Engineering **196** (2017) 012032 doi:10.1088/1757-899X/196/1/012032

using X-ray Diffraction (XRD). The catalytic activity of the Fe_3O_4/SnO_2 nanocomposites with various annealing temperature was analyzed using methylene blue as organic pollutant model. The catalytic activity of the Fe_3O_4/SnO_2 with 700°C annealing temperature showed best photocatalytic performance under UV-light irradiation.

2. Experimental

2.1. Materials
All reagents used were analytical grade and without further purification. Tin chloride anhydrous ($SnCl_2$), Iron (II) sulfate heptahydrate ($FeSO_4.7H_2O$), sodium hydroxide (NaOH), methylene blue (MB), ethanol, ethylene glycol (EG), ammonium oxalate, sodium sulfate, and tertbutyl alcohol (TBA) were purchased from Merck.

2.2. Synthesis of nanomaterials
The Fe_3O_4 nanoparticles were synthesized using the same method of our previous study [12]. The SnO_2 nanoparticles were synthesized using modification method reported by Yue li and friends [13]. First, $SnCl_2$ was dissolved into ethanol and the aqueous mixture then added into the NaOH solutions using magnetic stirring. Then, the mixture solutions were heated 180°C temperature for 3 hours and chilled in room temperature. The precipitated obtained was centrifuged and washed using aqueous and ethanol several times. The precipitated particles were dried in vacuum condition at 80°C temperature. The SnO_2 particles obtained after calcination process for 3 hours at various temperature (400°C, 500°C, 600°C, 700°C and 800°C).

Fe_3O_4/SnO_2 nanocomposites were synthesized using the sol-gel method. First, SnO_2 nanoparticles were mixed with Fe_3O_4 in aqueous and ethanol mixture. Each mixture was given ultrasonic for 2 hours then was centrifuged to obtain product on its precipitated. The product result obtained was dried in vacuum condition at 80°C temperature for 12 hours to obtain Fe_3O_4/SnO_2 nanoparticles.

2.3. Characterization
The sample was characterized using X-Ray Diffraction (XRD) measurement by using Rigaku Miniflex 600 with radiation source Cu K-α (λ =1,5406 Å).

2.4. Photocatalytic experiments
The photocatalytic experiment was done by mixing the sample Fe_3O_4/SnO_2 (400 °C,500 °C,600 °C,700 °C,800°C) into 100mL methylene blue (MB) solutions used as pollutant model with concentrations 20mg/L respectively, and pH solutions adjusted NaOH. The solution was allowed to stand in stirring state for adsorption and desorption balance. The photocatalytic test, the solution was given UV light irradiation with power 40W and wavelength of 320-400 nm. The solution was given UV light irradiation for 2 hours. Every 15 minutes span the concentration of MB solution was analyzed using UV-vis spectrometer.

3. Result and Discussion
Figure 1 shows the XRD spectrum of SnO_2 nanoparticles and Fe_3O_4/SnO_2 nanocomposites with various annealing temperature. The diffraction pattern of Fe_3O_4 sample observed at the value 2θ = 30.14°, 35.49°, 43.28°, 57.20°, 62.83° and 74.8° which shows each area (220), (311), (400), (511), (440) and (533) of the cubic spinel structure. Diffraction pattern of SnO_2 nanoparticle sample observed at value 2θ = 26,5°, 33.8°, 38°, 39°, 51.8°, 54.8°, 58°, 62°, 64.7°, 65.8°, 71.2°, 78.2°, 81.2°, 83.7° which shows the existence of areas (110), (101), (200), (111), (211), (220), (002), (310), (112), (301), (202), (321), (400), (222) of tetragonal structure of SnO_2 nanoparticle. The XRD pattern of Fe_3O_4/SnO_2 nanocomposites shows the existence of cubic spinel phase of Fe_3O_4, followed by the addition of tetragonal SnO_2 nanocomposites. As it is clear from this figure, there is no any impurity in the nanoparticle sample.

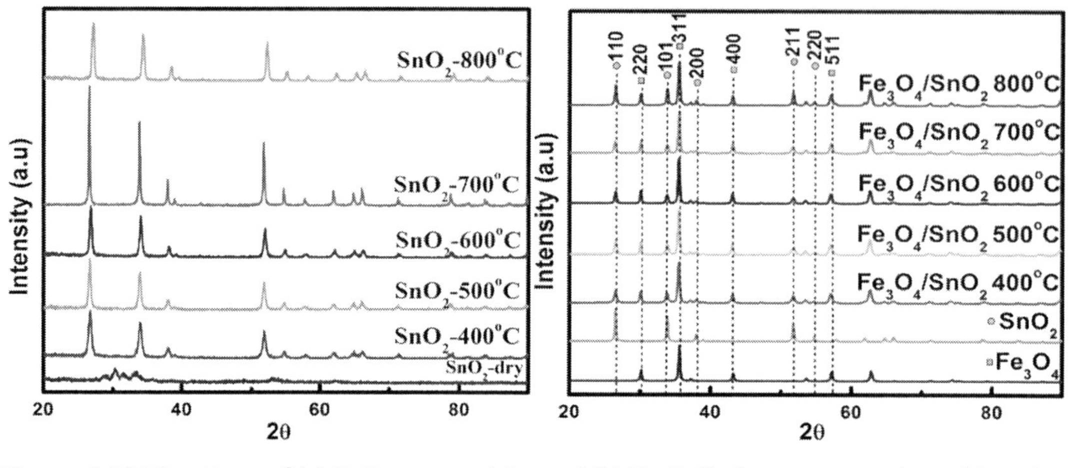

Figure. 1. XRD patterns of (a) SnO_2 nanoparticles and (b) Fe_3O_4/SnO_2 nanocomposites with various annealing temperature

Table 1. Lattice parameters and grain size of Fe_3O_4/SnO_2 nanoparticles with various annealing temperature

Sample	Lattice Parameter Fe_3O_4 a=b=c	SnO_2 a=b	SnO_2 c	<D> (nm) Fe_3O_4	<D> (nm) SnO_2
Fe_3O_4/SnO_2 400°C	8.384	4.738	3.193	24	24
Fe_3O_4/SnO_2 500 °C	8.391	4.739	3.190	27	22
Fe_3O_4/SnO_2 600 °C	8.388	4.740	3.196	29	23
Fe_3O_4/SnO_2 700 °C	8.380	4.760	3.190	28	21
Fe_3O_4/SnO_2 800 °C	8.387	4.741	3.193	30	28

Strong and sharp peaks can indicate that the prepared samples are highly crystalline [14]. The grain size measurement was calculated using Scherrer's equation and lattice parameter was calculated using Rietveld refinement method. The analysis result of grain size and lattice parameter is shown in Table 1.

Fe_3O_4/SnO_2 nanocomposites with various annealing temperature were subjected to methylene blue (MB) degradation process in the condition of UV at 320-400 nm. The catalytic degradation of methylene blue was observed (Fig. 2). The maximum degradation of photocatalytic activity is 43.5%, 53.2%, 60.6%, 68.1%, and 65% for Fe_3O_4/SnO_2 400°C, Fe_3O_4/SnO_2 500°C, Fe_3O_4/SnO_2 600°C, Fe_3O_4/SnO_2 700°C, and Fe_3O_4/SnO_2 800°C, respectively. As can be seen from the Fig. 2, the degradation rate of MB is increased with increasing calcination temperature until 700°C. But the photocatalytic activity of Fe_3O_4/SnO_2 calcined at 800°C is slightly lower than that of the photocatalyst calcinated 700°C. It may be due to the crystallite size of Fe_3O_4/SnO_2 are bigger than the other (Table. 1.), which resulting in a reduction of quantum size effect [15]. Fe_3O_4/SnO_2 700°C nanocomposite exhibits the best photocatalytic performance under UV light irradiation.

Figure 3 shows the effect of catalyst dosage addition, Fe_3O_4/SnO_2 calcinated at 700°C nanocomposites from 0.1 g/L until 0.4 g/L, to the efficiency of methylene blue degradation through a

IOP Conf. Series: Materials Science and Engineering **196** (2017) 012032 doi:10.1088/1757-899X/196/1/012032

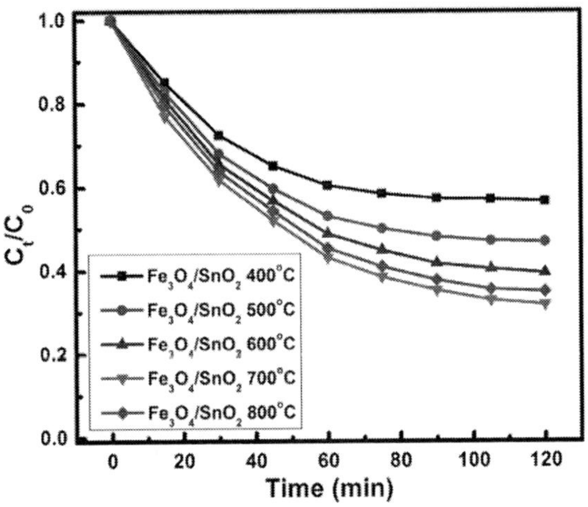

Figure 2. Photocatalytic activity of Fe_3O_4/SnO_2 nanocomposites with different annealing temperature.

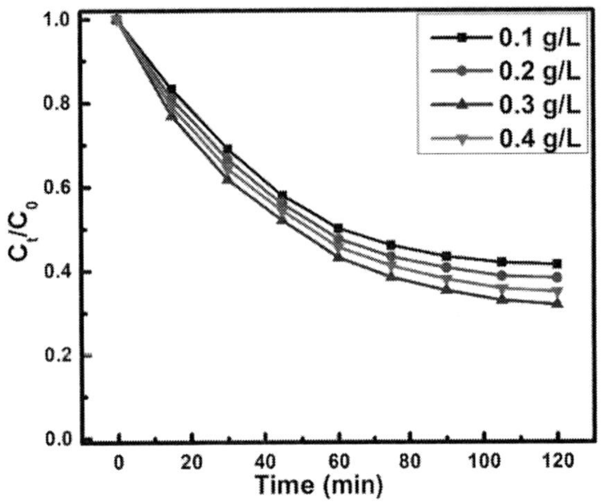

Figure 3. Effect of dosage on photocatalytic activity of Fe_3O_4/SnO_2 700°C nanocomposite.

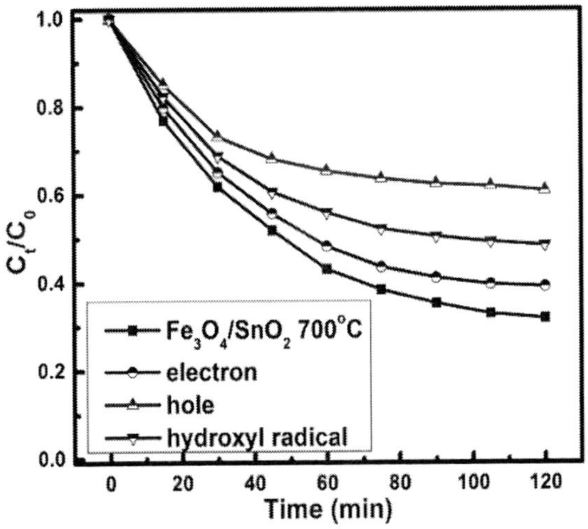

Figure 4. Effect of scavengers on photocatalytic activity of Fe_3O_4/SnO_2 700°C nanocomposite.

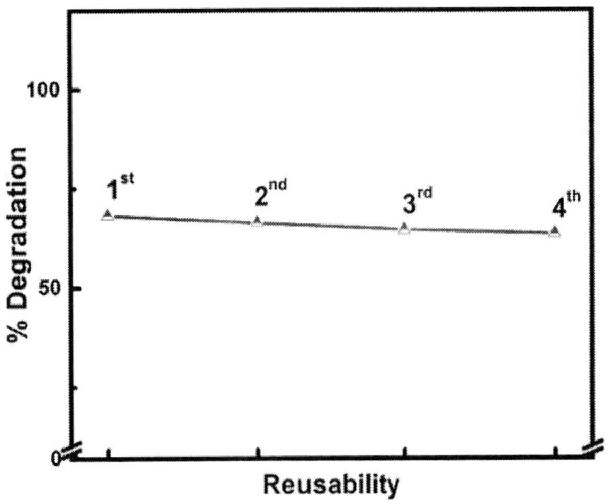

Figure 5. Reusability of Fe_3O_4/SnO_2 700°C nanocomposites.

photocatalytic process. As seen in Fig. 3, after 120 minutes, maximum degradation for Fe_3O_4/SnO_2 calcinated at 700°C with dosage 0.1 g/L, 0.2 g/L, 0.3 g/L and 0.4 g/L each are 58.6%, 61.8%, 68.1%, and 65%. The amount of sample will influence two things; the amount of active site on the sample and the amount of dyes adsorbed by the sample [16]. Catalyst dosage addition will also increase the amount of active site. The result in Fig. 3 indicates that the optimum catalyst dosage for Fe_3O_4/SnO_2 calcinated at 700°C nanocomposite sample is 0.3 g/L.

To know the most involved reactive species MB degradation to the all samples, the addition scavenger was done. In Fig. 4 shows the result of MB degradation in the photocatalytic process with scavenger addition influence. The scavenger addition serves to bind the reactive species to not take role in the photocatalytic process. The result obtained shows that the addition of hole scavenger shows the lowest degradation capacity. This indicates that the most influential reactive species for the samples is hole.

To analyze the catalyst stability, the cycling test was done. The cycling test was done by reusing catalyst four times for photocatalytic process. After photocatalytic process test, the sample was separated from its solutions using magnetic external as seen in Figure 5 to be reused as catalyst in next experiment. In figure 5 shows the result of reusing data of Fe_3O_4/SnO_2 calcinated at 700°C nanocomposites. It is shown that the sample showed good cycling capacity result with the decrease only about 5% after reusing three times. It proves that photocatalytic activity for the Fe_3O_4/SnO_2 of nancomposite is stable.

4. Conclusion

The Fe_3O_4/SnO_2 nanocomposites with different annealing (400°C, 500°C, 600°C, 700°C, 800°C) temperature have been successfully synthesized via sol-gel method. The maximum degradation of photocatalytic activity is 43.5%, 53.2%, 60.6%, 68.1%, and 65% for Fe_3O_4/SnO_2 400°C, Fe_3O_4/SnO_2 500°C, Fe_3O_4/SnO_2 600°C, Fe_3O_4/SnO_2 700°C, and Fe_3O_4/SnO_2 800°C, respectively. Fe_3O_4/SnO_2 700°C nanocomposite exhibits the best photocatalytic performance under UV light irradiation. The optimum catalyst dosage for Fe_3O_4/SnO_2 calcinated at 700°C nanocomposite sample is 0.3 g/L. Scavengers experiment shows that the most influential reactive species for the samples is hole. Sample showed good cycling capacity result with the decrease only about 5% after reuse three times.

References

[1] Jun Wang, Yanhui Lv, Liqun Zhang, Bin Liu, Renzheng Jiang, Guanxi Han, Rui, Xu, Xiangdong Zhang 2010 *Ultrasonics Sonochemistry* **17** 642-648
[2] Hongbo Zhang, Chunsheng Wei, Yingying Huang, Jun Wang, *Ultrasonics Sonochemistry* 2016 **30** 61-69
[3] Yiming He, Lihong Zhang, Maohong Fan, Xiaoxing Wang, Mikel L. Walbridge, Qingyan Nong, Ying Wu, Leihong Zhao 2015 *Solar Energy Materials and Solar Cells* **137** 175-184
[4] K. Ikehata, N.J. Naghashkar, M, Gamal El-Din 2006 *Ozone Science & Engineering* **28** 353-414
[5] D. Singh, S. Verma, R. K. Gautam, V. Krishna 2015 *Journal of Environmental Chemical Engineering* **3** 2161-2171
[6] N. Shanmugam, T. Sathya, G. Viruthagiri, C. Kalyanasundaram, R.Gobi, S. Ragupathy 2016 *Applied Surface Science* **360** 283-290
[7] Ali Fakhri, Sajjad Behrouz, Melika Pourmand 2015 *Journal of Photochemistry and Photobiology B: Biology* **149** 45-50
[8] Hui Liu, Tingting Liu, Xiaonan Dong, Yuwu Lv, Zhenfeng Zhu 2014 *Matterials Letters* **126** 36-38
[9] Hua-Yue Zhu, Ling Xiao, Ru Jiang, Guang-Ming Zeng, Li Liu 2011 *Chemical Engineering Journal* **172** 746-753
[10] Jianqi Ma, Shaobo Guo, Xiaohua Guo, Hongguang Ge 2015 *Applied Surface Science* **353** 1117-1125
[11] Huayun Liu, Huanhuan Wang, Shouzhu Li 2014 *Materials Science in Semiconductor Processing* **25** 294-300
[12] A. Taufik, I. K. Susanto, R. Saleh 2015 *Materials Science Forum* **827** 37-42
[13] Yue Li, Yanqun Guo, Ruiqin Tan, Ping Cui, Yong Li, Weijie Song 2009 *Materials Letters* **63** 2085-2088
[14] Gengxia Zhou, Xinglong Wu, Lizhe Liu, Xiaobin Zhu, Xiaoshu Zhu, Yanling Hao, Paul K. Chu 2015 *Applied Surface Science* **349** 798-804
[15] Maolin Zhang, Taicheng An, Xiaohong Hu, Cun Wang, Guoying Sheng, Jiamo Fu 2004 *Applied Catalysis A: General* **260** 215-222
[16] Yu-Ju Chiang, Chia-Chang Lin 2013 *Powder Technol.* **246** 137-143

Synthesis and Characterization of Composite UPR/Fe_3O_4 for Its Use as Electromagnetic Wave Absorber

Yusmaniar[1], W A Adi[2], Y Taryana[3], R Muzaki[1]

[1,4] Department of Chemistry, Faculty of Mathematics and Natural Science, Universitas Negeri Jakarta, Jl. Pemuda No 10, Rawamangun 13220, Jakarta, Indonesia
[2] Centre for Science and Technology of Advanced Materials, National Nuclear Energy Agency, Kawasan Puspiptek Serpong, Tangerang Selatan , Banten, Indonesia
[3] Centre for Research of Telecommunication and Electronic, Indonesia Institute of Science, Komplek LIPI Sangkuriang, Bandung, Jawa Barat, Indonesia

E-mail: ys_maniar@yahoo.com

Abstract. Synthesis and characterization of UPR/Fe_3O_4 composite were performed to evaluate its potency as a electromagnet absorbent. The composite was prepared from the mixture of unsaturated polyester resin and magnetite powder of iron oxide. Fe_3O_4 was used as filler and unsaturated polyester resins (UPR) was used as a matrix. Fe_3O_4 magnetite was synthesized from iron objects using electro synthesis method. The raw material was blended in the beaker glass for 120 min and then pressed at temperature of 60 °C for 30 min. The composite is in a semi-crystalline form that consists of amorphous matrix and the various crystalline fillers. The functional group's analysis of the composite showed that crosslink (bridges) was formed between the chains of individual polymer and between Fe_3O_4 magnetite filler and UPR matrix. The performance of microwave absorbent measured by VNA method showed the highest RL at the frequency of 10.44 GHz and11.74 dB. This value was achieved with a composition of 10wt% Fe_3O_4 and 90wt% UPR. We concluded that the composite of UPR/Fe_3O_4 has been successfully demonstrated as an electromagnetic wave absorber.

1. Introduction

In the past years, the use of wireless electronic devices grows rapidly. The wireless technology is based on transmission of energy through electromagnetic wave, which eventually may cause tissue damage from the radiation above the certain threshold. This danger can actually be avoided upon use of materials that can counteract the excess of electromagnetic wave radiation [1]. This material is called electromagnetic wave absorber.

Electromagnetic wave absorber can counteract transverse waves, which is the electromagnetic wave that propagates in a vacuum state. The wave can be transmitted, reflected and absorbed. The material absorber should be able to absorb the electromagnetic wave radiation, thereby it must have an optimum permeability (μ), permittivity (ε), and resistivity, and low coercivity (Hc) [2].

One of the materials used as an electromagnetic wave absorbent is magnetite or iron (II)(III) oxide (Fe_3O_4) [3][4]. The magnetite acts as a filler or filler matrix in the composite. It has also been used as catalyst, gas sensor, and data storage device materials. Magnetite is derived from iron objects by an electrochemical method [5], which is the most effective methods to improve the ability of a material to absorb wave. Magnetite has good permeability. Therefore, it is an excellent candidate for electromagnetic wave absorbent. In this study, the magnetite was derived from the iron nail, an iron object that is commonly used in construction works. The leftovers, wasted, and used iron nails can be utilized as a cheap source of good magnetic and solid structure materials for magnetite.

The other component of composite is the matrix polymer, which serves as the binder. In this research, Unsaturated Polyester Resin (UPR) was selected because it has good mechanical properties, i.e. dielectric constant, making it an optimum supporting material for the composite. Characterization of the composite was carried out with the electromagnetic wave type X band around the frequency of 8-12 GHz. This frequency was selected because it is used in applications such as RADAR, telecommunication, and wireless electronic device.

2. Materials and Method

Magnetite powder was made by electro-synthesis. At first, two iron nails prepared in a beaker filled with saline solution. Salt solution used is 0.9% NaCl solution. Iron nail that was submerged in a saline solution and then connect it to flow through the cathode and anode copper wire that fed the DC current from a battery or power supply. Electrolysis iron nails take about three hours to get precipitate magnetite (Fe_3O_4) blackish brown. After completion of the electrolysis process, the precipitate is filtered with Whatman paper 41. The precipitate has been filtered and after that allowed for about 15 minutes at the room temperature. The precipitate is then injected into the furnace at a temperature of approximately 400 °C for one hour. The next step was made of the composite between UPR and Fe_3O_4 with a weight ratio of 50: 50; 60: 40; 70: 30; 80: 20 and 90: 10. UPR as a matrix resin, hardener, and Fe_3O_4 were stirred for 10 minutes. After that in printing on a round plate with a thickness of 2 mm. Characterization of Fe_3O_4 synthesized is FTIR, SEM-EDS, and VSM. Phase composition analysis in the data used x-ray diffraction with a range 10-80° angle measurements; each specimen was analyzed by Vector Network Analyzer (VNA) to determine the reflection loss of the electromagnetic waves in the frequency range of 8 -12 GHZ.

3. Results and Discuss

Elemental and chemical property synthesized Fe_3O_4 was analyzed with EDS while the morphology was evaluated under the electron microscope.

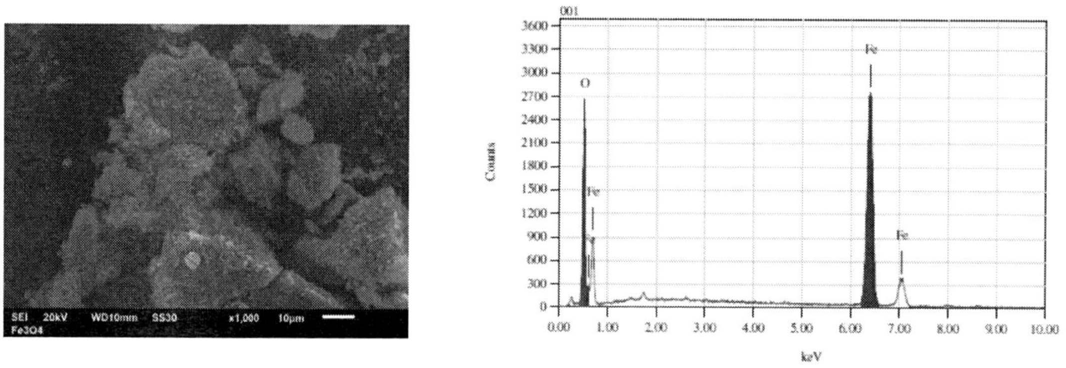

Figure1. The result of SEM-EDS from sample Fe_3O_4

The EDS spectra indicated the presence of Fe and O in the synthesized material. Under the microscope, the material appeared as a crystalline (Figure 1). This result suggests that the material is Fe_3O_4. However further analysis must be performed to confirm that hypothesis by means FTIR and XRD.

The identity of the synthesized material as to contain Fe_3O_4 was shown by the IR spectra (Figure 2). The peak at 576.64 cm-1 is characteristic for a Fe-O stretching [6]. Further confirmation of the material identity was provided by the X-ray diffraction experiment.

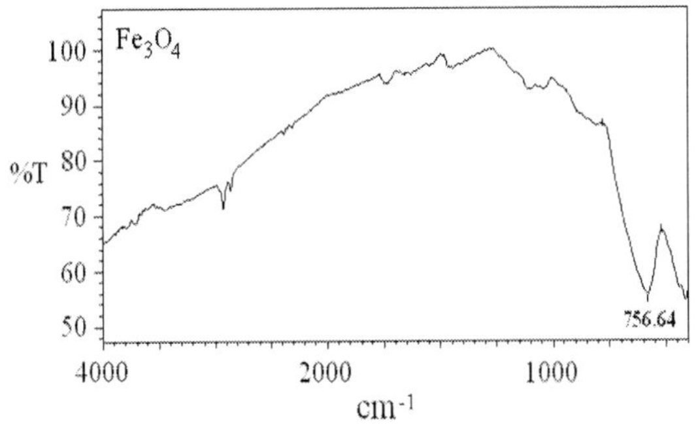

Figure 2. IR Spectra of the synthesized material Fe3O4 sample

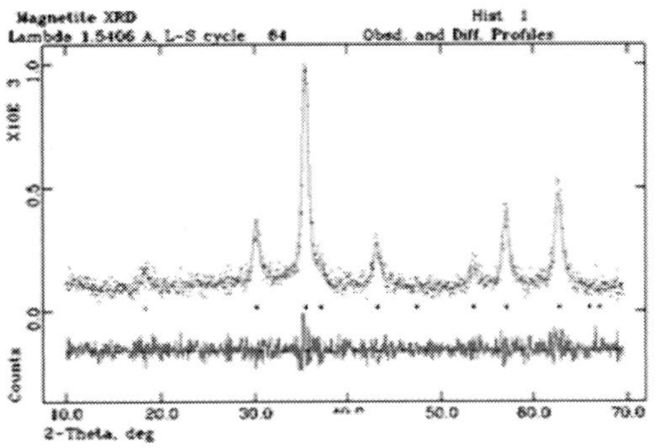

Figure 3. The refinement results of X-ray diffraction pattern Fe3O4 sample

Figure 3 shows XRD patterns of the Fe_3O_4. five characteristic peaks at 30, 35, 43, 53, and 63 were corresponding to the (220), (311), (400), (511) and (440) crystal planes of a pure Fe_3O_4 with a spinal structure (JCPDS file PDF no.65-3107) [7]. In Figure 3, the correction factor R_{wp} of 17:50% and the chi squared (χ^2) of 1.30 are presented. This latter result concludes that the material is composed of Fe_3O_4. Next, further experiment was performed to determine the magnetic properties of Fe_3O_4 using VSM.

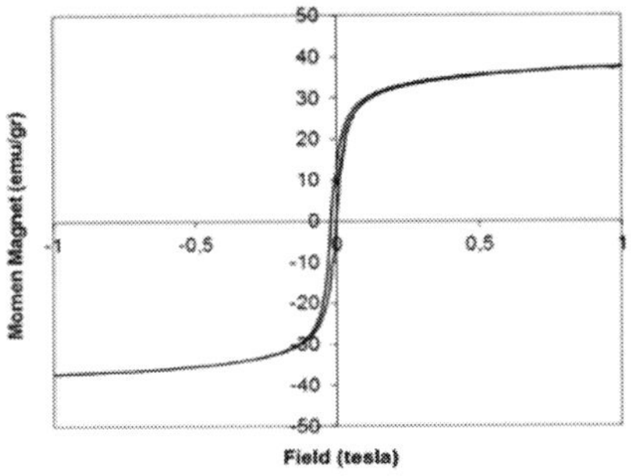

Figure 4. The curve Hysteresis of Fe3O4 sample

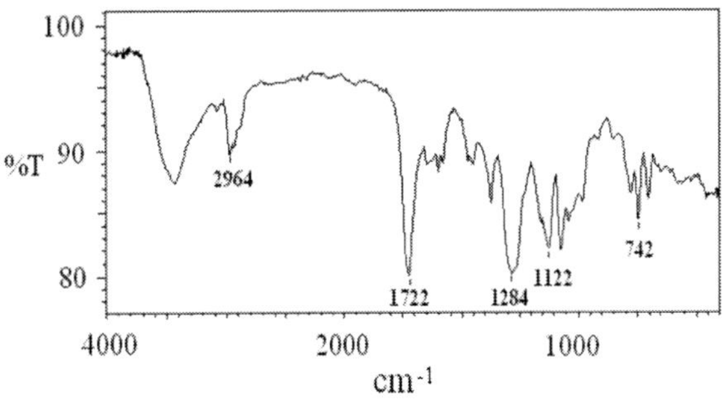

Figure 5. IR Spectra of the composite UPR/magnetite

The magnetic characteristic of Fe_3O_4 in the synthesized material was shown by the remanent magnetization (Mr), saturated magnetization (Ms), coercivity field (Hc) respective of 15.57 emu/g, 35.5 emu/g, and 90 Gauss. These values showed the Fe_3O_4 ferromagnetic properties. Thus the Fe_3O_4 in the material has successfully been identified and characterized.

Figure 5 shows an IR spectral of the UPR. In the spectral characteristics of UPR compounds with functional groups at wave numbers of 742cm^{-1}, 1122.57cm^{-1}, and 1284.59cm^{-1}, respective for benzene, polyester, and ester, while stretching -CH bond at wave number of 1722cm^{-1} and 2964.59cm^{-1} [8-10].

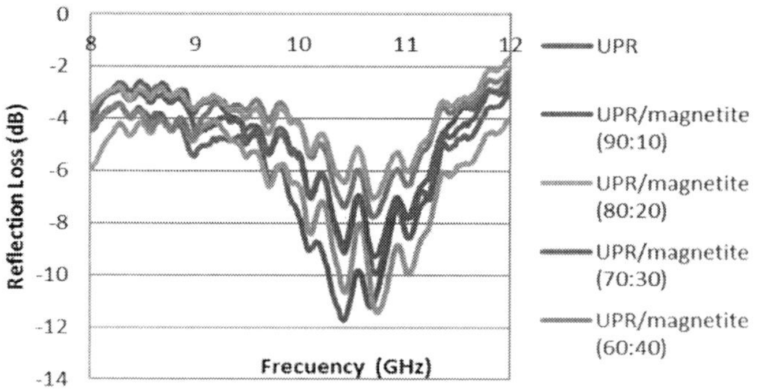

Figure 6. The result of Curve Reflection Loss composite UPR/magnetite

The composites with the UPR: Fe_3O_4 ratio of 90:10, 80:20, 70:30, 60:40, and 50:50, were analyzed with VNA method to determine the electromagnetic wave absorption power (Figure 6). Apparently, the proportion of 10-50% Fe_3O_4 to the UPR has no significant influence on the absorption capacity. The best absorption performance was achieved at proportion of 10% Fe_3O_4, reaching -11.73 dB at the frequency of 10.44 GHz.

4. Conclusion

We have successfully synthesized Fe3O4 and produced the composite that contains the magnetite and UPR. The identity of the composite was confirmed with SEM-EDS, FTIR, VSM, and XRD. The composite was also successfully produced with the proportion of 10% Fe_3O_4 to UPR to give the best electromagnetic wave absorption performance.

References

[1] Wu W 2008 *Nanoscale Res Lett.* **3** 397
[2] Cabrera L 2008 *Electrochimica Acta.* **53** 3436
[3] Jazirehpour M 2015 *Journal of Alloys and Compounds* **638** 188
[4] Cao J and Zhitomirsky I 2005 *Materials Chemistry and Physics* **96** 289
[5] Donya Ramimoghadam S B 2014 *Journal of Magnetism and Magnetic Materials* **368** 207
[6] Patil R M 2014 *Royal Society of Chemistry* **4** 4515
[7] Xiaotai Zhang J K 2013 *Journal of Crystal Growth* **372** 170
[8] Adeodu A O, Anyaeche C O and Oluwole O O 2014 *J. of Advancement in Engineering and Technology* 1
[9] Nariyal R K, Kothari P and Bisht B 2014 *Chemical Science Transactions* **3** 1064
[10] Darmawan A, Smart S, Julbe A and Diniz da Costa 2011 *J.C. Iron oxide silica derived from sol-gel synthesis. Materials* **4** 448

Characteristic of Thermally Reduced Graphene Oxide as Supercapacitors Electrode Materials

Vika Marcelina[1], Norman Syakir[1], Santhy Wyantuti[2], Yeni W Hartati[2], Rahmat Hidayat[3] and Fitrilawati[1]

[1]Department of Physics, University of Padjadjaran, Jalan Raya Jatinangor Km 21, Sumedang 45363, West Jawa, Indonesia
[2]Department of Chemistry, University of Padjadjaran, Jalan Raya Jatinangor Km 21, Sumedang 45363, West Jawa, Indonesia
[3]Department of Physics, Faculty of Mathematics and Natural Sciences, Institut Teknologi Bandung, Jalan Ganesha 10, Bandung 40132, Indonesia

*fitrilawati@phys.unpad.ac.id

Abstract. We investigated graphene like material named reduced graphene oxide (RGO) as an electrode material by employed graphene oxide (GO). Thin film of GO was prepared on the indium thin oxide (ITO) substrate by spin-coating method using varied concentration of GO that dispersed in water. In order to remove its oxygen contained, GO film was thermally reduced at 200 ^{0}C for 1 hour. We used cyclic voltammetry to measure its CV characteristic and estimated its specific capacitance. We obtained the highest specific capacitance of 6.53 mF g^{-1} that measured from 4 mg ml^{-1} RGO thin film at scan rate 25 mVs^{-1}.

1. Introduction

Supercapacitor is known as a complement of other power sources such as batteries and fuel cells, and it can have long life cycles and can be rapidly charged and discharged at high power densities due to its highly reversible charge storage process [1]. Therefore supercapacitor is a potential candidate as an alternative energy storage devices.

Graphene has a potential application as an electrode material for electrochemical energy storage since it has very high electrical conductivity and a high specific surface area (2630 m^2 g^{-1}) [2]. It is reported an intrinsic capacitance of graphene of 21 μF cm^{-2} [3]. Graphene oxide (GO) is a most common starting material for graphene based applications since it can be produced in large quantities [4]. Since GO contains various oxygen functional groups, it is necessary to remove them in order to recover its conjugated structure [5, 6]. Thermal reduction using heat treatment is a common way to remove oxygen contained groups of GO [7].

In this paper, we report some results on measurement of specific capacitance value of reduced graphene oxide (RGO) films using cyclic voltammetry at varied scan rate in 1 M H_2SO_4 electrolyte.

2. Experiments

We used a commercial GO suspension (Graphenea) that dispersed in water. The GO dispersion was sonicated using ultrasonic cleaner (Branson 1800) for 1 hour at ambient temperature to exfoliate GO before used for film preparation. Indium Tin Oxide glass (ITO glass) was employed as a conductive substrate. The substrate was cleaned with teepol, rinsed with milli-Q water, then sonicated in ethanol

IOP Conf. Series: Materials Science and Engineering **196** (2017) 012034 doi:10.1088/1757-899X/196/1/012034

at 50 ^{0}C and dried. The GO films was prepared on ITO glass using spin-coating method (Chemat Technology Spin-coater KW-4A) at 1000 rpm and dried at 30 ^{0}C. The GO films were prepared using varied concentration from 1 to 4 mg ml^{-1}. In order to obtain RGO films, the GO films were thermally reduced by heating at 200 ^{0}C for 1 hour.

Cyclic voltammetry (CV) measurements were carried out using a potentiostatic (MetrOhm Autolab) with a platinum as a counter electrode, an Ag/AgCl as a reference electrode, and RGO film as a working electrode. The measurements were done at ambient temperature in 1 M H_2SO_4 aqueous electrolyte. Cyclic voltammetry scans were recorded from -0.2 V to 0.8 V in varied scan rates.

3. Results and Discussions

The cyclic voltammetry curves (CV) of RGO samples that prepared using varied concentration of 1, 2, 3 and 4 mg ml^{-1} on ITO glass substrates that were measured in a three-electrode configuration using 1 M H_2SO_4 as the electrolyte at varied scan speed are showed in Figure 1. The CV curves indicate oxidation and reduction behavior of RGO samples that the oxidation and oxidation currents depend on the film thickness, the thicker sample leads to higher oxidation and reduction currents.

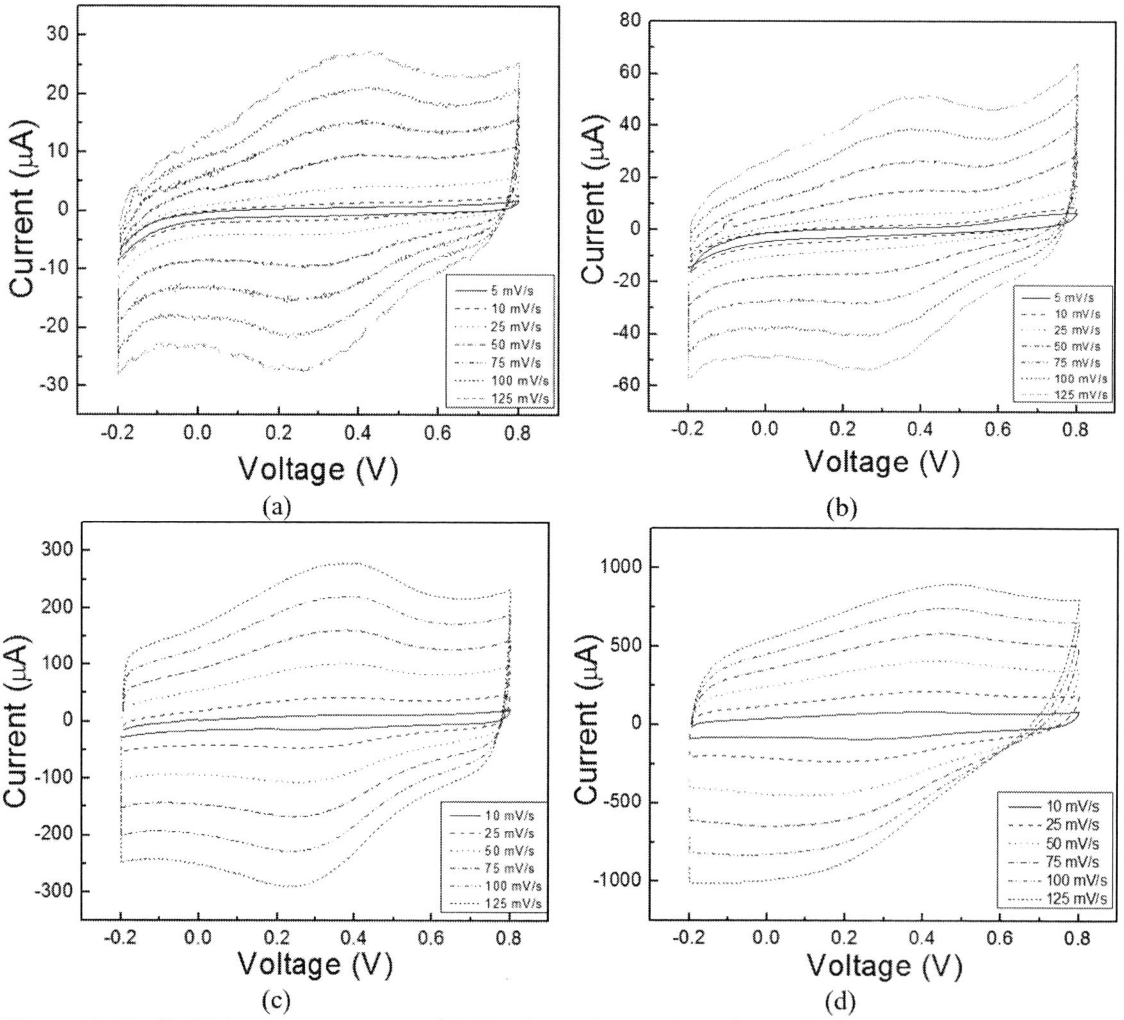

Figure 1. Cyclic Voltammetry curves of RGO electrodes on ITO glass substrate at various scan rates of 5, 10, 25, 50, 75, 100 and 125 mVs^{-1} measured from -0.2 V to 0.8 V of samples that prepared using varied GO concentration, 1 mg ml^{-1} (a), 2 mg ml^{-1} (b), 3 mg ml^{-1} (c), and 4 mg ml^{-1} (d).

From the Figure 1, it is obvious that all CV curves of RGO electrode measured from -0.2 V to 0.8 V at various scan rates of 5, 10, 25, 50, 75, 100 and 125 mVs^{-1}, respectively show a capacitive behavior. The increasing of a scan rate is only slightly distorted rectangular-like shape of CV curves, that is similar as previously reported [8]. The rectangular-like shape of the CV curves of prepared RGO films clearly indicated electrochemical double-layer capacitance characteristics, similar as previously reported [8]. The figures also show that each similar rectangular shape of CV has one pair of redox peaks. It is assigned as a typical transition between quinone or hydroquinone groups of carbon materials [9]. Electrochemical behavior of RGO electrode in 1 M H_2SO_4 has similar shape as reported previously that interpreted as a hybrid type of both a pseudocapacitive and an electric double layer capacitive [10].

Effect of increasing a scan rates from 5 to 125 mVs^{-1} on CV curves of RGO electrode at four different RGO concentration are also shown in Figure 1. The curves are not deformed due to increment of scan rate, but proportionally enlarged as scan rates increased up to 125 mVs^{-1}. We can see an obvious hysteresis of the CV curves.

The CV hysteresis area for these samples are increased due to increasing ofthe scan rates. In order to get a quantitative value of an increment of the hysteresis area, we applied equation (1) [11] on the data of Figure 1. The symbol P is a CV hysteresis area (in μW), i is a current respond, and dV is step potential that was 0.00244 V in this measurement.

$$P = \left(\int_{-0.2}^{0.8} i \, dV\right)_{ox} + \left(\int_{0.8}^{-0.2} i \, dV\right)_{red} \quad (1)$$

The hysteresis area of each samples that calculated using equation (1) are shown in Figure 2. We can see that the CV hysteresis area of each samples are increased with the rise of scan rate. It is related with current response of RGO electrode, which at the higher scan rate the RGO electrode can give the higher current response. This characteristics indicate a stability of capacitance of the RGO electrode. In order to obtain a specific capacitance value of the RGO electrodes from CV curves, we employed equation (2) [11] and using previous calculation results of hysteresis area. The symbol C is specific capacitance of RGO (F g^{-1}), m is mass of electro-active materials (g), ΔV is potential window (V), s is scan rate (mV s^{-1}), and i is current response in a given potential V.

$$C = \frac{1}{2m\,\Delta V\,s}\left(\int_{-0.2}^{0.8} i \, dV\right)_{ox} + \left(\int_{0.8}^{-0.2} i \, dV\right)_{red} \quad (2)$$

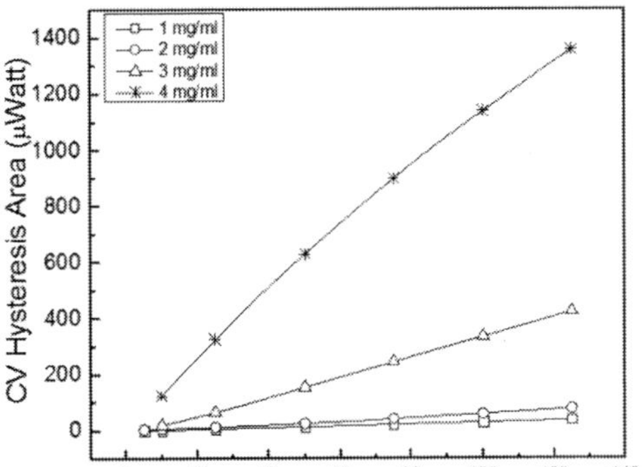

Figure 2. The CV hysteresis area of various RGO electrodes that prepared from varied concentration of GO dispersion that measured at various scan rates.

The calculation results of specific capacitance of RGO electrodes that prepared from varied concentration obtained from varied scan rates measurements are shown in Figure 3 (a) and (b). The specific capacitance of RGO electrode that prepared from certain concentration is remain the same when measured at varied scan rates. Further for RGO electrodes prepared from 1, 2, and 4 mg ml^{-1}, increasing the scan rates can result a decreasing of the specific capacitance. It can happen because an increment of CV hysteresis area is inversely proportional with scan rates increment as shown by equation (2). The highest specific capacitance that can be reached in this experiment is 6.53 mF g^{-1}. It is obtained from RGO electrode which was prepared using 4 mg ml^{-1} and scan rate 25 mV s^{-1}.

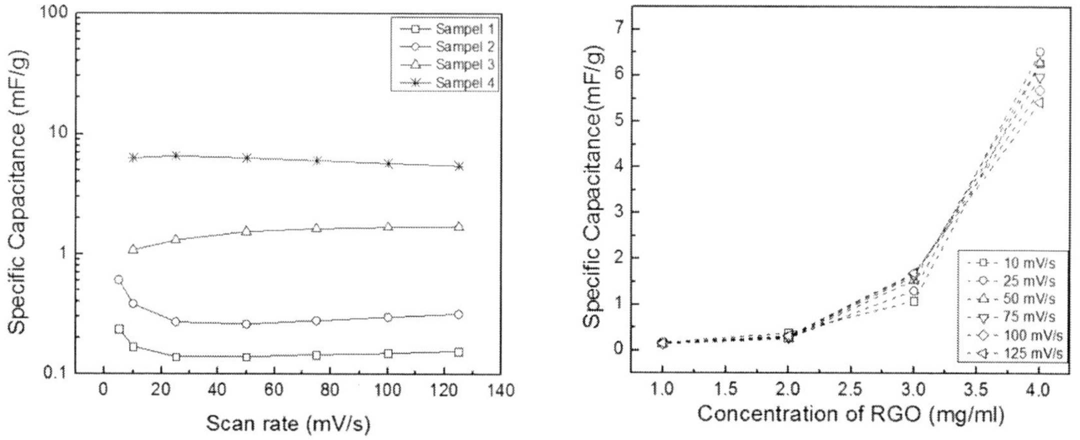

(a) (b)

Figure 3. The calculated specific capacitance of RGO electrodes at various scan rates that prepared from varied concentration of GO dispersion (a), and at various concentration that measured at varied scan rates (b).

4. Conclusions

The thermally RGO film is potentially used as an electrode material for supercapacitor as confirmed by cyclic voltammetry measurement. The capacitance of RGO film will be increased by increasing film thickness through using higher concentration of GO dispersion and will be decreased by increment a scan rate. The highest specific capacitance obtained in this work is 6.53 mF g^{-1} from RGO film that prepared using 4 mg ml^{-1} GO dispersion and measured at scan rate of 25 mVs^{-1}.

Acknowledgements

This work was partially funded by Academic Leadership Grant (ALG) contract no 480/UN6.RKT/Kep/HK/2016.

References

[1] Shukla A K 2001 *Resonance*, **6**(7) 72
[2] Zhao B, Liu P, Jiang Y, Pan D, Tao H, Song J, Fang T, Xu W 2012 *J. Power Sources* **198** 423
[3] Xia J, Chen F, Li J, and Tao N 2009 *Nat. Nanotechnol.*, **4**(8) 505
[4] Dreyer D R, Park S, Bielawski W, and Ruoff R S 2010 *Chem. Soc. Rev.* **39** 228
[5] Kauppila J, Kunnas P, damlin P, Viinikanoja A, Kyamström 2013 *Electrochim. Acta* **89** 84
[6] Pei S and Cheng H 2011 *Carbon* **11** 1
[7] Mcallister M.J, Li J L, Adamson D H, Schniepp H C, Abdala A A, Liu J, Alonso M H, Milius D L, Car R, Prudhomme R K and Aksay I A 2007 *Chem. Mater*, **19**(18) 4396
[8] Ramadoss A, Saravanakumar B, and Kim S J 2015 *Nano Energy* **15** 587
[9] Wang D W, Li F, Zhao J, Ren W, Ghen Z G, Tan J, Wu Z S, Gentle I, Lu G Q and Cheng H M 2009 *ACS Nano* **3**(7) 1745
[10] Oh Y J, Yoo J J, kim Y I, Yoon J K, Yoon H N, Kim J H, Park S B 2014 *Electrochim. Acta* **116** 118
[11] Chen Y and Ma Y 2010 *Carbon*, **49**(2) 573

The Optimization of CMC Concentration as Graphite Binder on the Anode of $LiFePO_4$ Battery

S Hidayat[1], T Cahyono[1], J Y Mindara[1], N Riveli[1], W Alamsyah[1] and I Rahayu[2]

[1] Department of Physics, Faculty Mathematic and Natural Sciences, University of Padjadjaran, Jl. Raya Bandung-Sumedang Km 21 Jatinangor 45363, West Jawa Indonesia
[2]Department of Chemistry, Faculty Mathematic and Natural Sciences, University of Padjadjaran, Jl. Raya Bandung-Sumedang Km 21 Jatinangor 45363, West Jawa Indonesia

E-mail: sahrul.hidayat@phys.unpad.ac.id

Abstract. Recently, the most dominating power supply on the mobile electronics market are rechargeable Lithium-ion batteries. This is because of a higher energy density and a longer lifetime compared to similar rechargeable battery systems. Graphite is commonly used as anode material in the Lithium-ion batteries, because of its excellent electrochemical characteristics and low cost fabrication. In this paper, we reported the optimization of the concentration of the CMC (carboxymethyl cellulose), that acts as the binder for graphite anode. Based on our experimental results, the best composition of graphite : C : CMC is 90 : 8 : 2 in weight %. Anode with such composition has, based on SEM measurement, a relatively good surface morphology, while it also has relatively high conductivity, about 2.68 S/cm. The result of cyclic voltammogram with a scan rate of 10 mV/s in the voltage range of 0 to 1 Volt, shows the peak of reduction voltage at 0.85 Volts and the peak voltage of oxidation is at −1.5 Volt. The performance of the battery system with $LiFePO_4$ set as the cathode, shows that the working voltage is about 2.67 Volts at 1 mA current-loading, with the efficiency around 47%.

1. Introduction

Environmental pollution and energy crisis have been severely accumulated due to the excessive utilization of fossil fuel resources. To overcome these problems, novel clean energy resources and it's related energy conversion and storage materials and devices are highly demanded. Among the diverse energy storage devices, lithium-ion batteries (LIBs) have been studied overwhelmingly, and certain kinds of LIBs have been commercialized already. $LiCoO_2$ is one of the main LIB cathode materials that are used in industrial scale. But $LiCoO_2$ have several drawbacks, such as that it could pollute the environment during production process, overcharge during usage thus causing potential safety hazard, and that cobalt is expensive for its limited storage [1]. Therefore, search for alternative materials is always a research hotspot. Having similar crystal structure as $LiCoO_2$, $LiNiO_2$ has the advantage of a lower price, but it also has difficulties in its synthesis, it has poor structure, thermal, and cycling stability [2].

Therefore, the exploits of high-performance electrode materials, electrolytes, and membrane for LIBs have attracted great attention during the last decade. Lithium iron phosphate ($LiFePO_4$, LFP) with olivine structure is one of the most promising cathode materials for LIBs, owing to its high theoretical capacity (170 mAh g^{-1}), acceptable operating voltage (3.4 V vs. Li^+/Li), good cycling stability, low toxicity, good thermal stability, and low cost [3].

Lithium ion batteries are made up of one or more generating compartments called cells. Each cell is composed of three components: a positive electrode, a negative electrode, and an electrolyte between them. The positive electrode is made from chemical compound named lithium iron phosphate ($LiFePO_4$). The negative electrode is made up of carbon (graphite) and the electrolyte varies from one type of battery to another. All lithium ion batteries more or less work in the same manner. During the charging of the battery, lithium based positive electrode expel some of its lithium ions, which move through the electrolyte to reach to the negative electrode and remain there. The battery stores energy during this process. When the battery is discharging, the lithium ions move back across the electrolyte to the positive electrode, producing the energy that powers the battery. In both cases, electrons flow around the outer circuits in the opposite direction to the ions. Electrons do not flow through the electrolyte as it tends to be an effective insulating barrier. The movement of ions (through the electrolyte) and electrons (around the external circuit, in the opposite direction) are interconnected to each other; if any one of them stops, the other will also stops. If the ions are no longer move through the electrolyte because the battery is completely discharged, the electrons can't move through the outer circuit either, then the battery is no longer powered.

In commercial Li-ion batteries, PVDF has been widely used as the binder for both the negative and the positive electrodes, because of its good electrochemical stability and high binding power. PVDF binder is expensive, not easy to recycle, and involves the use of volatile organic compounds such as the toxic and environmentally harmful N-methyl-2-pyrrolidone (NMP), in its processing.

As an alternative binder material, we use CMC that are cheaper and environmentally friendlier than PVDF [5]. CMC is a familiar subtance in the food industry, used as emulsifiers. We use local produced CMC, ordered from the PUDAC Scientific. CMC, which is produced from the insertion of carboxymethyl groups into natural cellulose, is the commonly used binder for anodes and cathodes of Li-ion batteries. CMC has a strong shear-thinning behavior that adjusts to the slurry rheology.

2. Experiment

The main interest of our experiment is the optimization of CMC as binder for the graphite-Carbon cathode. In the experiment, a layer of CMC-graphite-C is deposited on the glass substrate. Study of characterization is then conducted, that consists of conductivity measurement and photo micro. Several different composition of the cathode are tested. Once we found the composition of best quality, we use it to replace the anode in the commercial $LiFePO_4$ battery. As the electrolyte, we use 1M Lithium hexafloro phosphate in mix with Ethylene carbonate, Di-methylene Carbonate and Di-ethylene carbonate. Finally, we perform battery performance test, that includes charge-discharge process, energy density and efficiency.

In the fabrication for anode layer, graphite, conductive carbon and the CMC are mixed in distilled water using magnetic stirrer. The layer is the constructed using the Doctor Blade methode. Figure 1 shows the steps of the battery packaging. First, the anode and cathode layer are prepared. Then, components of the battery are stacked as follows: cathode, separator and anode. The battery with ordered components is then placed into the packaging cell. Finally, the electrolyte is filled to the cell via the glove box with argon environment. The battery is then ready for characterizations.

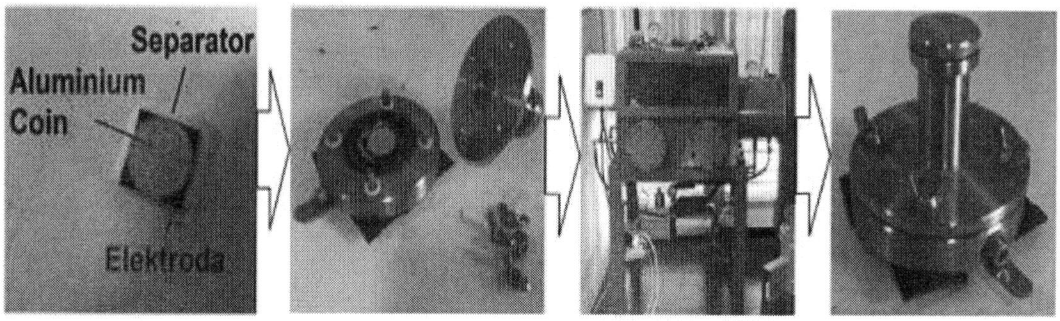

Figure 1. The steps of battery packaging.

In the battery characterization step, we used voltage/current source for charging process, automatic battery loader for discharging process, and multimeter for measuring the voltage or current of battery. The system is integrated with the personal computer.

3. Results and Discussion

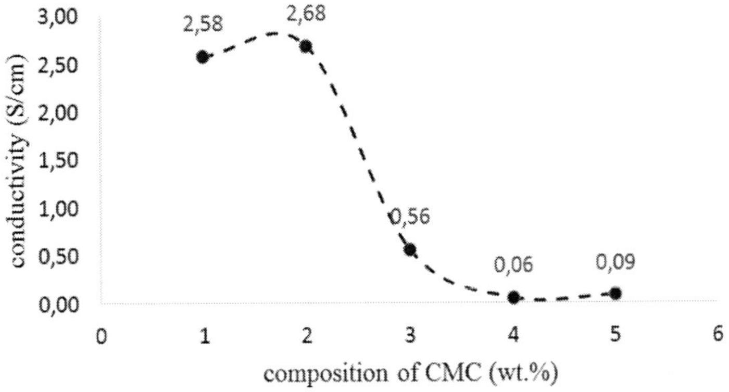

Figure 2. Conductivity of CMC-graphite-C.

Figure 2 shows the conductivity of CMC:graphite:C layer with five different compositions of CMC. The conductivity values are obtained through the four line probe method. As the basis, we start the measurement with the layer when the CMC composition is of 5%. We then reduce the CMC composition and obtained an increasing tendency of the conductivity. Based on our experiment, the maximum conductivity is obtained when the CMC composition is of 2%. The layer with such composition is then used in the battery system.

Cyclic voltammogram is used to studies of oxidation and reduction processes of CMC-graphite-C layer in electrolyte media. In this test, we use CMC-graphite-C deposition on ITO substrate as working electrode and platinum as counter electrode. In Fig. 4 is shown the reduction peak and the oxidation peak. When the scan voltage increased, the lithium ion move intercalation into the graphite layer and the opposite happen when scan is reversed. In Figure 3 we can also observe the reduction and oxidation in CMC-graphite-C layer occurs in reversibly.

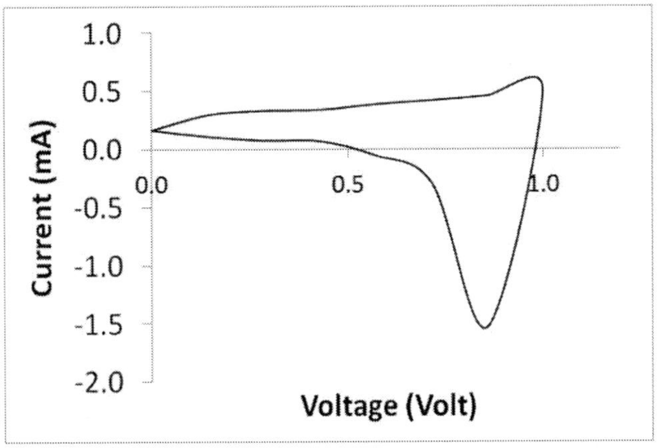

Figure 3. Cyclic voltammogram of the CMC-graphite-C layer.

The battery is made in the small cell model, where the area of electrode is 4 cm^2 only, causing the capacity of battery is limited. When the charging time is less than 20 minute, the working voltage of battery goes below 2.5 Volt. While as the charging time is above 25 minute, the working voltage does not change significantly. We found that 30 minutes is the optimal charging time. Figure 4 shows the optimization of current discharging process. We can observe that the optimal discharge current is about 1 mA, and that corresponds to the maximum battery efficiency of about 47%. We can increase the discharge current by making the battery with wider electrode.

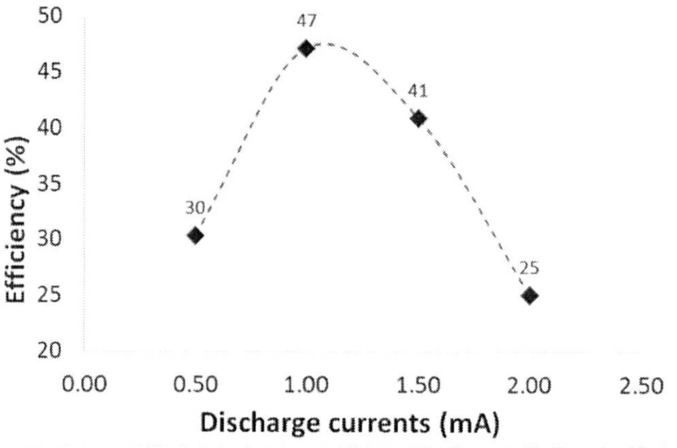

Figure 4. Optimization of discharging currents.

4. Summary

In this study we vary the composition of CMC which act as the binder in CMC-graphite-C cathode in a LIB battery system. We found that the layer with CMC composition of 2 % (%wt) is the one that gives best perfomnce. It produces good surface profile and high conductivity of about 2.68 S/cm. This shows that the use of CMC increases the binding power of the active cathode components. CMC composition of less than 2% produces less binding power, while the composition of higher than 2% will start to to slows down the charge movement, does reduces it's conductivity. That is due to the fact that CMC is a cellulose with the properties of an isolator. For the use of the cathode layer in the battery system, the optimal charging voltage is at 4 Volt for about 30 minutes, The working voltage is about 2.7 volts and the maximum efficiency is 47%, at a discharge current of 1 mA.

References

[1] Lee J H, Paik U, Hackley V A, Choi Y M 2005 *J. of The Electrochemical Society* **9** A1763
[2] Hamid N.A., Wennig S., Hardt S., Heinzel A., Schulz C., Wiggers H 2012 *Journal of Power Sources* **216** 76
[3] Pei B, Yao H, Zhang W, Yang Z 2012 *Journal of Power Sources* **220** 317
[4] Hidayat S, Amelia O, Rahayu I, Fitrilawati 2015 *Materials Science Forum* **827** 125
[5] Ebrahimzadeh S, Ghanbarzadeh B, Hamishehkar H 2016 *International Journal of Biological Macromolecules* **84** 16

IOP Conf. Series: Materials Science and Engineering **196** (2017) 012036 doi:10.1088/1757-899X/196/1/012036

Synthesis and Photoluminesence Study of Reduced Graphene Oxide (rGO)/ZnO for Solar Energy Absorbing Materials

Christina Martha, Malik Anjelh Baqiya, Yoyok Cahyono, Darminto

Advanced Materials Research Group, Department of Physics, Faculty of Mathematics and Natural Science, Institut Teknologi Sepuluh Nopember, Kampus ITS Sukolilo, Surabaya 60111, Indonesia

E-mail: christinmartha45@gmail.com

Abstract. Materials combining reduced graphene oxide (rGO) from coconut shells and commercial ZnO have been synthesized by dry-mixing in weight ratio of 1:2, 2:2, and 3:2. For photoluminesence (PL) characterization, the solutions with concentration of rGO/ZnO in aquadest up to 0.003 mg/mL were prepared. The absorbing photon energy by the samples at wavelength 280-426 nm (ultraviolet-purple) has induced electron transition to conduction band. Further, the returning electron to valence band was followed by photon emission at wavelength of 530-880 nm (green-infrared). The PL intensity was observed to drastically enhance with increasing content of rGO in the solution by 252.5%, 285.0% and 291.3% for the corresponding samples compared to the solution containing pure ZnO. The rGO/ZnO materials exhibit higher absorbance with wider wavelength range, and therefore can potentially be used as solar energy absorbing materials to enhance the efficiency of solar cell.

1. Introduction

A solar cell converts solar energy becomes electricity. The main issue of solar cells up to the present is about the low efficiency. One of the problems may possibly arise from the fact that the low absorption of solar energy in material [1]. ZnO is a semiconductor that has high transparancy and can generally be used as solar energy converter material [2]. It has gap energy about 3.3 eV.

Meanwhile, reduced graphene oxide (rGO) has good characteristics in electrics and can be applied as solar energy absorbing material [3].The starting material which can be used to produce rGO is old coconut shell which is as a waste, since it contains lignin, hemicellulose, and cellulose which are constituted by carbon elements. As we know that Indonesia as tropical archipelago is a huge habitat of coconut trees as natural resources. Referred to recent research by Nugraheni (2014) and Ananta (2015), the coconut shell has successfully been converted by calcination at 400°C to be material with main molecule bonding like C=C and C-C, and other impurities such as C-H, C-O, C=O and O-H, that indicates rGO phase [4]-[5]. This rGO has an optical bandgap about 0.07 - 1.2 eV, which was measured by Nasrullah (2014) [4].

Photoluminescence (PL) is a phenomenon of light spontaneous emission from a material that is excited by photon energy [7]. The PL spectroscopy, which is a non destructive method, is a selective electrical characterization which is very sensitive for discrete electronic states. Emission spectrum feature can be used to identify surface, interface bonding, and impurity level. According to Chien (2012), if oxygen is reduced in the substance, the PL curve will shift to a lower wavelength [3]. Meanwhile, if ZnO material is added by rGO material, the emission intensity will increase. By receiving photon energy at 426 nm, the electrons inside material can be excited and then return to ground state by emitting electromagnetic wave in range of green-red visible light. This shifting can occur because of oxygen vacancy defect, having energy formation of about 1.7 eV- 2.4 eV [6]. So, if more rGO is added to ZnO, the gap energy will decrease, to be between ZnO and rGO [2]. Thus, the rGO/ZnO can be applied as solar energy absorbing materials (SEAM), which will be studied in this paper.

2. Materials and Experiments

Materials used in this research are old coconut shell waste, ZnO (SAP Chemicals), alcohol, and aquadest. The old coconut shell waste was dried under the sunlight for several days, and then cleaned. To convert it further to carbon compound, it was burned in a carbonization box. It was then crushed by mortar to be granulated shape. Afterwards, it was calcined in a furnace at temperature 400°C for 5 hours. The rGO powder was dissolved in aquadest for ultrasonication for 2 hours, stirring for 1 hour, and centrifugation at 2000 rpm for 30 minutes.

The synthesized rGO was then mixed with ZnO in the weight ratio of 1:2; 2:2; and 3:2, and dissolved in water to be solution, where ZnO 0.002 mg/mL was mixed by each of rGO: 0.001, 0.002, and 0.003 mg/mL for PL (Perkin Elmer LS-55) and particle size (PSA) measurements. Besides, the rGO and ZnO were also characterized by XRD (D8 Advance Bruker) and scanning electron microscopy (SEM) characterizations.

3. Results and Discussions

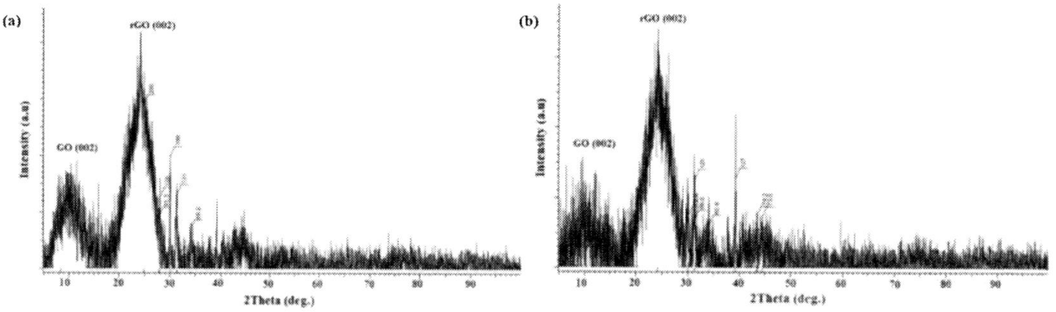

Figure 1. XRD characterization of: (a) old coconut shell charcoal, (d) rGO material.

Presented in Fig. 1 are the XRD patterns of the old coconut shell charcoal powders resulted from the burning process (a), and that after furnace heating at 400°C for 5 hours (b). In Fig. 1 (b), the diffraction angle at 11° decreases, which means that the graphene oxide (GO) phase has partially transformed to rGO. The reduction of GO phase in the sample is caused by heating that makes oxygen atoms release from GO structure. Fig. 2 (a) shows the rGO layers and Fig. 2 (b) shows the ZnO powders with heterogeneous wurtzite structure. Further, characterization by PSA shows the ZnO particle size distribution having two peaks at 685.3 nm and 193.4 nm. It means that the ZnO particle size is heterogeneous. Meanwhile rGO exhibits the particle size average about 300 nm.

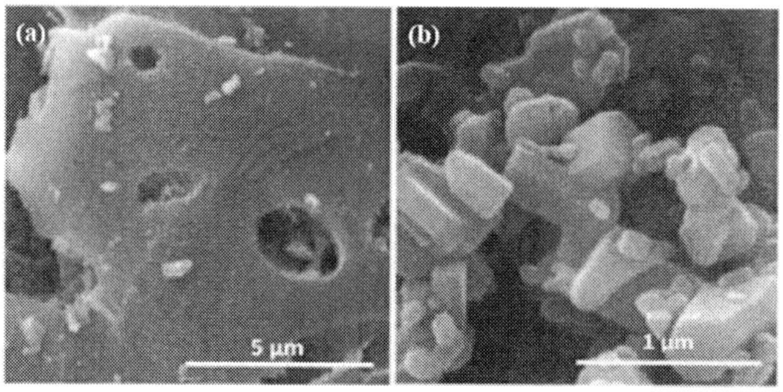

Figure 2. SEM characterization results: (a) rGO, (b) ZnO.

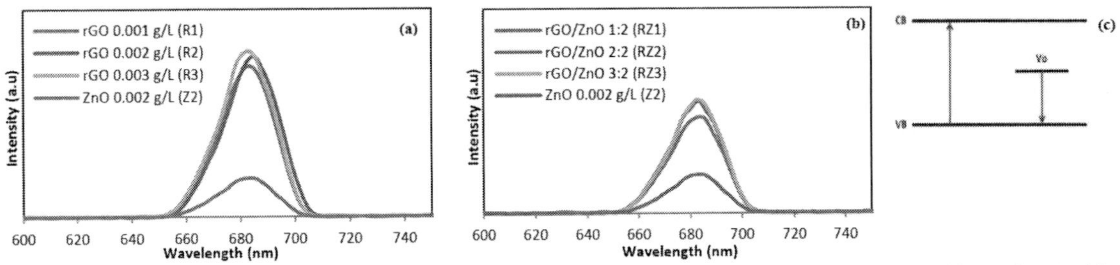

Figure 3. Emission spectra under excitation wavelength of 341 nm: (a) rGO and ZnO, (b)rGO/ZnO, (c) emission mechanism of red visible light.

The emission spectra with the main peak of about 682 nm resulted by 341 nm excitation wavelength are shown in Fig. 3. Fig. 3 (a) shows that emission spectrum of ZnO which is lower than that of rGO. The higher content of rGO will result in higher intensity of emission. It is seen that samples emit red visible light under the excitation using wavelength at 341 nm. Fig.3 (b) shows that the emission spectrum of rGO/ZnO having higher intensity than that of pure ZnO. More rGO that was added could increase the emission spectrum of ZnO, as shown in Fig. 3 (b). Fig. 3 (c) shows the process and mechanism of how PL is happened. Excitation spectrum of rGO/ZnO material shifts from 341 nm to 682 nm. Since more oxygen atoms still exist in the samples, so the emission has shifted to higher wavelength. This result is supported by the XRD pattern in Fig. 1 (b), showing that there is still GO phase having peak at diffraction angle of around 11°.

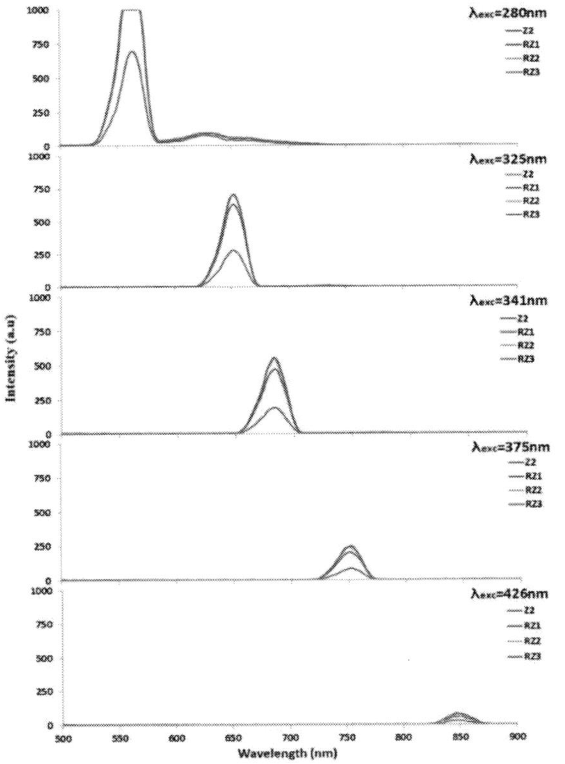

Figure 4. Emission spectrums of some excitation spectrums for rGO/ZnO and ZnO.

In Fig. 4, the emission spectra of the whole sampels using excitation wavelengths of 280 nm, 325 nm, 375 nm, and 426 nm are presented. This figure also shows the same characteristic for 4 samples in the range of 530-590 nm (form green to yellow of the visible light). From these 5 excitation wavelength parameters, it is shown that the emission wavelengths are twice of its excitation wavelength or a half of its excitation energy. The emission spectrum intensity decreases when the excitation wavelength increases because of the decreasing energy given to material. This result is in nice agreement with previous reports [2]-[3]. This composite is a material system consisting of 2 components, rGO and ZnO. This condition makes the light intensity which be absorbed by this composite system is divided by both of rGO and ZnO. Because of that, the emission intensity of rGO/ZnO is between rGO and ZnO.

4. Summary

The absorbance of ZnO material can be enhanced by introduction of rGO to form a composite. The rGO/ZnO system was able to absorb photon energy of 280-426 nm (ultraviolet-purple visible light) and then emit electromagnetic wave at 530-880 nm (green visible light-infrared). The shift is possible because of energy used by electrons to vibrate as a non-radiative transition. Thus, the rGO/ZnO composite has been developed to have higher absorbance with wider range and higher intensity of light.

Acknowledgements
Partial support by Hibah Kompetensi, Kemenristekdikti, 2016, for this research would thankfully be appreciated.

References

Reference to articles:
[1] Huang X 2013 *Chem. Soc. Rev.* **42** 173
[2] Narayanan G N 2015 *International Journal of ChemTech Research* **7** 1065
[3] Chien C T 2012 *Angewandte Communications* DOI: 10.1002/anie.201200474
[4] Nugraheni A Y, Nasrulloh M, Fandi A P, Astuti F, Darminto 2014 *Material Science Forum* **827** 285
[5] Mas'udah K W, Nugraha I M A, Abidin S, Astuti F, Darminto 2016 *AIP Conference Proceedings* **1725** 020045
[6] Ding J 2014 *Journal of Colloid and Interface Science* **416** 289
[7] P. Atkins, *Physical Chemistry Ninth Edition*, W.H. Freeman and Company, New York, 2010.

Pin-Hole Free Perovskite Film for Solar Cells Application Prepared by Controlled Two-Step Spin-Coating Method

A Bahtiar[1], S Rahmanita[1] and Y D Inayatie[1]

[1]Department of Physics, Universitas Padjadjaran
Jl. Raya Bandung-Sumedang km. 21 Jatinangor, Sumedang Indonesia 45363

E-mail: ayi.bahtiar@phys.unpad.ac.id

Abstract. Morphology of perovskite film is a key important for achieving high performance perovskite solar cells. Perovskite films are commonly prepared by two-step spin-coating method. However, pin-holes are frequently formed in perovskite films due to incomplete conversion of lead-iodide (PbI_2) into perovskite $CH_3NH_3PbI_3$. Pin-holes in perovskite film cause large hysteresis in current-voltage curve of solar cells due to large series resistance between perovskite layer-hole transport material. Moreover, crystal structure and grain size of perovskite crystal are also other important parameters for achieving high performance solar cells, which are significantly affected by preparation of perovskite film. We studied the effect of preparation of perovskite film using controlled spin-coating parameters on crystal structure and morphological properties of perovskite film. We used two-step spin-coating method for preparation of perovskite film with varied spinning speed, spinning time and temperature of spin-coating process to control growth of perovskite crystal aimed to produce high quality perovskite crystal with pin-hole free and large grain size. All experiment was performed in air with high humidity (larger than 80%). The best crystal structure, pin-hole free with large grain crystal size of perovskite film was obtained from film prepared at room temperature with spinning speed 1000 rpm for 20 seconds and annealed at 100°C for 300 seconds.

1. Introduction

Recently, perovskite materials have been intensively studied as active material for high power conversion efficiency (PCE) solar cells due to their wide light absorption spectrum and can be processed into thin films using low-cost techniques, therefore, the perovskite materials are good candidate for large scale production and mass utilization of solar cells replacing high-cost conventional solar cells, such as silicon and germanium arsenide. Today, state of-the-art PCE of perovskite solar cell is reached more than 21% which is comparable to conventional and market available silicon based solar cell [1]. However, reproducibility issue is a main problem to fabricate this solar cell in large scale. The difficulty to control morphology of perovskite films is a cause of large variation of the PCE of solar cells. The main strategic issues for improving the PCE and its reproducibility are control morphology of perovskite films and to understand charges carrier transport in active material.

Methylamonium lead (III) iodide ($CH_3NH_3PbI_3$) material is the most popular material used as active material for perovskite solar cells because it has a direct bandgap of ~ 1.55 eV, corresponding to an absorption onset at 800 nm and intense absorption of the entire visible spectrum [2-3]. These properties are expected to produce solar with PCE comparable with high-cost conventional inorganic solar cells such as silicon and germanium arsenide. Moreover, the exciton binding energy of this material is very small and has long exciton diffusion length in the order of 100 -1000 nm with lifetime of ~ 100 ns [4-5], so this perovskite material promises great potential for high performance solar cells with low-cost and large scale production.

Perovskite material was firstly used as light absorber for liquid dye-sensitized solar cell by Miyasaka et al with a PCE of only 3.8% [6]. However, a significant improvement of PCE was obtained by replacing the liquid hole transport material (HTM) with the solid 2,2',7,7'-tetrakis(N,N-di-p-methoxyphenylamine)-9,9;spirobifluorene (Spiro-OMeTAD) to hinder the dissolution of perovskite material [7-10]. Snaith et al [11] has successfully prepared a solid-state Al_2O_3 mesoscopic solar cells using $CH_3NH_3PbI_3$ as active material with PCE of 10.9%. They also successfully fabricated the perovskite solar cells without mesoscopic layer through control morphology of perovskite film [12]. Graetzel et al used two-step sequential deposition method for producing uniform morphology of perovskite film and a PCE higher than 15% was achieved [13]. Recently, Yang et al reported planar perovskite solar cells with PCE of 19.3% by efficient interface engineering [14]. The world record of more than 21% PCE of perovskite solar cell is already reported [15].

Although, the 20% psychological barrier of PCE solar cells for large scale production is already exceeded, but there are some problem must be solved. First, the problem of hysteresis curve in solar cells characteristic, due to some problem in high quality perovskite film with pin-hole free and large grains size, the interface between perovskite material, hole transport layer and electron transport layer which causes the inaccuracy in evaluating the PCE. Second, the most crucial issue for PCE is to control morphology of perovskite film as active material of solar cells. Many photo-physical properties, such as light absorption, charge carrier transport and diffusion length can be dramatically influenced by the crystallization of the perovskite film. The defects and the crystal grains act as charges carrier traps which would encourage the electron recombination and hence would affect the PCE of solar cells. Third, the problem related to the reproducibility of the measured PCE of solar cells due to the difficulty to control perovskite film morphology. Therefore, the device performance strongly depends on the morphology of the perovskite film [12,15-18].

Here, we reported our recent study on preparation perovskite $CH_3NH_3PbI_3$ films with high quality crystal and large scale grains (larger than 500 nm) through controlled parameters of two-step spin coating method in air with high humidity. We varied spinning speed, spinning time and temperature of spin-coating process to control growth of perovskite crystal aimed to produce high quality perovskite crystal with pin-hole free and large grain size. We obtained the best crystal structure, pin-hole free with large grain crystal size of perovskite film was obtained from film prepared at room temperature with spinning speed 1000 rpm for 20 seconds and annealed at 100°C for 300 seconds. These parameters can be used as standard method for preparing high quality perovskite crystal with large grains size in air with high humidity (> 80%) to produce high PCE of solar cells.

2. Experimental Method

Lead(II) iodide or PbI_2 and methylammonium iodide (CH_3NH_3I, MAI), poly(3,4-ethylenedioxythiophene):poly-(styrenesulfonate) (PEDOT:PSS) solution were purchased from Luminescence Technology Inc. Taiwan and used as received. Dimethylformamide (DMF) and 2-isopropanol (IPA) solvents were purchased from Sigma Aldrich and used without further purification. The PbI_2 solution was prepared by dissolving of 900 mg PbI_2 in 2 ml DMF. The solution was heated while stirred at 70°C for 24 hours. After heating and stirring process, the solution was centrifuged with a speed of 4000 rpm for 15 minutes to obtain homogeneous solution. The homogeneous PbI_2 solution was confirmed by yellow transparent solution. Meanwhile, the MAI solution was prepared by putting 90 mg MAI in 2 ml IPA and heated at 70°C for 30 minutes while stirred using magnetic stirrer.

Perovskite films were prepared with two-step spin-coating method on PEDOT:PSS film spin-coated on glass substrate. PEDOT:PSS solution was spin-coated on cleaned glass substrate with spinning speed 1000 rpm for 20 seconds. The PEDOT:PSS film was annealed at 110°C for 20 minutes. In two-step spin-coating method, PbI_2 film was first spin-coated onto PEDOT:PSS film and subsequently followed by spin-coated of MAI solution. Three perovskite films were prepared with different spin-coating

IOP Conf. Series: Materials Science and Engineering **196** (2017) 012037 doi:10.1088/1757-899X/196/1/012037

parameters and different treatment of annealing temperatures after spin-coating films, as shown in Table 1. We used two-step annealing temperature at variation-1 and variation-3. All solutions and films preparation were conducted in air with high humidity (> 80%). Crystal structure perovskite films were measured by XRD and SEM was used to analyze perovskite film morphology.

Table 1. Spin-coating and annealing parameters of preparation thin films perovskite

Variation	Solutions	Temperature of PbI_2 before spin-coating of MAI (°C)	Spin-coating parameters		Annealing temperature (°C)	Annealing time (s)
			speed (rpm)	time (s)		
1	PbI_2	100	1000	20	40	180
					100	300
	MAI	Room Temperature	1000	20	100	300
2	PbI_2	Room Temperature	6000	35	70	300
	MAI	Room Temperature	6000	35	100	300
3	PbI_2	Room Temperature	1000	20	40	180
					100	300
	MAI	Room Temperature	1000	20	100	300

3. Results and Discussion

Figure 1 shows XRD patterns of PbI_2 and perovskite films spin-coated on PEDOT:PSS coated glass. PbI_2 film was oriented in c-axis only as indicated by four main peaks at 12.7°, 24.6°, 38.7°, and at 52.4° that shows crystal orientation at (001), (002), (003), and (004), respectively [13]. PbI_2 films prepared at 70°C and room temperature have similar XRD pattern which shows that PbI_2 crystal is not affected by spin-coating temperature from homogeneous and yellow transparent solution of PbI_2. In perovskite variation-1 film shows diffraction peaks at 14.2°, 24.6°, 28.5°, 31.8°, 40.5°, 44.0°, and at 50.3° which correlation to crystal plane of (110), (202), (220), (310), (224), (134) and (404), respectively [19-20]. This pattern confirms that perovskite was formed with tetragonal crystal structure. However, a peak at 12.7° originated from PbI_2 peak is still observed. It shows that PbI_2 was not fully converted into perovskite $CH_3NH_3PbI_3$. We presumed it was caused by lack of temperature control during spin-coating process. The film should be gradually cooled to room temperature to give a time for crystal growing.

In variation-2, we used the parameters in literature [21], due to similar materials that used in our experiment. A significant different XRD pattern was observed in perovskite film variation-2 as compared to variation-1. Only five main peaks are observed at 14.2°, 28.5°, 31.8°, 40.5°, and at 44.0° which correlated to the crystal orientation at (110), (220), (310), (224), and (134), respectively. Two peaks at 12.7°, and 28.5° are much higher intensity as compared to other peaks, which shows that perovskite crystal variation-2 is mainly oriented at x-y axis. Moreover, no peak originated from PbI_2 is observed. It shows that all PbI_2 is fully converted to perovskite $CH_3NH_3PbI_3$ crystal.

XRD pattern of perovskite variation-3 shows similar pattern as variation-1, except no peak at 12.7° is observed. It shows that spin-coating of PbI_2 and MAI solutions at room temperature can obtain fully perovskite crystal. Room temperature spin-coating process is more preferable for large area and mass production and also easier spin-coating control process of perovskite film.

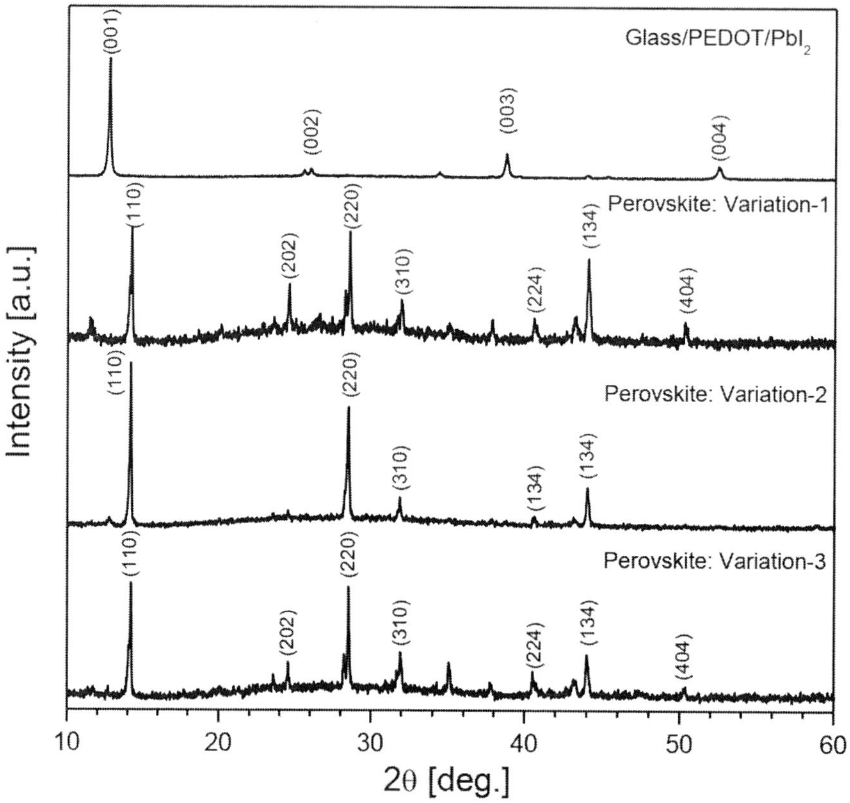

Figure 1. XRD pattern of PbI_2, and perovskite films prepared with different spin-coating parameters

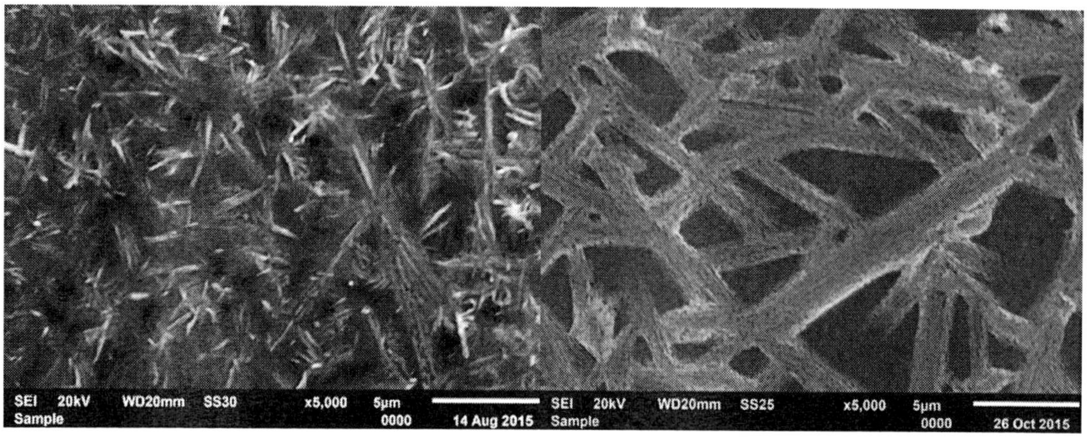

Figure 2. SEM photographs of PbI_2 films spin-coated at 70°C (left) and at room temperature (right)

SEM of PbI_2 films prepared at 70°C and at room temperature are shown in Figure 2. Both pictures show similar fibril or needle like structure which mean that no significant different of PbI_2 crystal spin-coated from homogenous solution at 70°C or at room temperature. It means that spin-coating method at room temperature can be used for standard preparation of high quality PbI_2 film.

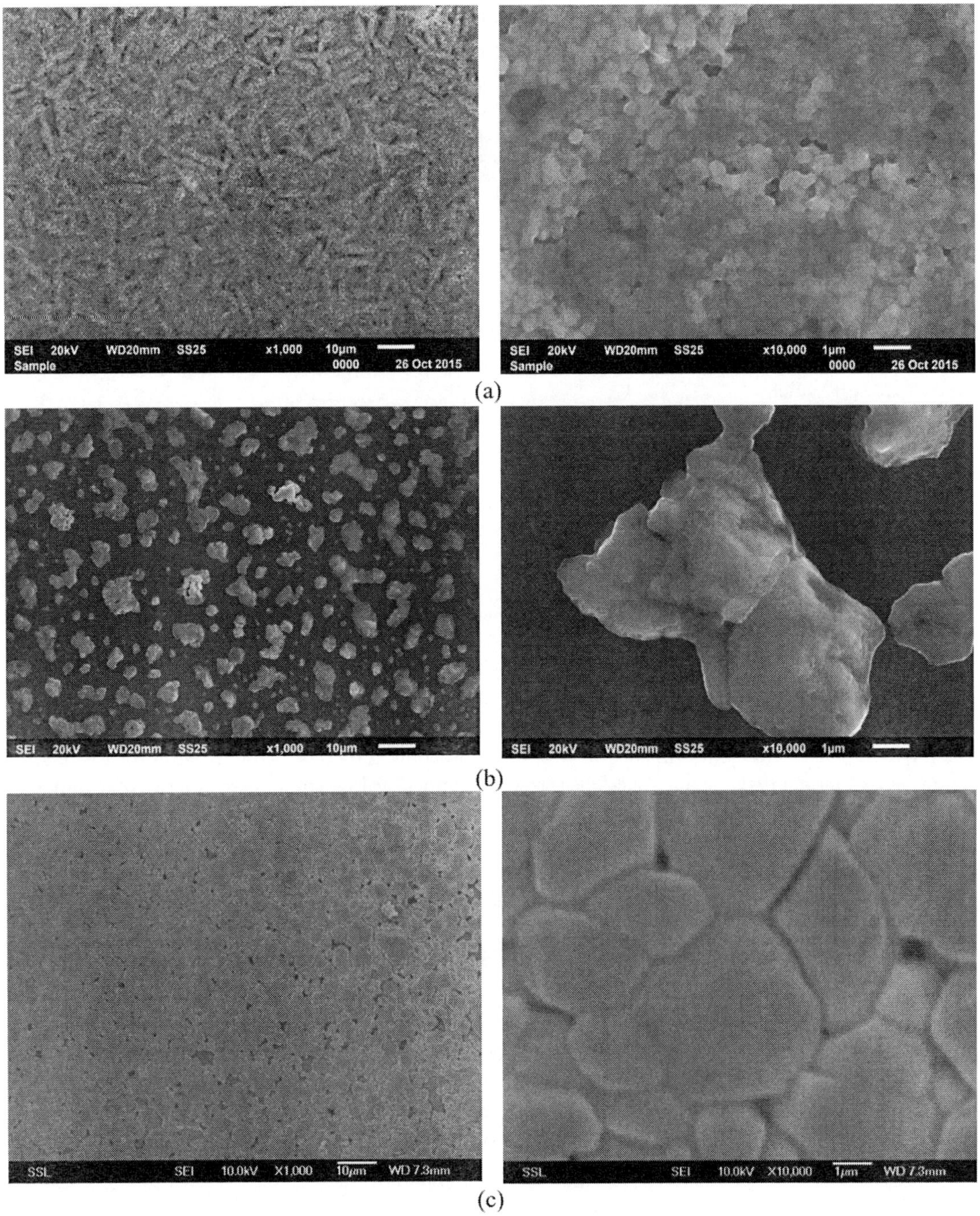

Figure 3. SEM of perovskite films prepared with different spin-coating parameters. (a). Variation-1, (b). Variation-2, and (c). Variation-3.

Figure 3 shows SEM images of perovskite films with different magnification prepared by using two-step spin-coating method with different spin-coating parameters and different treatment after spin-coating. Needle-like of PbI_2 structures are still clearly observed in the film. It confirmed that PbI_2 was not fully converted into perovskite as approved from PbI_2 peak in XRD pattern as discussed above. Incomplete conversion of PbI_2 into perovskite lead to large hysteresis of J-V curve and low PCE in perovskite solar cells [22]. Small grain size of perovskite crystal with average size of 200 nm is obtained and also some pin-holes or voids between grains of perovskite are still clearly observed. These pin-holes must be reduced to avoid direct contact of other layers deposited onto perovskite film which lead to inferior performance and large hysteresis in J-V curve of perovskite solar cells [23].

In case of perovskite variation-2, large empty spaces between perovskite grains or clusters are clearly observed. It might be caused by too high of spinning speed of spin-coated PbI_2 and MAI. Although, PbI_2 was fully converted into perovskite crystal as shown in its XRD pattern, this perovskite film is not suitable for perovskite solar cells due to large pin-holes between perovskite crystals.

In contrast, small and only few pin-holes are observed in perovskite film variation-3. Moreover large grain size of perovskite crystal with average size above 500 nm is obtained by spin-coating parameter of 1000 rpm for 20 seconds and at room temperature. Therefore, we recommend these parameters can be used as standard parameters of two-step spin-coating method in air with high humidity for preparing high quality perovskite film with pin-hole free and large grain size to obtain high performance of perovskite solar cells.

Recent study of Cohen et al [23] shows that pin-hole free of perovskite film can produce high PCE perovskite solar cells with reduced hysteresis, due to small series resistance of devices and reduced charge carrier recombination in active layer of solar cells. Kim & Park [24] and Kim et al [25] studied perovskite solar cells with different grains size of perovskite. They found that the PCE of solar cells is improved when grains crystal size is also improved from 100 nm to 500 nm. The improvement of PCE with increasing of grain size is caused by the increased of External Quantum Efficiency (EQE) at longer wavelengths which related to negligible losses in the charge generation and collection in solar cells with large perovskite grains size [25]. The same trend was also found in the absorption spectra measured in the reflectance mode. They found that the photocurrent improvement is primarily attributed to the enhanced absorption due to the thicker active layer with larger perovskite crystal grains size [25].

4. Conclusion

We have prepared perovskite films using two-step spin-coating method with different parameters. Pin-hole free and large crystal size above 500 nm was obtained at low spinning-speed at 1000 rpm and at room temperature spin-coating process. These parameters can be used as a standard process of two-step spin-coating method in air with high humidity (> 80%) for producing high crystal quality with large grains size perovskite films.

Acknowledgments

This work was financially supported by International Collaboration and Scientific Publication Grant from Ministry of Research, Technology and Higher Education (Kemenristekdikti), Contract No. 431/UN6.3.1/PL/2016 dated on February 19, 2016.

References

[1] http://www.nrel.gov/ncpv/images/efficiency_chart.jpg (accessed on September 25, 2016).
[2] Baikie T, Fang Y, Kadro J M, Schreyer M, Wei F, Mhaisalkar S G, Gräztel M and White T J 2013 *J. Mater. Chem A.* **1** 5628.
[3] Bretscheider S A, Weickert J, Dorman J A and Schmidt-Mende L 2014 *App. Phys. Lett.* **2** 040701.
[4] Hao F, Stoumpos C C, Cao D H, Chang R P H and Kanatzidis M G 2014 *Nat. Photon.* **8** 489.
[5] Noel N K, Stranks S D, Abate A, Wehrenfennig C, Guarmera S, Haghighirad A A, Sadhanala A, Eperson G E, Johnston M B, Petrozza A M, Herz L M and Snaith H J 2014 *Energy Environ. Sci.* **7** 3061.
[6] Kojima A, Teshima K, Shirai Y and Miyasaka T 2009 *J. Am. Chem. Soc.* **131** 6050.
[7] Boix P P, Nomomura K, Mathews N and Mhaisalkar S G 2014 *Materials Today* **17** 16.
[8] Wang B, Xiao X and Chen T 2014 *Nanoscale* **6** 12287.
[9] Kim H S, Im S H and Park N G 2014 *J. Phys. Chem. C.* **118** 5615.
[10] Green M A, Ho-Baillie A and Snaith H J 2014 *Nat. Photon.* **8** 506.

[11] Lee M M, Teuscher J, Miyasaka T, Murakami T N and Snaith H J 2012 *Science* **338** 643.
[12] Eperon G E, Burlakov V M, Docampo P, Goriely A and Snaith H J 2014 *Adv. Funct. Mater.* **24** 151.
[13] Burschka J, Pellet N, Moon S J, Humphry-Baker R, Gao P, Nazeeruddin M K and Gräztel M 2013 *Nature* **499** 316.
[14] Zhou H, Chen Q, Li G, Luo S, Song T, Duan H S, Hong Z, You J, Liu Y and Yang Y 2014 *Science* **345** 542.
[15] Bi D, Yi C, Luo J, Décoppet J D, Zhang F, Zakeeruddin S M, Li X, Hagfeldt A and Grätzel M 2016 *Nat. Energy* (in press), DOI: 10.1038/NENERGY.2016.142.
[16] Zhang W, Saliba M, Moore D T, Pathak S K, Hörantner M T, Stergiopoulos T, Stranks S D, Eperon G E, Alexander-Webber J A, Abate A, Sadhanala A, Yao S, Chen Y, Friend R H, Estroff L A, Wiesner U and Snaith H J 2015 *Nat Commun.* **6** 6142.
[17] Chang C Y, Chu C Y, Huang Y C, Huang C W, Chang S Y, Chen C A, Chao C Y and Su W F 2015 *ACS Appl. Mater. Interfaces* **7** 4955.
[18] Cohen B E and Etgar L 2016, *Front. Optoelectron.* **9** 44.
[19] Cao D H, Stoumpos C C, Malliakas C D, Katz M J, Farha O K, Hupp J T, Kanatzidis M G 2014 *APL Mater* **2**, 091101.
[20] Zheng E, Wang X F, Song J, Yan L, Tian W, Miyasaka T 2015 *ACS Appl. Mater. Interfaces* **7** 18156
[21] Shao Y, Xiao Z, Bi C, Yuan Y, Huang J 2014 *Nat Commun.* **5** 5784.
[22] Jacobsson T J, Baena J P C, Anaraki E H, Philippe B, Stranks S D, Bouduban M E F, Tress W, Schenk K, Teuscher J, Moser J E, Rensmo H, Hagfeldt A 2016 *J. Am. Chem. Soc.* **138** 10331.
[23] Cohen B E, Aharon S, Dymshits A and Etgar L 2016, *J. Phys. Chem. C* **120** 142.
[24] Kim H S and Park N G 2014 *J. Phys. Chem. Lett.* **5** 2927.
[25] Kim H D, Ohkita H, Benten H, Ito S 2016 *Adv. Mater.* **28** 917.

IOP Conf. Series: Materials Science and Engineering **196** (2017) 012038 doi:10.1088/1757-899X/196/1/012038

Annealing treatment of a-Si:H films deposited by PECVD and their properties

Yoyok Cahyono, Fuad Darul Muttaqin, Umi Maslakah and Darminto

Department of Physics, Faculty of mathematics and Natural Sciences, Institut Teknologi Sepuluh Nopember (ITS), Sukolilo ITS Campus, Surabaya 60111, Indonesia

Email : yoyok@physics.its.ac.id

Abstract. The hydrogenated amorphour silicon (a-Si:H) films have been grown on the glass substrates by plasma enhanced chemical vapor deposition (PE-CVD) employing silane (SiH_4) with hydrogen (H_2) dilution. The as – deposited films were then annealed at various temperatures of 200, 300 and 400°C for 30 minutes. Annealing process at 300°C was also performed for 60 and 90 minutes. Examinations using X-ray diffractometry, infrared and UV-Vis spectroscopy demonstrated that the annealed films show an increasing crystalinity of 3.26 – 6.80% and reduced dangling bond content down to more than one order of magnitude (from $2.3x10^{19}$ to $1.2x10^{18}$ cm^{-3}). Meanwhile, the energy bandgap and Urbach energy of the films are around 1.71 – 1.75 eV and 0.21 – 0.27 eV respectively.

1. Introduction

Among the solar cell material, hydrogenated amorphous silicon is a promising material in photovoltaic technology today. Initial efficiency ranges between 9-10% achieved in the laboratory for a single connection, then the value rises to 13% with multiple structure and silicon alloy (mostly with Ge). Crystalline silicon (c-Si or mc-Si) are more stable and have higher efficiency, but the production costs for both are very expensive [1]. Therefore, although the efficiency of a-Si: H is lower compared to c-Si (and mc-Si), a-Si: H has the potential to be cheaper so that it will be able to replace crystalline silicon technology which has a more stable efficiency. However, the Si-H bond formed easily separated by the effect of irradiation with high intensity, which is then left dangling bond defect. Dangling bond is a prime candidate causes a decrease in the efficiency of solar cells based on a-Si: H, which is known as the Staebler-Wronski effect [2]. Another defect is the addition of the bond length deviation of 1.9% and a bond angle of about 10° from its crystal structure [3].

Although it has been shown clearly that hydrogen is the trigger of the state of the defect within a thin layer of amorphous silicon, however, the study of the fundamental properties of defects and their interaction with hydrogen is still being conducted in a serious and profound. For example, how the contributing role of dangling bond defects that arise and changes in particle size on properties of thin layers. This paper discusses the process of anneal after deposition, aims to determine the effect of heat treatment on the bond between the silicon and hydrogen (SiH, SiH2), the value of the energy gap, the size of the particle, and its relationship with the phase change. This is related to the stability and density of states bonds are made. When the bonds between silicon and hydrogen is lost due to the heat treatment, it will form a dangling bond in the structure of the film. Dangling bonds is a major defect that resulted in losses on the solar cell material based a-Si: H, and it is one factor contributing to the Stabler-Wronski effect.

2. Experiment Methods

Layer deposition of a-Si: H is done by using RF-PECVD, by diluting the silane gas with hydrogen, namely $R=H_2/SiH_4=36$. Other deposition parameters are P= 2 torr, T=150°C, RF=5 watt, and t=60 minutes. Anneal is one of the heat treatment process in which the sample is heated and held at a given temperature, then allowed to cool at room temperature. This anneal process aims to determine the effect of heat treatment on the energy gap, defects in thin layers, particle size, and its relationship with the phase change. At the time of the annealing process, the sample is placed in a furnace with a thin layer facing up, to avoid the deposition layer is not damaged. In this anneal process, used furnace apparatus, with variations in temperature, T = 200°C, 300°C, 400°C and variations of time, t = 30 minutes, 60 minutes and 90 minutes.

To determine the character of the results of deposition and annealing, performed three types of characterization, ie UV-Vis spectrometer, X-ray diffraction (XRD), and Fourier-Transform Infra-Red (FTIR). XRD characterization is done by firing angle 3° over a range of angles (2θ) = 5°-60°. XRD characterization is used to determine the level of crystallinity layer. Determination of the crystallinity of the films using X'Pert HighScore Plus software (HSP). FTIR characterization performed at wave number 400–4000 cm^{-1}. While UV-Vis performed at a wavelength of 250 – 800 nm.

3. Results and Discussion

Figure 1 shows the transmittance of the wave function for the effect of temperature anneal 200°C, 300°C, and 400°C with anneal time is 30 minutes. Seems that the anneal temperature of 200°C, 300°C, and 400°C, a thin layer that transmits visible light respectively by 0,29% - 86,73%, 0,05% - 86,99%, and 0,13% - 76,23% in the wavelength range 500nm – 800 nm. In the anneal temperature of 300°C, transmittance increased and decreased in temperature anneal 400°C. Layers with anneal time variation of 30, 60 and 90 minutes, transmit light at 0.05% - 92.22%, 0.09% - 87.28%, and 0.002% - 69.05% in the wavelength range 500nm - 800 nm. The longer the anneal time, the transmittance decreases. The increase and decrease in transmittance certainly is not caused by the thickness of the layers, because of the interference fringe pattern seen relatively the same layer thickness. The increase and decrease in transmittance be linked to changes in crystallinity, energy gap, Urbach energy and particle size.

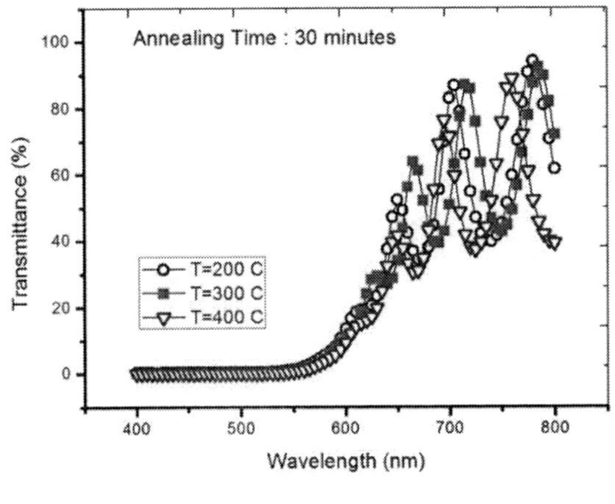

Figure 1. The transmittance of a thin layer of Si:H as a function of wavelength and anneal temperature.

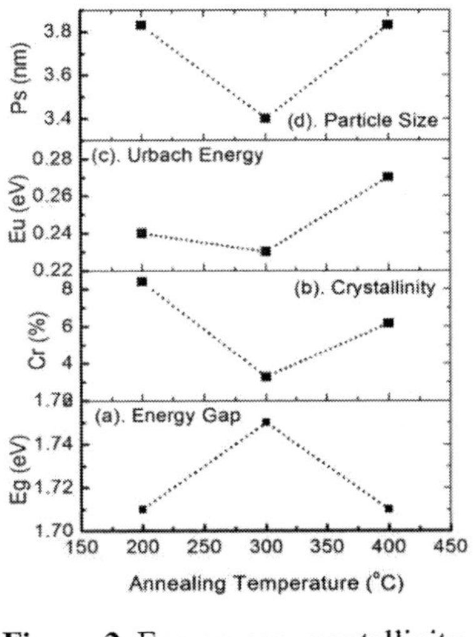

Figure 2. Energy gap, crystallinity, Urbach energy, and the particle size.

IOP Conf. Series: Materials Science and Engineering **196** (2017) 012038 doi:10.1088/1757-899X/196/1/012038

Figure 2 shows the crystallinity, energy gap, Urbach energy, and the particle size under the influence of temperature anneal. The increase in transmittance at the anneal temperature 300oC due to increased energy gap. Increased energy gap resulting in a reduction of crystallinity, and the particle size of the thin film, compared the effect of a decrease in energy Urbach. The influence of the presence of localized state is less sensitive when compared with the effect of particle size reduction (quantum confinement) [4,5]. When the anneal temperature is raised to 400°C, energy gap decreases and the crystallinity increases, so that the transmittance of the layer was decreased. Usually anneal temperature increases can result in the release of hydrogen bonds that result in increased localized state, because the dangling bond activation and increased the randomness due to changes in the length and angle of the chain bonding, and this is indicated by the increase in energy Urbach. However, due to changes in particle size is getting bigger, the energy gap decreases, although the Urbach energy is increasing.

At the time of annealing process to anneal time of 60 minutes, the resulting increase in energy Urbach and crystallinity in the layer. These two contradictory consequences that impact, due to rising energy Urbach will raise the energy gap, otherwise the increase in crystallinity lowers energy gap. Because two things are contradictory, then a large energy gap layer unchanged. Annealing process is carried out with anneal time of 90 minutes, resulting in increased crystallinity layer, Urbach energy decreases, and the particle size enlarged, so that the energy gap decreases. In general, when the dangling bond in the layer increases, the energy gap will increase. This is due to defect or localized state increased. But the energy gap was also strongly influenced by the particle size (grain).

Dangling bond density of states can be approximated by the density of electron spin state of unpaired. The density of states of the electron spin can be determined using the absorption coefficient values at hv = 1.2 eV [6]. Because UV-Vis characterization is only done for the value of hv between 1:55 eV to 3.1 eV, then the value of the absorption coefficient is determined by a mathematical equation based on Urbach energy trendline. Figure 3 shows the results of calculation of changes in the density of states of the dangling bond, after treatment anneal temperature of 200°C, 300°C and 400°C, using electron spin state density approach. At 300°C temperature, electron density of states or the dangling bond decreased compared with a thin layer on anneal temperature of 200°C, and increased in layers with anneal temperature of 400°C. This is due to the release of hydrogen atoms in SiH. Theoretically changes in the density of states Si-H, inversely proportional to the dangling bond. This is consistent with that seen from Figure 3, where the density of states of the SiH tends to fall when the dangling bond density of states increases.

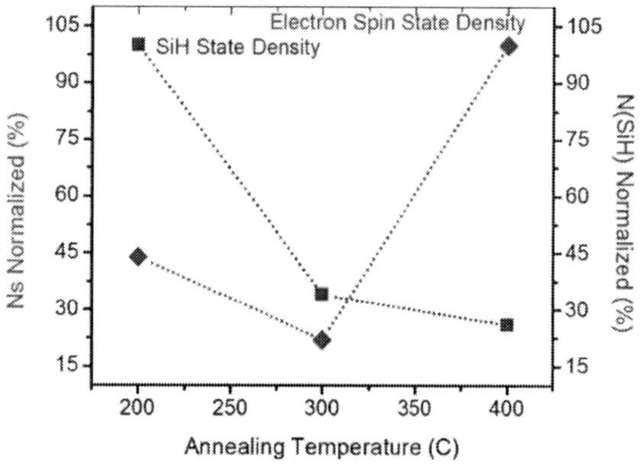

Figure 3. Normalized dangling bond (electron spin) and SiH (hydrogen) density of states.

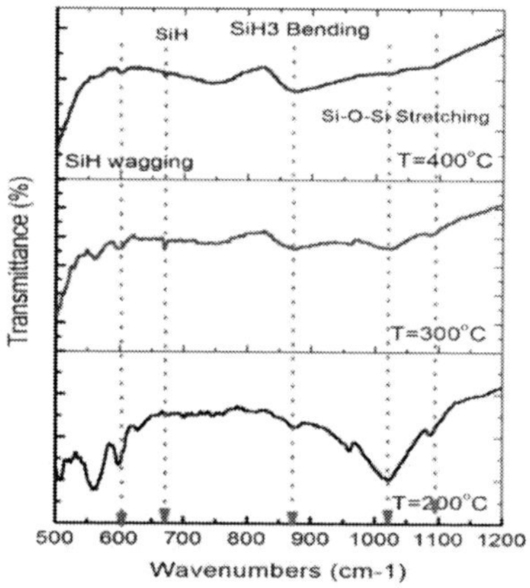

Figure 4. FTIR absorption spectra of Si:H thin layers for some anneal temperatures.

However, Si-H and dangling bond has a relationship with the complex interplay. This is because during the anneal process will many possibilities happen. For example, For example, when SiH bond breaking and H atoms is released, it is possible to form Si-H_2 and left dangling bond. Thus, the density of states of SiH is inversely proportional to the density of states of dangling bond and Si-H_2. However, when one of the H atoms in SiH_2 is released, it will form SiH and dangling bond, which is formed HSi-(db). Because it can be understood that the change in the density of dangling bond (electron spin state density) and SiH in layers with anneal temperature to 300°C decreased, compared to the anneal temperature of 200°C. It was expected because some of the dangling bond is released, most of the SiH and not all of them form a radical SiH. Others form a radical such as SiH_2, SH_3, and (SiH_2)n.

Figure 4 shows the FTIR spectrum in the wave number interval 500-1200 cm-1, for layers with anneal temperature of 200°C, 300°C, and 400°C, and anneal time is 30 minutes. It is seen that the absorption peaks of silicon hydride compounds that are in wave numbers 598 cm^{-1} (600/SiH wagging), 669 cm^{-1} (667/SiH), 887 cm^{-1} (890/SiH_3 bending) [7]. Aside from that shown in Figure 4, the absorption peaks are also on the wave number 1999 cm^{-1} (2000/SiH stretching), and 2090 cm^{-1} (2090/SiH_2 Stretching The peak intensity at around 600 cm^{-1} decreases with rising temperatures and anneal time. This indicates that the density of SiH bond at the position of the wave number decreases with increasing temperature and anneal time. The opposite occurs in the top position at wave number 669 cm^{-1} and 887 cm^{-1}, wherein the peak intensity increases on rising anneal temperature. An increase in the density of SiH bond at the peak position. Nevertheless, SiH bond formation is dominated by changes in the position of the peak of 600 cm^{-1}, and is in accordance with the decreasing density of SiH in Figure 3. Shifting the top point of absorption also occurs around the wave number of 2000 cm^{-1} to changes in anneal temperature, ie [i] peak at 1997 cm^{-1}, [ii] at 1999 cm^{-1}, and [iii] the position of the wave number 1998 cm^{-1} for layers with anneal temperature of 400°C. Shifting positions absorption peak can be associated with increasing regularity in the network silicon, and an increase in the fraction of crystalline [8].

4. Conclusions

Based on research that has been done can be concluded that the higher the anneal temperature and the longer the anneal time, there is generally a tendency to decrease transmittance. Decrease in transmittance due to changes in crystallinity, energy gap, Urbach energy and particle size. Theoretically changes in the density of states of the Si-H is inversely proportional to the density of dangling bonds. However, Si-H and dangling bond has a relationship with the complex interplay. Many possibilities occur.

References

[1] Vilamitjana D S, "Amorphous Silicon Solar Cells Obtained By Hot-Wire Chemical Vapour Deposition," Universitat de Barcelona, Barcelona, 2004.
[2] Street R, *Hydrogenated amorphous silicon*. Cambridge University Press, 1991.
[3] Munyeme G., "Experimental and Computer Modelling Studies of Metastability of Amorphous Silicon Based Solar Cells," Universiteit Utrecht, Netherland, 2003.
[4] Kim T W, Cho C H, Kim B H, and Park S J 2006 *Applied Physics Letters* **88** 123102–1
[5] Ameer N M A and Abdulrida M C 2011 *Journal of Modern Physics* **2** 1530
[6] Kre R, Moussé M, Tchétché Y, Bella F, Aka B, and Thomas P 2010 *International Journal of the Physical Sciences* **5** 675
[7] Ali A, Kobayashi H, Inokuma T, and Hajry A 2013 *Materials Research Bulletin* **48** 1027
[8] Park Y B and Rhee S W 2001 *Journal Of Applied Physics* **90** 217

IOP Conf. Series: Materials Science and Engineering **196** (2017) 012039 doi:10.1088/1757-899X/196/1/012039

The Effect of Variations in the Ratio of Matrix/Solvent on the Physical and Mechanical Properties of Chitosan Biopolymer Membranes

N N Rupiasih[1, 2], M Sumadiyasa [1] and I K Putra [1]

[1] Department of Physics, Faculty of Mathematics and Natural Sciences, Udayana University, Denpasar, Bali 80361, Indonesia

[2] Group Research Material Sciences and Technology-Polymer and Biomaterial, Udayana University, Denpasar, Bali 80361, Indonesia

E-mail: rupiasih@gmail.com

Abstract. The membranes with various ratio of matrix/solvent such as 1%, 2%, 3%, and 4% were prepared using casting method. Chitosan powder was used as a matrix and acetic acid 1% (v/v) solution as solvent. The physical properties of membranes have been characterized by different techniques including FTIR, XRD, swelling index (SI), BJH method, and filtration, while the mechanical properties have been characterized by tensile strength measurements. The results showed that increasing the amount of matrix resulted in an increased in thicknesses, pores size, the elongation-at-break, elastic moduli, tensile strength and the pure water flux of membranes significantly, whereas the pores density, swelling index and crystallinity index decreased. This study can be simplified to be even faster and straightforward in designing polymer membrane structures for filtration purposes.

1. Introduction

Chitosan is a natural polysaccharide biopolymer derived by deacetylation of chitin. Chitin is a major component of the shells of crustacean such as crab, shrimp and crawfish [1-3]. Chitin is insoluble in water and common organic solvents but it dissolve only in solvents such as *N,N*dimethylacetamide, hexafluoroacetone or hexafluoro-2-propanol; whereas chitosan is soluble in aqueous acidic solutions (pH < 6.0), therefore it is largely used in different applications as gels, films and fibers [4-6].

Chitosan has attracted great interest due to its biocompatibility, high charge density, non-toxicity and adhesion. It has antibacterial effect, heavy metal adsorption, antioxidation, film and membranes formability [5, 6]. The film properties of chitosan depend on its morphology, which is effected by molecular weight, degree of N-acetylation, solvent evaporation, and free amine regenerating mechanism [7]. Chitosan film has a potential to be employed for packaging, particularly as an edible packaging. Chitosan based membranes have been used in reverse osmosis, gas separation, dialysis, and pervaporation [8]. It is used as a composite, polyelectrolyte complex or a blend with another polymer [7, 8]. Chitosan as a membrane alone has high mechanical strength, permeable to urea, and able to reject high molecular weight compounds [7, 9].

Many studies have been carried out in order to understand the mechanism of the basic processes of forming chitosan polymer membranes. The use of organic acid solution such as glycolic acid, malic acid and ascorbic acid solutions for chitosan dissolution produces chitosan membranes with improved mechanical properties and enhanced hydrophilicity [10]. Le *et al* reported that the Young's Modulus of the hydroxyapatite (HA) chitosan composites is lower than pure chitosan [11]. The Young's Modulus of the composites decreases with HA content, while the failure strength and strain increase with the HA content. From ionic interaction between the chitosan and concentration of sulfuric acid, protonation and crosslinking takes place for a typical concentration of 0.1-0.5 M of 25 μm thick membranes. Beyond 0.5 M concentration of sulfuric acid, ion diffusion is favored and degree of crosslinking increases [1].

Based on those backgrounds, it becomes clear that the studies carried out so far are more related to the mechanism of forming composites chitosan membranes and their physical characteristics. Therefore, it will be interesting to investigate the effect of variations in the ratio of matrix/solvent on the physical and mechanical properties of chitosan polymeric membranes. This study reports the effect of variations in the ratio of chitosan (as a matrix) to acetic acids (as a solvent) on physical and mechanical properties of chitosan membranes include FTIR, XRD, pore size, pores density, swelling index, tensile strength, and pure water flux (PWF).

2. Materials and Methods

2.1. Materials

Chitosan with a degree of deacetylation (DD) of 87.9%, solubility in acetic acid of 99.4% and an average molecular weight of 900,000 was used as membrane materials. The chemicals used in this study were analytical grade and demineralized water was used in preparing the solutions.

2.2. Chitosan membranes preparation

Various chitosan membranes such as 1%, 2%, 3%, and 4% have been prepared by casting method. Chitosan is used as matrix and acetic acid 1% (v/v) solution as solvent. 1% (w/v) membrane solution was prepared by dissolving 2.5 g of chitosan powder in 250 ml of acetic acid 1%. The mixture was stirred for 8 hours at room temperature to obtain membrane solution. The membrane solution was poured onto a glass plate sized of 18.5 cm x 24.0 cm then dry at room temperature for 6 days. The dried membrane was immersed into 1 M NaOH for 12 minutes, washing repeatedly with distilled water and finally drying again at room temperature. The membrane obtained was chitosan membrane 1%. Other membranes have made by the same method, namely by varying the amount of chitosan in the same volume of acetic acid 1%. Finally, it obtained dry chitosan membranes, which are ready to characterize or used.

2.3. Characterization methods

The FTIR spectra of all membranes were obtained using an IR Prestige-21 FTIR spectrophotometer, Shimadzu. Spectra were taken with a resolution of 4 cm^{-1} and averaged over 10 scans. The spectra were recorded in the range of 400 to 4000 cm^{-1}.

The X-ray diffraction spectra of dry membranes have been recorded at room temperature using X-ray Difractometer (XRD). The X-ray source used was Cu-Kα radiation wavelength of 1.54 Å, with irradiation conditions were 30 kV and 40 mA and the diffraction angle, 2θ between 5° to 60°. From the spectra obtained, we can calculate the crystallinity index (CI) of the membranes using the XRD peak height method developed by Segal and coworkers, formula (1) [12]:

$$CI = \frac{(I_{110} - I_{am})}{I_{110}} \quad (1)$$

Where I_{110} is the highest crystalline peak (110) and I_{am} is the amorphous peak which is at about 12° in Figure 4.

The pore size and pores density of membranes were analyzed by BJH (*Barrett, Joynerand Halenda*) method using Nova 1200e instrument. Around 1.0 g of membrane was cut into small pieces and put in sample chamber. Degassing process was done at temperature of 60 °C for 7 hours. The analyses of pore size and pores density were done using ASiQ software.

For filtration purposes, membrane are usually used in an aqueous environment, so the water uptake capability which called as swelling index (SI) is an important property of membranes. The membranes were cut in circulars form with diameter 3.5 cm, weighed, and they were kept in beaker with 20 ml of distilled water at 28°C for 5 minutes. Then the membranes were taken out after 5 minutes carefully, removed the excess water from the surface of the membranes with filter paper and weighed [4]. The swelling index of the membrane was calculated using the following formula:

$$SI\,(\%) = \frac{(m_{wet} - m_{dry})}{m_{dry}}\, x\ 100\% \qquad (2)$$

where m_{wet} and m_{dry} are the mass of the wet and dry membranes, respectively.

The tensile strength (TS) and elongation-at-break (ε) of membranes were measured using Screw Test Stand Machine ALX S following the guidelines of the ASTM Standard Method D638-90. The measurements have done in dry condition, at room temperature. The membranes were cut into "dogbone-shaped" specimens with the length of narrow section was 12.7-14.2 mm and width of 2-9 mm. Tensile strength was expressed in MPa and ε in percent. The measurements were replicated five times for each membrane and the average values are reported. The elastic moduli (E) of membranes were determined from the stress-strain graphs.

The measurement of pure water flux (PWF) was performed using *dead-end* filtration method with pressure 50-60 kPa. The effective surface area of membranes (A) used is $1.048x10^{-3}$ m^2 approximately. The membranes soaked in distilled water for around 10-15 minutes before being used as filter. The PWF was calculated using formula (3) [13].

$$PWF\,(J) = \frac{V}{At} \qquad (3)$$

where V is volume of the permeate (l), A is effective surface area of the membrane (m^2) and t is time (h).

3. Results and Discussion

The thicknesses of chitosan membranes obtained were 0.041, 0.134, 0.204, and 0.236 mm, respectively, as shown in Figure 1. They were increased as increased the amount of chitosan from 1%, 2%, 3%, and 4%.

Figure 2 shows FTIR spectra of chitosan membranes 1%, 2%, 3%, and 4%. The spectra obtained were similar to the spectra that have been reported by other researchers [14-15]. All spectra showed similar pattern and in the figure only spectrum of membranes 3% is labeled, other spectra are written in Table 1.

Based on Figure 2 can be described the main bands observed in chitosan polymer membranes. The broad band at around 3692 cm^{-1} is due to overlapping of the O-H and N-H stretching vibrations of functional groups engaged in hydrogen bonds [15]. The band at around 1568 cm^{-1} is assigned to the stretching vibration of amino group of chitosan. The bands which assigned to the stretching vibration of C–O–C linkages in the saccharide structure (glucosamine rings) appear at 1181, 1044 and 899 cm^{-1} [15]. Very weak bands at around 1431 and 1609 cm^{-1} corresponding to the C–H bond in the methyl and stretching vibrations of the carbonyl group (amide I), respectively, that reveal the high degree of deacetylation of the chitosan used [16].

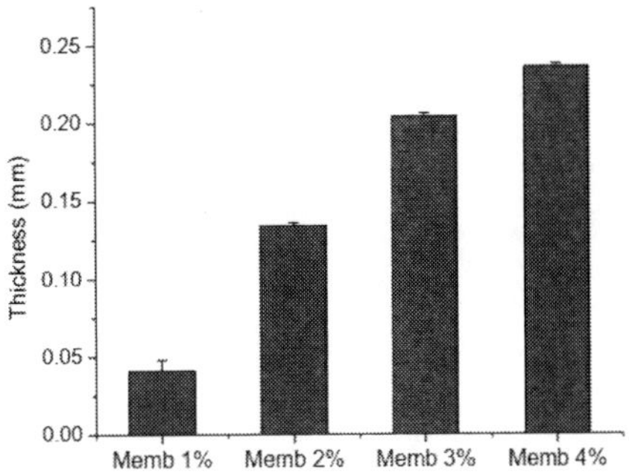

Figure 1. Thickness of chitosan membranes: 1%, 2%, 3%, and 4%.

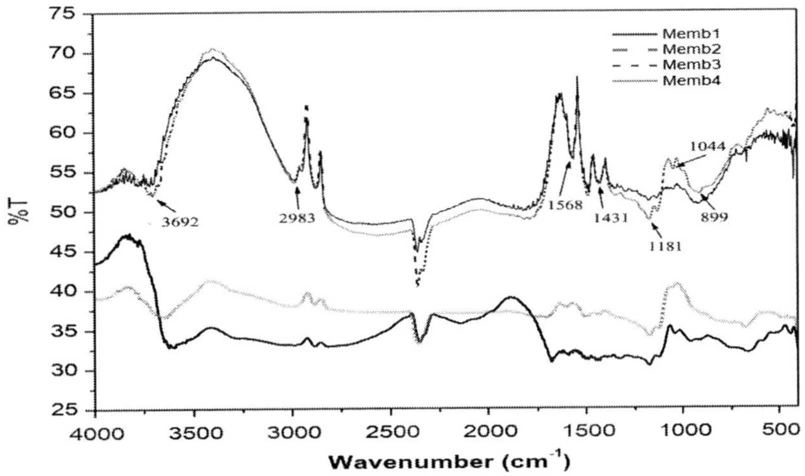

Figure 2. FTIR spectra of chitosan membranes: 1%, 2%, 3%, and 4%.

The data in Table 1 indicate that there are no new bands were observed, which means no new molecules are formed on the membranes by increased the amount of chitosan. Some bands have been slightly shifted to the higher wavenumbers or lower, but they are still within their group ranges.

Figure 3a shows a typical stress-strain curve of various membranes, which are similar to the characteristics of the tensile modulus plastic films [17]. The tensile strength (TS), elongation-at-break (ε) and modulus of elasticity (E) was determined from stress-strain curves (Figure 3a) and the results are shown in Figure 3b. Tensile strength is the maximum tension that can be supported by the membranes until it break. Elongation-at-break is a measure of flexibility of the membranes that can be considered as characteristic defects of the membranes before breaking [18].

Figure 3b shows membrane 1% has very small value of E and TS compared with the other membranes (2%, 3%, and 4%). The increased of the amount of matrix also resulted in an increased in elongation-at-break (ε) of the membrane gradually from 191% to 341% for membrane 1 % to 3% and decreased slightly to 329% for membrane 4%. This suggests that the increased the amount of matrix more than 3% decreased the elongation of the membrane. It reduced the ability of stretching as shown by Fig. 3a, where the graph looked steeper [18].

Table 1. The main bands observed in chitosan membranes.

Functional Groups	Wavenumbers Observed (cm^{-1})			
	Membrane 1%	Membrane 2%	Membrane 3%	Membrane 4%
NH_2 and –OH stretch	3619	3660	3692	3700
Symmetric CH_2 stretching	2973	2989	2983	2983
C=O (carbonyl group (amide I))	1609	1609	1609	1609
NH_2 in amino group (amide II)	1588	1588	1568	1568
C-H stretch	1431	1431	1431	1431
C–O–C stretch (saccharide structure)	1181, 1044 and 907	1181, 1044 and 907	1181, 1044 and 899	1181, 1044 and 907

Ref. 15-16.

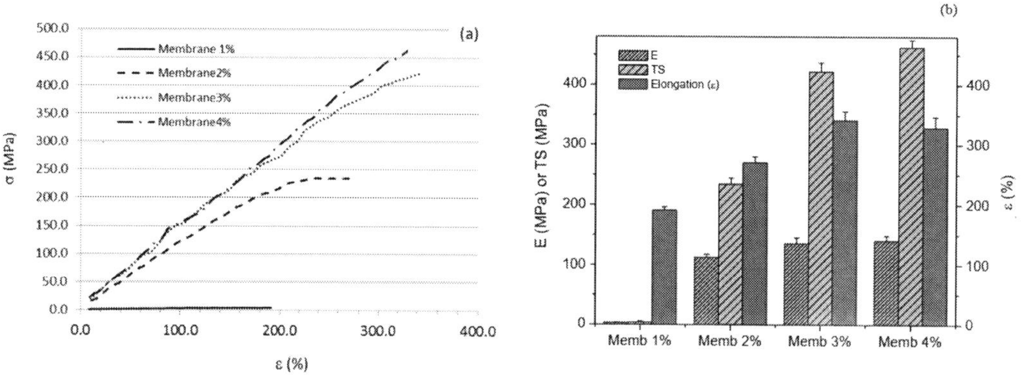

Figure 3. Mechanical properties of chitosan membranes in dry condition: (a) stress-strain curves (representative result is shown), (b) Tensile strength (TS), elongation-at-break (ε) and elastic moduli (E).

The X-ray diffraction spectra of all membranes are displayed in Figure 4. It showed two peaks located at 2θ at about 10° and 20°. This result is in agreement with that have been reported by other researchers [18]. It is known that chitosan always contains bound water (5%) even if it has been extensively dried [18]. The incorporation of bound water molecules into the crystal lattice, commonly termed hydrated crystals, generally gives rise to a more dominated amorphous part which can be normally detected by a broad crystalline peak in the corresponding X-ray pattern. Therefore, the crystalline peak I centred at around 10° is attributed to the hydrated crystalline structure of chitosan and the crystalline peak II centred at around 20° is crystalline 110 of chitosan [18].

By using formula 1 have been calculated the crystallinity index (CI) of the membranes as shows in Figure 5. The highest crystallinity index is obtained for membrane 1% of about 87%, while the other membranes (2%, 3%, and 4%) have almost same values of about 68%. This result is strengthened by the modulus of elasticity (E, Fig. 3b) of membranes where membrane 1% has very small of about 3 MPa and the other membranes have very high, about 130 MPa. This can described that the crystallinity index of membrane 1% is high so its flexibility which expressed by E be very small. Also, it is in accordance with the value of ε which is increased gradually with increasing the amount of matrix from 1% to 4%. These results suggest that the increased the amount of matrix in the membrane resulted in the less ordered structure of the membranes e.g. decreased in crystallinity, thereby increased it tensile strength (TS) [18].

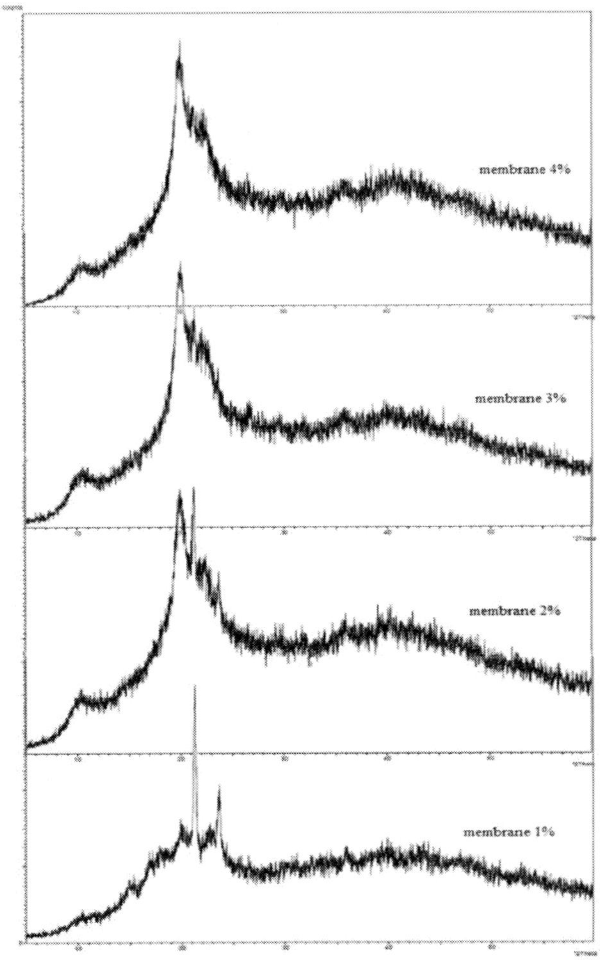

Figure 4. X-ray diffraction spectra of chitosan membranes 1%, 2%, 3%, and 4%.

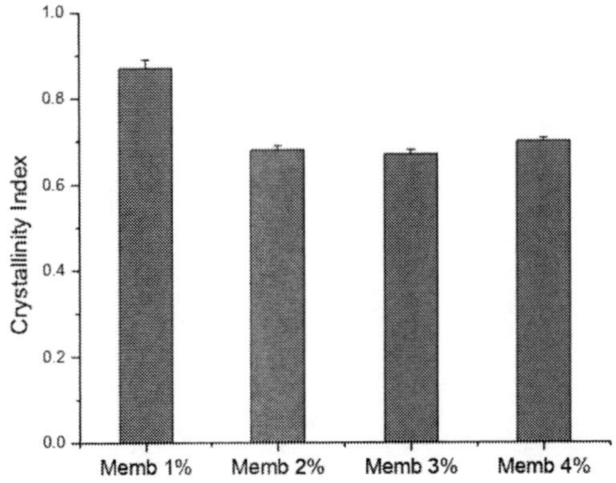

Figure 5. The crystallinity index (CI) of chitosan membranes: 1%, 2%, 3%, and 4%.

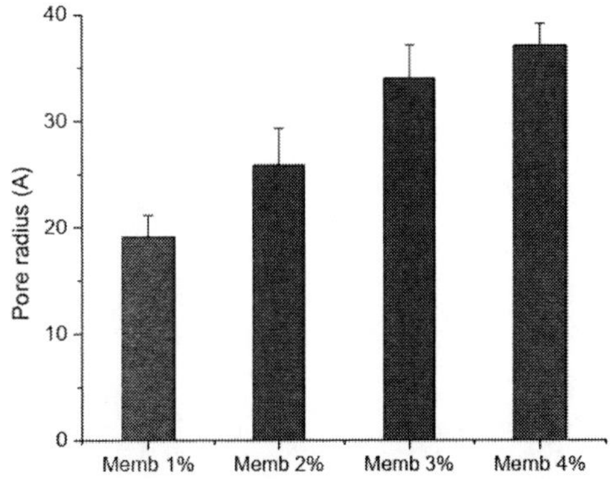

Figure 6. The pore radius of chitosan membranes: 1%, 2%, 3%, and 4%

The pore size and pore density of various membranes obtained using BJH method are shown in Figure 6 and 7. Figure 6 shows that the pore size (means radius of the pores) increased from 19.12 Å to 37.09 Å with increasing the amount of matrix from 1% to 4%, while the pores density decreased from $2.12 x 10^9$ to $0.98 x 10^9$ pores/g as shown in Figure 7. Based on the IUPAC convention for nomenclature porosity, 1972 [19], all membranes are belong to mesoporous group with the pore radius ranges from 19.12 Å to 37.09 Å, which can be used in microfiltration process.

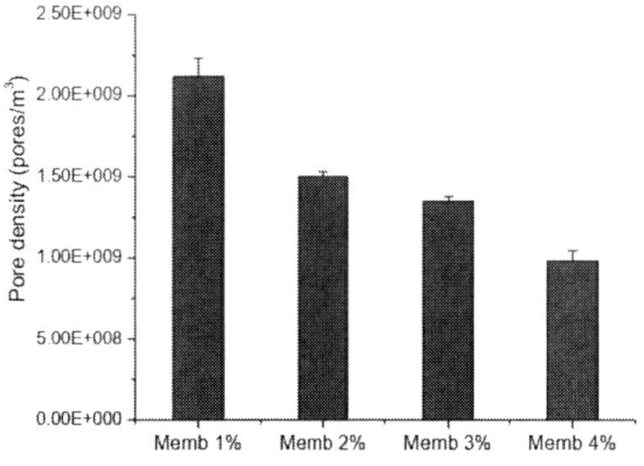

Figure 7. The pores density of chitosan membranes: 1%, 2%, 3%, and 4%.

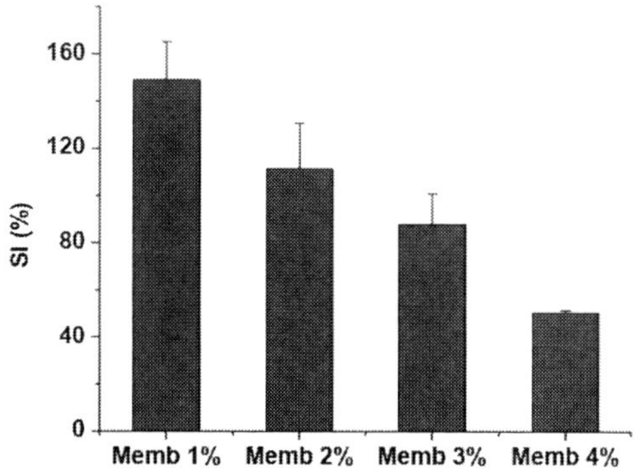

Figure 8. The swelling index (SI) of chitosan membranes: 1%, 2%, 3%, and 4%.

Figure 8 shows the swelling index (SI) of chitosan membranes 1%, 2%, 3%, and 4%. This shows the decreased of the swelling index as increasing the amount of the matrix. These results are in accordance with the pores density of the membranes (Fig. 7) which is increased as increasing the amount of matrix from 1% to 4%. The higher the pores density of the membrane will increase their water uptake capability.

Figure 9 shows the pure water flux (PWF) of all membranes. The PWF values showed significant increased with increasing the concentration of matrix from 1% to 4%. These results are consistent with the pore size of membrane samples which increased as increasing the amount of matrix. From those results (pore size, pores density and filtration), we can conclude that PWF mainly depend on the pores size of membranes. It is an important characteristic in filtration case.

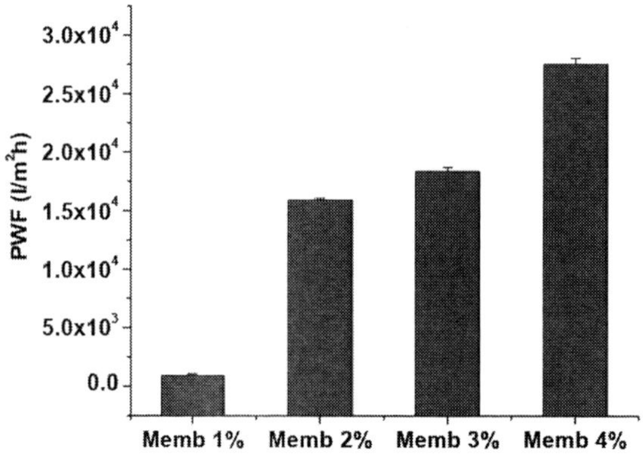

Figure 9. The pure water flux (*PWF*) of chitosan membranes: 1%, 2%, 3%, and 4%.

4. Conclusion

We have demonstrated that the variation in the ratio of matrix/solvent can affect the physical and mechanical properties of chitosan polymeric membranes. This type of study can give structural information which is correlated with performance related information, when analysing membranes designed for filtration purposes. This work allows predicting possible ratio of matrix/solvent in designing polymer membranes, without the necessity of performing time consuming (especially in the case of filtration membranes). In this study, we have shown that increasing the amount of matrix resulted in an increased in thicknesses, pores size, the elongation-at-break, elastic moduli, tensile strength and the pure water flux of membranes, significantly, whereas the pores density, swelling index and crystallinity index decreased.

Acknowledgements

This work was financially supported by the Ministry of Research, Technology and Higher Education, Republic of Indonesia, under Fundamental Research Grant of Udayana University, Year Budget 2016 are highly acknowledged.

References

[1] Begum A A, Radhakrishnan R and Prem Nazeer K 2011 *Malaysian Polymer Journal* **6** 27
[2] Rupiasih N N, Purnomo R R and Sumadiyasa M 2016 *Journal of Physics: Conference Series* **710** (012009)
[3] Jayakumar R., Prabaharan M., Sudheesh Kumar P T, Nair S V , Furuike T and Tamura H. 2011 *Biomedical Engineering Trends in Materials Science Edited by Mr Anthony Laskovski* 3-24 www.intechopen.com.
[4] Rinaudo M 2006 *Prog Polym Sci*, **31** 603
[5] Khorrami M., Najafpour G D, Younesi H and Hosseinpour M N 2012 *Chem Biochem Eng Q* **26** 217
[6] Kumirska J, Mirko X W, Czerwicka M, Bychowska Z K A, Brzozowski K, Thöming J and Stepnowski P 2011 *Biomedical Engineering, Trends in Materials Science in Mr Anthony Laskovski (Ed.)* 25www.intechopen.com.
[7] Rupiasih N N, Puspita Y E and Sumadiyasa M 2015 *Journal of Physics: Conference Series* **622** (012004)
[8] Rupiasih N N 2016 *Journal of Physics: Conference Series* **776** (012045)
[9] Nasir N F M, Zain N M, Raha M G and Kadri N A 2005 *American Journal of Applied Sciences* **2** 1578
[10] Po-Hui Chen, Ya-Shi Hwang, Ting-Yun Kuo, Fang-Hsuan Liu, Juin-Yih Lai and Hsyue-Jen Hsieh 2007 *Journal of Medical and Biological Engineering* **27** 23
[11] Le H.R., Qu, S, Mackay R E and Rothwell R 2012 *Journal of Advanced Ceramics* **1** 66
[12] Srinivasa P C, Ramesh M N, Kumar K R and Tharanathan R N 2004 *Journal of Food Engineering* **63** 79
[13] Rupiasih N N and Vidyasagar P B 2008 *Polymer Degradation and Stability* **93** 1300
[14] Eun-Jung Lee, Du-Sik Shin, Hyoun-Ee Kim, Hae-Won Kim, Young-Hag Koh and Jun-Hyeog Jang 2009 *Biomaterials* **30** 743
[15] Monarul Islam Md, Shah Md Masum, Rahman M M, Ashraful Islam Molla Md, Shaikh A A and Roy S K 2011 *International Journal of Basic & Applied Sciences IJBAS-IJENS* **11** 77
[16] Paula M. Lo´pez-Pe´rez, Marques A P, da Silva M R P, Pashkuleva I Reis R L 2007 *J. Mater. Chem* **17** 4064
[17] Souza B W S, Cerqueira M A, Martins J T, Casariego A J, Teixeira A and Vicente A A 2010 *Food Hydrocolloids* **24** 330
[18] Kirkwan M and Strawbridge J 2003 *In Coles R, McDowell D, Kirkwan M (ed) Food packaging technology, Blakwell Publishing, CRC-Press* 174
[19] IUPAC 1972 2005 *Convention for Nomenclature of Porosity* (Cambridge (UK):IUPAC Recommendations 2005)

The Effect of Pressure and Temperature on Separation of Free Gadolinium(III) From Gd-DTPA Complex by Nanofiltration-Complexation Method

Iman Rahayu, Anni Anggraeni, Ukun MSS, Husein H Bahti

Departement of Chemistry, Faculty of Mathematics and Natural Science, Universitas Padjadjaran, Jl. Raya Bandung Sumedang km. 21, Jatinangor, West Java 45363, Indonesia

E-mail: iman.rahayu@unpad.ac.id

ABSTRACT. Nowdays, the utilization of rare earth elements has been carried out widely in industry and medicine, one of them is gadolinium in Gd-DTPA complex is used as a contrast agent in a magnetic resonance imaging (MRI) diagnostic to increase the visual contrast between normal tissue and diseased. Although the stability of a given complex may be high enough, the complexation step couldnot have been completed, so there is possible to gadolinium(III) in the complex compound. Therefore, the function of that compounds should be dangerous because of the toxicity of gadolinium(III) in human body. So, it is necessarry to separate free gadolinium(III) from Gd-DTPA complex by nanofiltration-complexation. The method of this study is complexing of Gd_2O_3 with DTPA ligand by reflux and separation of Gd-DTPA complex from gadolinium(III) with a nanofiltration membrane on the variation of pressures(2, 3, 4, 5, 6 bars) and temperature (25, 30, 35, 40 °C) and determined the flux and rejection. The results of this study are the higher of pressures and temperatures, permeation flux are increasing and ion rejections are decreasing and gave the free gadolinium(III) rejection until 86.26%.

1. INTRODUCTION

Rare earth elements (REE's) have been used increasingly in the field of industry. In the midle of 2008, about 120,000 tons of REE have been widely used in the some of industry sector, where 60% of them used in the field of chemical engineering, metalurgy, catalyst, glasses, catode tubes and fluoresences. Meanwhile, about 40% of them used in the high technology materials fabrication, such as baterry alloy, ceramic, guns military, satellite system, and magnet [1]. Beside that, rare earth elements is also used as supporting materials such as in the medical field which is using gadolinium in Gd-DTPA complex form as contrasting agent in a magnetic resonace imaging (MRI) diagnostic equipment to increase visual contrast between normal and abnormal tissues [1,2].

The separation and purification of rare earth elements which has been done by scientists on the 18th - 19th centuries is the hardest challenge. Eventually, until the 20th century all of REEs had not been identified due to their similarity in physical and chemical characteristics [1,3]. The separation of REEs could be increase their purity actually involves liquid-liquid extraction method. This method is a great important due to its ability to produced highly purified lanthanides which commonly used in material fabrication, such as electronic, ceramic, optic, and catalyst. However, liquid-liquid extraction required additional solvent, which is problem in a matter of environmental preservation [4]. Another method for REEs separation has been investigated. Solvent extraction with chelating agent di-n-dibutyldithiocarbamate (DBDTC) could attract gadolinium metal and formed gadolinium complex. However, only 20,8% purified gadolinium would be produced [2,4]. The separation of REEs has been accomplished by nanofiltration (NF) process with associating a metal-ion complexation step on charge organic membrane: effect of pH, pressure, flux, ionic strength, and temperature and gave 95% yields gadolinium [2]. Based on those previous research, the separation of REEs (gadolinium) will be achieve a great result by nanofiltration membrane process.

Among the different available separative methods (liqui-liquid extraction and solvent extraction), membran technology has many advantages such as high efficiency, no exchange phase, low energy, produce very few waste, and easy to operate. Nanofiltration (NF) is one of the most promising membran process to separate rare earth elements due to high permeability and permselectivity, suited for rejecting organics with weight molecular > 200 g/mol and metal ions with moderate pressure. Due to these characteristics, the NF process show a great interest for REE ion rejection [2,4]. The REE ion rejection by nanofiltration complexation can be explained by two mechanism. The first is the steric effect based on MWCO membrane which is related to pore size or pore radius of the membrane. The charge effect is the second mechanism which refers to electrostatic interactions between solute and charge surface membrane. Most membrane acquire an electric surface charge when brought into contact with an aqueous solution. Solutes having the same charge than the membrane are excluded, and solutes having the opposite charge can passed through it. Solutes having different valences can be separated based on the charge effect.

In this work, the rejection of free gadolinium(III) by nanofiltration process assisted by complexation using DTPA ligand was studied according to the pressure (2, 3, 4, 5, 6 bars) and temperature (25, 30, 35, 40 °C).

2. EXPERIMENT

Aqua miliqu, glacial acetic acid (p.a., 100%), buffer acetate buffer pH 5.8, diethylene triamine pentaacetic acid (p.a., Sigma Aldrich), gadolinium chloride hexahydrate (p.a., Sigma Aldrich), gadolinium oxide (p.a., Sigma Aldrich), sodium hydroxide (p.a.) and xylenol orange.

Device the reflux tool set, then as much as 5.4375 g Gd_2O_3 weighted, put into a 1000 mL reflux flask and added with 6.4902 g of DTPA. Then added 540.00 mL aqua miliqu to the mixture and then refluxed for 3 hours on the oil bath until the solution is clear and colorless. Allowed the solution to stand at room temperature. Furthermore, the pH of the solution is measured by a pH meter. If the solution is more acid, added 3 N sodium hydroxide gradually until reaching a pH of 7.0-7.5. Then the solution was filtered using filter paper [9].

Amount of 500.00 mL of Gd-DTPA complex solution diluted with aqua miliqu until 3000.00 mL in the dilution flask . A certain amount of aqua miliqu passed into the membrane and nanofiltration process is run for 60 minutes until the membrane reached a stable condition, and then permeates were collected and the flux values were calculated. A solution of Gd-DTPA complex which acts as the feed solution is passed through the membrane and carried out the separation process of Gd-DTPA complex solution at pressure of 2, 3, 4, 5, and 6 bar and temperature of 25, 30, 35, 40 ° C for 135 minutes with cross-flow system. Permeate was collected, and then permeate fluxs were calculated for each variation of pressure and temperature by using the formula $Jv = V / (A.t)$. The concentration of free gadolinium(III) in the feed and permeate determined by using a visible spectrophotometer and the rejections of free gadolinium(III) were calculated by using the formula $R = (1\text{-}Cp / Cf) \times 100\%$.

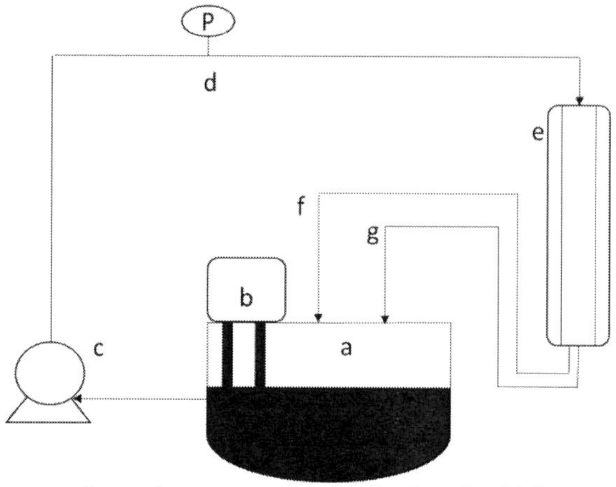

Figure 1. Schematic flow diagram of membrane system: (a) feed tank, (b) heat exchanger, (c) pump, (d) pressure gauge, (e) nanofiltration membrane, (f) permeate, (g) retentate

3. RESULT AND DISCUSSION

An optimuml condition of membrane performance is generally expressed by the permeation flux and rejection of free gadolinium(III) in the feed solution. The higher the value of fluxes and rejections, the membrane has a better performance. But in fact, in a separation process by nanofiltration membrane discovered a phenomenon that the membrane flux is inversely proportional to its rejection. If the membrane flux was increased, the rejection will decreasd, and vice versa. If the membrane flux was decreased then rejeksinya will increased. In this study, an optimum conditions of membrane will be obtained by influenced of several parameters, such as the effects of pressure and temperature.

a. Effect of Pressure

Figure 2 shows the effect of pressure on the permeation fluxes and free gadolinium(III) rejections. In that figure, it can be seen the flux values at a pressure of 2, 3, 4, 5, and 6 bars respectively 237.530 L / m^2.h, 371.061 L / m^2.h, 523.100 L / m^2.h, 700.930 L / m^2. h, and 864.552 L / m^2.h. The higher of applied pressure, the permeation fluxes increase linearly. This is consistent with the driving force on the membrane operation, where the higher of pressure can make the propulsion that given to the feed solution across the membrane getting bigger so those chemical species that contained in the feed solution will be quickly moving into the membrane pores and out as permeates. In addition, the higher pressure caused by high feed diffusion rate. The high diffusion rate may causing interactions between a feed solution to the membrane surface occured faster thus the speed of the feed solution to through the membrane pores getting bigger, hard to resist the membrane feed solution and a lot of free gadolinium(III) which can diffuse through the membrane.

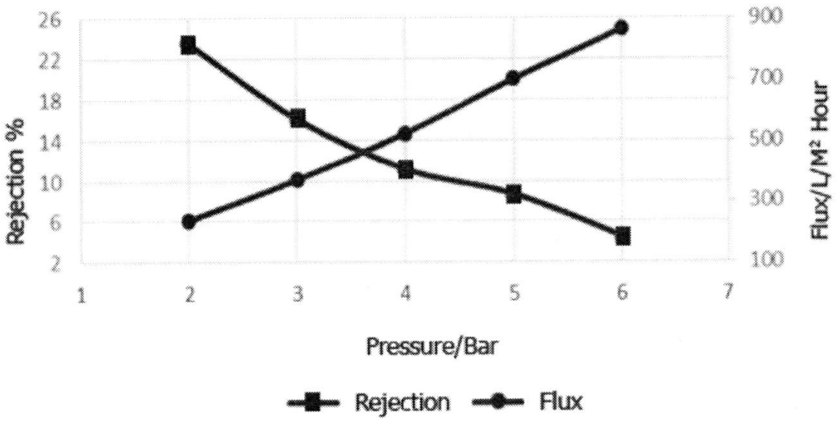

Figure 2. The effect of pressure on permeation fluxes and free gadolinium(III) rejections

The rejections of free gadolinium(III) at a pressure of 2, 3, 4, 5, and 6 bars respectively 23.56%; 16.12%; 11.16%; 8.68% and 4.49%. In this study, the percentages of rejection explained the ability of membrane to be able to reject the free gadolinium(III) across the membrane pores. In Figure 2 it can be seen that the rejections of free gadolinium(III) are decreasing with the higher pressure. It happened because the more species of free gadolinium(III) which passed through the membrane pores due to the propulsion on the feed solution is greater so that those smaller particle (in this case free gadolinium(III)) can passed through the membrane, while Gd-DTPA complex which have larger size will be retained. So, the membrane's ability to reject free gadolinium(III) are decreasing. 6 bar pressure is the most optimum pressure on the separation of free gadolinium(III) from Gd-DTPA complex because it produced the smallest rejection about 4.49%. In this study, the decreasing rejections did not indicate the concentration polarization on the membrane.

b. Effect of Temperature

Figure 3 shows the permeation fluxes as a function of a feed temperature. The permeate flux values at 25, 30, 35, and 40 ° C respectively 782.268 L/m^2.h, 847.458 L/m^2.h, 995.022 L/m^2.h, and 1129.938 L/m^2.h. In Figure 3 it can be seen that the higher of feed temperature, the permeation fluxes are increasing. This is due to the higher temperature causes the viscosity solution is decreasing [12]. The higher temperature causes the distance between the particles in the feed solution increasingly stretched so that the friction force getting smaller, consequently viscosity solution is decreasing. Decreasing viscosity solution may causing the speed of the particles (in this case free gadolinium(III)) to pass through the membrane pores become larger thus permeation fluxes are increasing.

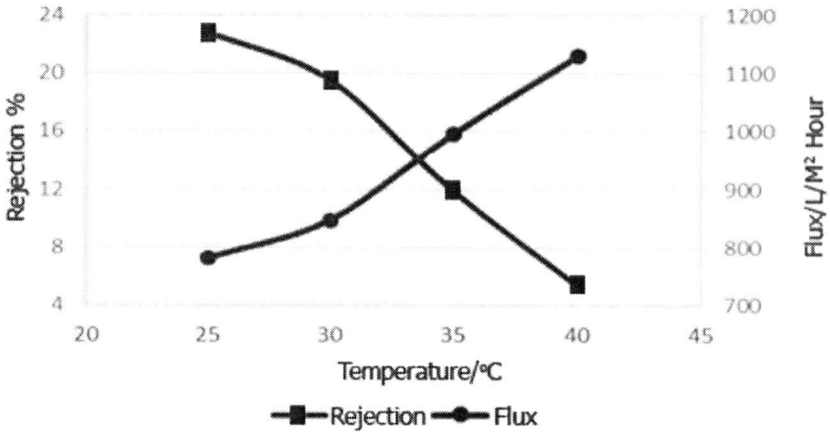

Figure 3. The effect of feed temperature on permeation fluxes and free gadolinium(III) rejections

Rejections of free gadolinium(III) at temperature of 25, 30, 35 and 40°C, respectively 22.79%; 19.53%; 11.93% and 5.42%. Figure 3 shows that the higher of feed temperature, the rejection of free gadolinium(III) are decreasing. It happened due to the pore size of the membrane getting bigger so that the more free gadolinium(III) in the feed solution that passed through the membrane pores and out as permeates thus the rejection of free gadolinium(III) are decreasing. 45°C is an optimum temperature for separation of free gadolinium(III) from Gd-DTPA complex because it produces the lowest rejection about 5.42%.

4. CONCLUSION

The higher of applied pressure and feed temperature cause permeation fluxes are increasing and free gadolinium(III) rejections are decreasing. An optimum condition for separation of free gadolinium(III) from Gd-DTPA complex have been done at pressure 6 bar and temperature 40°C where the rejection values are about 4,49% and 5,42%.

REFERENCES

[1] Chitry F, Pellet-Rostaing S, Gozzi, C, & Lemaire M 2001 *Journal of Separation Science and Technology* **36** 605
[2] Sorin A, Favre-Reguillon A, Pellet-Rostaing S, Sbai M, Szymczyk A, Fievet P, & Lemaire M 2005 *Journal of Membrane Science* **267** 41
[3] Murthy Z V P, and Choudhary A 2011 *Journal of rare earths* **29** 297
[4] Choi S, Yun Z, Hong S and Ahn K 2001 *Desalination* **133** 53

IOP Conf. Series: Materials Science and Engineering **196** (2017) 012041 doi:10.1088/1757-899X/196/1/012041

Raman scattering study of the effect of *A*- and *B*-site substitution on the room-temperature structure of $ABi_4Ti_4O_{15}$

A Prasetyo[1,2], B Mihailova[3], V Suendo[1], A A Nugroho[4], Zulhadjri[5] and Ismunandar[1]

[1] Inorganic and Physical Chemistry Research Group, Faculty Mathematics and Natural Sciences, Institut Teknologi Bandung, Jl. Ganesha 10, Bandung 40132, Indonesia
[2] Department of Chemistry, Universitas Islam Negeri Maulana Malik Ibrahim, Jl. Gajayana 50, Malang 65144, Indonesia
[3] Department Geowissenschaften, Universität Hamburg, Grindelallee 48, Hamburg 20146, Germany
[4] Physics of Magnetism and Photonic Group, Faculty Mathematics and Natural Sciences, Institut Teknologi Bandung, Jl. Ganesha 10, Bandung 40132, Indonesia
[5] Department of Chemistry, Faculty of Mathematics and Natural Sciences, Universitas Andalas. Kampus Limau Manis, Padang 25163, Indonesia

E-mail: ismu@chem.itb.ac.id; antoniaprasetyo@gmail.com

Abstract. Aurivillius-type materials exhibit promising ferroelectric and multiferroic properties that can be tailored via chemistry variations in the perovskite block. Hence, it is important to clarify the relations composition-structure, also on a local-scale level. The aim of this contribution is to give further insights into the effect of *A*- and *B*-site cations to the room-temperature local structure of Aurivillius four-layered$ABi_4Ti_4O_{15}$ (A = Sr, Pb, Ba) and $Pb_{1-x}Bi_{4+x}Ti_{4-x}Mn_xO_{15}$ (x = 0, 0.2, 0.4) by Raman scattering. The effect of *A*-site cation to the local structure of perovskite block was identified by the phonon mode near 750 and 870 cm^{-1}arising from BO_6 stretching. *A*-site Ba^{2+}, having the largest ionic radius among the considered elements, significantly stiffens the TiO_6 octahedra, as derived from the fact that the TiO_6 stretching modes have the highest wavenumber for $BaBi_4Ti_4O_{15}$, i.e. the Ti-O bond strength is strongest for this compound. The replacement of Ti^{4+} by Mn^{3+} cation at the *B*- site also influences the *B*-O bond. The comparison of the phonon modes near 700 and 870 cm^{-1} in $Pb_{1-x}Bi_{4+x}Ti_{4-x}Mn_xO_{15}$ with x = 0, 0.2, and 0.4 shows that the lowest wavenumber, which is due to the elongation of Ti-O bonds is observed for x = 0.4.

1. Introduction

Aurivillius oxides ($A_{n-1}Bi_2B_nO_{3n+3}$ with n = the number of BO_6 layers in the perovksite (ABO_3) block, $A = Ca^{2+}, Sr^{2+}, Ba^{2+}, Pb^{2+}, Bi^{+3}, Na^{+1}$, $B = Ti^{4+}, Nb^{5+}, Ta^{5+}, W^{6+}$ or Mo^{6+}) have a great potential to be used in diverse technological applications, due to their ferroelectric, magnetoelectric, thermoelectric, luminescence and photocatalytic properties [1-3]. On cooling four-layered Aurivillius compounds $ABi_4Ti_4O_{15}$ (A = Ba, Sr, and Pb) undergo a paraelectric-ferroelectric phase transition from tetragonal *I4/mmm* to orthorhombic $A2_1am$ symmetry [4]. The perovskite block of Aurivillius compounds exhibit structural flexibility, allowing for cationic substitution at both the *A* site and *B* site, which in turn gives opportunity to improve or tune the physical properties. Recently double-doped $PbBi_4Ti_4O_{15}$ with Mn^{3+}

and Bi^{3+} substituting for B-site Ti^{4+} and A-site Pb^{2+}, respectively, has been synthesized, aiming to introduce magnetic properties via partial B-site substitution of Ti^{4+} by Mn^{3+} and indeed paramagnetic properties have been reported for $Pb_{1-x}Bi_{4+x}Ti_{4-x}Mn_xO_{15}$ with x = 0.2 and 0.4 [5]. The substitution partially of Ti^{4+} by Mn^{3+} also affected to local structure of BO_6. The neutron diffraction of $Pb_{1-x}Bi_{4+x}Ti_{4-x}Mn_xO_{15}$ (x = 0, 0.2 and 0.4) showed that the introduction of Mn^{3+} to B site cation changes the bond length of B-O [5].

The X-ray diffraction (XRD) technique is commonly used to reveal the effect of A- and B-site cations on $ABi_4Ti_4O_{15}$ (A = Ba, Sr, and Pb) structure, albeit Bragg diffraction delivers information only on the long-range order, i.e. the average over ~ 10 unit cells and therefore subtle structural changes on the local level of a few unit cells can hardly be detected. Raman spectroscopy is an ideal tool for detecting local distortions in the crystal structure such as weak layer octahedron tilting, cation disorder in a fine-scale range, or local structural deviations from the global symmetry [6,7]. In this work, the local structure of $A_{1-x}Bi_{4+x}(Ti_{4-x}B''_x)O_{15}$ the different types of cations at the A and B sites (A = Ba, Sr, and Pb, B'' = Mn) is studied using Raman spectroscopy. The peak positions (ω) of all compounds are compared to deduce the chemically induced changes in bond lengths and bond angles.

2. Experiment

Polycrystalline $ABi_4Ti_4O_{15}$ (A = Ba, Sr) were synthesized via a solid state reaction of Bi_2O_3, ACO_3 (A = Ba, Ca) and TiO_2 [4]. Polycrystalline $PbBi_4Ti_4O_{15}$ (PBT), $Pb_{0.8}Bi_{4.2}Ti_{3.8}Mn_{0.2}O_{15}$ (PBTM0.2), and $Pb_{0.6}Bi_{4.4}Ti_{3.6}Mn_{0.4}O_{15}$ (PBTM0.4) were synthesized by the molten salt method [5]. Raman spectra were collected with a Horiba Jobin-Yvon T64000 triple-grating spectrometer with spectral resolution of ~2 cm^{-1} and peak-position precision of 0.35 cm^{-1}, using the 514.5-nm line of an Ar^+ laser. The spectra were temperature-reduced by the Bose-Einstein occupation factor and fitted with Lorenztians to obtain peak positions, full widths at half maximum (FWHMs), and integrated intensities.

3. Results and Discussion

According to group theory, the Raman active modes in the room-temperature phase (space group $A2_1am$) are $36A_1 + 35A_2 + 34B_1 + 36B_2$ [6]. Figure 1 shows the Raman spectra of $ABi_4Ti_4O_{15}$ (A = Pb, Ba, Sr) and $Pb_{1-x}Bi_{4+x}Ti_{4-x}Mn_xO_{15}$ (x = 0.2, and 0.4) at room temperature.

3.1 $ABi_4Ti_4O_{15}$ (A = Ba, Sr, Pb)

It is well known that the ionic-radius size of the A-site cation influences the BO_6 configuration in terms of internal octahedral distortion and octahedral tilts, which is mirrored by the tolerance factor $t=(r_A+r_O)/[\sqrt{2}(r_B+r_O)]$. The size of A-cation ionic radius in $ABi_4Ti_4O_{15}$ decreases in order Ba >Pb>Sr, which causes a decrease in t and a change in T_c [4]. Figure 2 shows the dependence of phonon wavenumbers on the type A-site cation for $ABi_4Ti_4O_{15}$ (A = Ba, Sr, Pb). The mode near 58 cm^{-1}, arising from displacements of the Bi_2O_2 fluorite-like planes relative to the perovskite-like blocks, shows no difference in the peak position. However, the modes near 40 and 100 cm^{-1}, involving displacements of A-site cations, shift towards higher wavenumbers. It is due to Sr^{2+} being lighter than Ba^{2+}. For these modes the Pb compound exhibits higher peak positions, although Pb^{2+} is heavier than Ba^{2+}. This indicates a considerably stronger Pb-O bond strength as compared to Ba-O and Sr-O, which is consistent with the fact Ba-O and Sr-O interactions have more ionic character than Pb-O and show no lone pairs in contrast to in Pb^{2+}. The influence of the type of A-site cation on the local structure of BO_6 can be deduced from the modes near 750 and 870 cm^{-1}, having the highest wavenumber for A = Ba, which means the energy of B-O stretching is highest. The symmetrical stretching BO_6 mode near 870 cm^{-1} consists predominantly of O vibrations and therefore the trend of its wavenumber with chemistry is directly related with the B-O bond strength. It is apparent that the B-O bond strength decreases with the decrease in the A-cation ionic radius and correspondingly the tolerance factor.

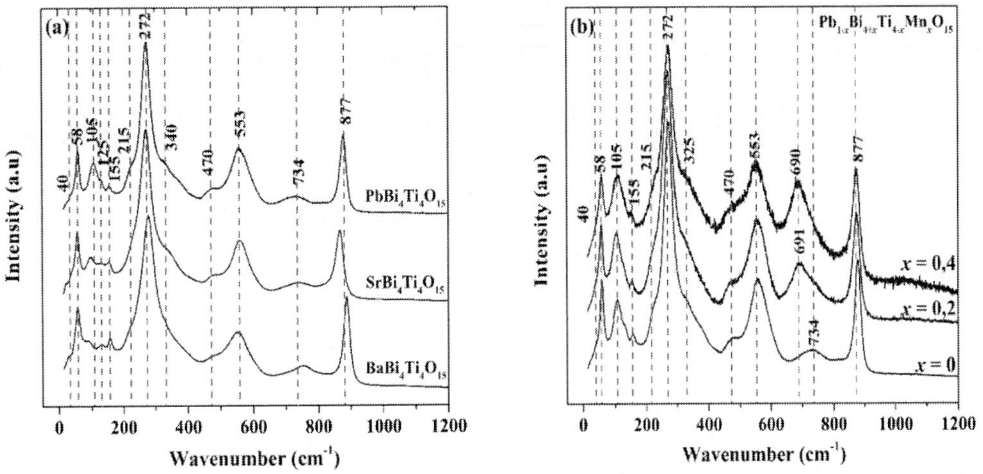

Figure 1. Raman scattering of (a) $ABi_4Ti_4O_{15}$ (A = Ba, Sr, Pb), and (b) $Pb_{1-x}Bi_{4+x}Ti_{4-x}Mn_xO_{15}$ (x = 0, 0.2, and 0.4) at room temperature.

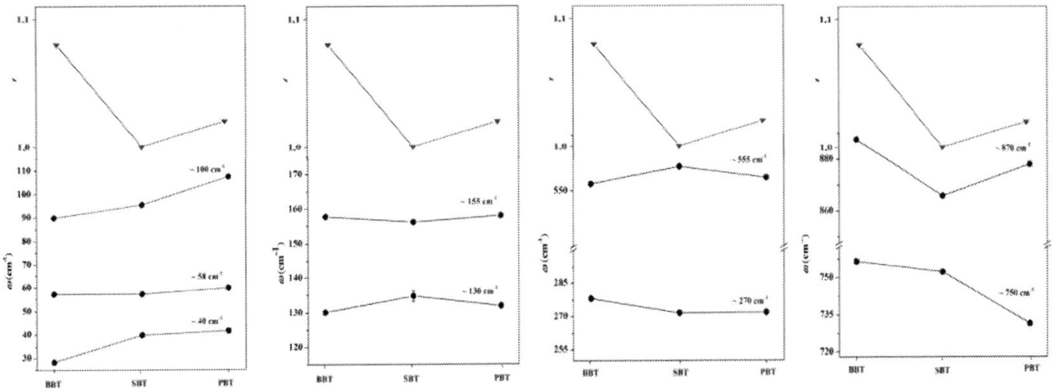

Figure 2. Peak position of phonon modes in $ABi_4Ti_4O_{15}$ (A = Ba, Sr, Pb). The red triangles in the upper part of the corresponding plot represent the trend of the tolerance factor t. (The data tolerance factor from Kennedy, *et al.* (2008)).

3.2 $Pb_{1-x}Bi_{4+x}Ti_{4-x}Mn_xO_{15}$ (x = 0, 0.2, and 0.4)

Figure 3 showed the dependence of phonon modes position on the Mn^{3+} concentration for $Pb_{1-x}Bi_{4+x}Ti_{4-x}Mn_xO_{15}$ (x = 0, 0.2, and 0.4). The peak positions of the modes near 40 and 100 cm^{-1}, which are related to Pb^{2+}/Bi^{3+} the displacements, are only subtly affected by the double doping. It is due to the fact that both the ionic radii and mass of 12-coordinated Pb^{2+} and Bi^{3+} are almost the same. The mode near 58 cm^{-1} has the same position and indicated that there is no Pb^{2+} occupied to Bi_2O_2 layer, similar to the case of $ABi_4Ti_4O_{15}$. The mode near 150 cm^{-1}, which is related to A-cation and BO_3 translation vibrations, shifts to the lower wavenumbers with the increase of doping. It is due to the Mn^{3+} with heavier mass replaced to Ti^{4+} at the B site. The results of neutron diffraction analysis of $Pb_{1-x}Bi_{4+x}Ti_{4-x}Mn_xO_{15}$ (x = 0, 0.2, and 0.4) showed that Mn^{3+} influenced mainly the local structure of BO_6 especially in the changes of bond length of B-O but there is no clear trend established for the Mn^{3+} concentration to B-O length [5].The change in the local structure of BO_6 octahedral was best mirrored by the BO_6 stretching modes near 740 and 870 cm^{-1}. Both modes shift towards lower wavenumbers, indicating a decrease in the B-O bond strength and consequent elongation of B-O bond length as a result of Jahn-Teller effect in MnO_6 octahedra.

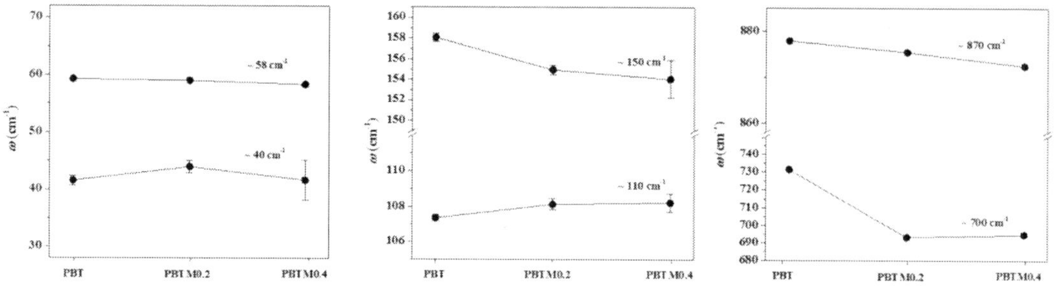

Figure 3. The peak position of external modes BO_6 on $Pb_{1-x}Bi_{4+x}Ti_{4-x}Mn_xO_{15}$ (x = 0, 0.2, and 0.4): (a) modes below 100 cm^{-1}, (b) modes at 100-200 cm^{-1}, and (c) modes at ~750, ~870 cm^{-1}.

4. Conclusion

For $ABi_4Ti_4O_{15}$ (A = Ba, Sr, and Pb), the size of ionic radius of the A cation influences strongly the BO_6 stretching modes near 750 and 870 cm^{-1}.The A-site cation with a largest ionicradius (Ba^{2+}) stiffens the BO_6 octahedra. For $Pb_{1-x}Bi_{4+x}Ti_{4-x}Mn_xO_{15}$ (x = 0, 0.2, and 0.4), the partial of Ti^{4+} by Mn^{3+} at the B site results in elongation of B-O bond.

Acknowledgement

This work was financially supported by the Desentralisasi Research Project 2014, Ministry of Research, Technology and Higher Education, Republic of Indonesia. A. A. Nugroho thanks to Ministry of Research, Technology and Higher Education, Republic of Indonesia through Riset Kerjasama Luar Negeri dan Publikasi Internasional 2016. Veinardi Suendo thanks to Penelitian Unggulan Perguruan Tinggi ITB 2016, Ministry of Research, Technology and Higher Education, Republic of Indonesia

References

[1] Eerenstein W, Mathur N D and Scott, J F 2006 Nature 442 759-765
[2] Zheng F, Li J, Chai X, Liu D, Wang X, Li Y and Yao X 2016 J. Mater. Sci. Mater. Electron. 27 7994-8000
[3] Zhou L, Jin C, Yu Y, Chi F, Ran S and Lv Y 2016 J. Alloy Compd. 680 301–308
[4] Kennedy B J, Zhou Q, Ismunandar, Kubota, Y and Kato K 2008 J. Solid State Chem. 181 1377-1386
[5] Zulhadjri, Prijamboedi B, Nugroho A A, Mufti N, Fajar A, Palstra T T M, and Ismunandar 2011 J. Solid State Chem. 184 1318-1323
[6] Kojima S, Imaizumi R, Hamazaki S, and Takashige M 1995 J. Mol. Struct. 348 37-40
[7] Prasetyo A, Mihailova B, Suendo V, Nugroho A A and Ismunandar 2015 J. Appl. Phys. 117 064102/1-6

Mechanical and molecular studies of biocomposites filled with oil palm empty fruit bunches microfibers

S Nikmatin[1], D R Saepulloh[1], Irmansyah[1], A Syafiuddin[2]

[1]Departmennt of Physics, Faculty of Mathematics and Natural Sciences, Bogor Agricultural University, 16680 Bogor, Indonesia
[2]Department of Geophysics Engineering, Faculty of Civil Engineering and Planning Sepuluh November Institute of Technology, 60111 Surabaya, Indonesia

E-mail: snikmatin@apps.ipb.ac.id

Abstract. The present work aims to investigate mechanical and molecular characteristics of acrylonitrile butadiene styrene (ABS) composites filled with oil palm empty fruit bunches (OPEFB) microfibers. OPEFB microfibers were produced using mechanical milling. Composite granules were fabricated using single screw extruder. These composites were then used for fabricating helmet according to the Indonesian National Standard (SNI). Mechanical testing confirms that the helmet produced using this biocomposites are suitable to the SNI. Molecular interaction between matrix with OPEFB can be described using orbital hybridization theory. In general, this study has successfully investigated mechanical and molecular properties of the biocomposites.

1. Introduction

Indonesia becomes the biggest producer oil palm in the world. It is estimated that Indonesia contributes 48% of the total volume of palm oil production. In 2014, its production reaches 31.5 million tons, where the quantity has stably increased for the last 20 years with 11% every year. In 2020, Indonesia's Crude Palm Oil (CPO) is estimated to reach 40 million ton. Every ton of the processed oil palm fresh fruit bunches (OPFFB) produces about 50% of the total waste, while the processed oil palm empty fruit bunches (OPEFB) produces about 23% of the total waste. These wastes became a serious problem since they were directly discharged to the environment. Utilization of these materials for various purposes is an interesting study to be explored [1-4]. Moreover, several works have been carried out to investigate their physical, chemical, and biological properties [5-9].

Furthermore, the present work aims to investigate mechanical and molecular characteristics of acrylonitrile butadiene styrene (ABS) composites filled with OPEFB microfibers. These knowledges are essential for comprehensive evaluation their composite properties in terms of macroscopic and microscopic scales.

OPEFB are widely known to contain a lot of fibers and celluloses that are potential to be explored as filler for producing biocomposite materials [2]. OPEFB's fiber is able to bind with ABS $((C_8H_8 \cdot C_4H_6 \cdot C_3H_3N)_x)$ polymer through compatibilizer in a composite structure. Binding between OPEFB fiber and ABS polymer can be explained by the molecular orbital theory. The molecular orbital theory stated that the binding molecules occured was due to the hibidisation of the orbitals for each molecules [10, 11]. When OPEFB fiber molecules placed near with the ABS molecules, the wave functions of each molecule are supepositioned linearly, forming hibrid binding having the hibrid energy levels characteristics, which principally can be found by solving the multiatomic Schrodinger equation.

2. Experimental

2.1. Materials

In this work, OPEFB were obtained from PT Perkebunan Nusantara VIII Cikasungka, Bogor. Recycled/virgin ABS was obtained from PT MUB Jaya Cibinong. In addition, additives used for manufacturing composites granules are antioxidants primer (butylated hydroxytoluene) from Mumbai, India, scavenger acid (calcium stearate) from Mumbai, India, and compatibilzer (maleic anhydride) from Darmastadt, Germany.

2.2. Filler production

OPEFB were washed to be cleaned of impurities and enumerations and then cut into chip size. Then, they were dried at 100 °C for 8 hours to reduce their water content. OPEFB were milled using milling machine at a speed of 25.000 rpm for 5 min. The results of the first milling were sieved to 20 mesh sieve in order to separate the large fibers and tiny fibers. Large fibers (retained 20 mesh) were then milled until their size were able to pass through the sieve. For the fibers passing 20 mesh were continued with a 60-mesh sieve and followed by 100 mesh sieve. While the fiber cannot pass through the sieve are reprocessed as aforementioned treatments. In the present study, fibers passing a 100 mesh, which are about 150 μm in size, were used for biocomposite fillers.

2.3. Manufacture of composite granules

Biocomposite granules were fabricated by combining filler, recycled/virgin polymer (matrix) with additives using industrial scale machine. The biocomposites were produced by extrusion using single-screw extruder machine (Model HXSJ-125/125, Kai Xin, China) with 15 wt% filler, 81.7 wt% recycled ABS polymer, and 1 wt% primary antioxidant (China), 0.3 wt% acid scavanger (Germany), and 2 wt% coupling agent (Germany). In the barrel, the samples were blended with gradient temperatures of 195, 215, 220, 220, 220, 225, 225, and 225 °C. The composition of each material is then listed in Table 1.

Table 1. Composition of biocomposites.

Code	OPEFB	Polymer	Material aditive
SN-1	Microfiber 15%	Recycled ABS	Antioxidan 1% Acid Scavanger 0.3% Maleic Acid 2%
SN-2	Microfiber 15%	Virgin ABS	Antioxidan 1% Acid Scavanger 0.3% Maleic Acid 2%

2.4. Test

ABB FTIR MB3000 model equipped with MB3000 software, Québec, Canada were employed for FTIR analysis. In this investigation, wave numbers from 450 to 4000 cm^{-1} at a resolution of 4 cm^{-1} was used. Pellets were produced by mixing biocomposites and KBr with a ratio of 1:100 using a hydraulic press at a pressure of 10 tons. The pellets were then placed in the FTIR holder and FTIR spectra were recorded. Mechanical properties of the helmet produced using this biocomposite materials were tested according to the Indonesian National Standard (SNI).

2.5. Molecular orbital theory

Molecular orbital theory states that the bonding molecules are due to hybridization of their orbital. It is obtained from linier superposition of orbital wave function. Wave function ψ can be developed from linier combination of basis function ϕ_i as follows:

$$\psi = \sum_i^n c_i \phi_i \tag{1}$$

where c_i is optimum using variation principle as a result its energy is minimum. Function ϕ_i is defined as orbital molecular.

The strength of the bond between the two molecular orbitals is affected by its molecular symmetry and type of bonding orbitals hybridize. Two types of bonding are σ and π bonds. In the bond σ, its line symmetrical electron density is concentrated along the line connecting the two atoms. For instance, Figure 1 shows the formation of the σ bond between two molecules. Moreover, the π bond is a bond where its superposition area is above and below the internuclear axis (see Figure 2).

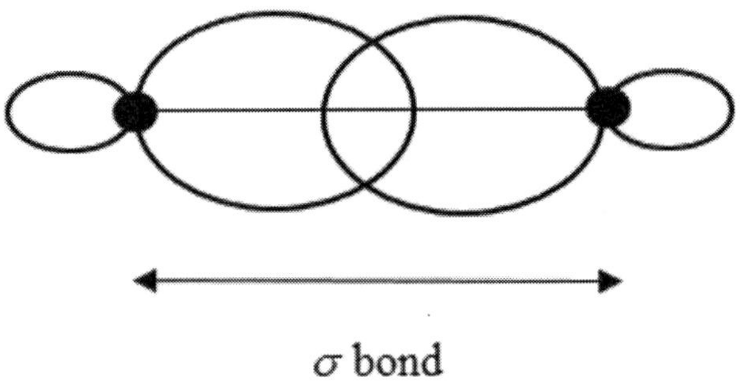

Figure 1. σ bond

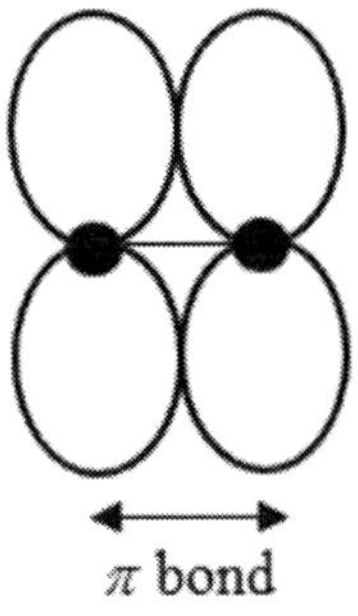

Figure 2. π bond

The number of bond that can be formed is crucially affected by the number of the valence electron at level bonding and antibonding and then can be estimated using the following equation:

$$N_{bond} = \frac{1}{2}\left(n_{eb} - n_{eab}\right) \tag{2}$$

where n_{eb} is the number of valence electron at bonding level and n_{eab} is the number of valence electron at antibonding level.

3. Results and Discussions

3.1 Biocomposite production

OPEFB fibers, recycled polymer and additives were prepared according to Table 1. Then the materials inserted into the mixing machine. Mixing was carried out to bind the material. The use of additives in this research is useful to maintain condition of the plastic/polymer used. In addition, antioxidant additive serves to prevent the oxidation reaction by oxygen that can cause degradation of the polymer. Acid scavenger acts to neutralize the acid formed during the extrusion process so that the polymer chains are not degraded. Moreover, the coupling agent serves to increase the degree of dispersion, where the coupling agent acts as the interface that affect the interfacial adhesion and interfacial tension. The Interfacial adhesion increases while the interfacial tension decreases with addition of the coupling agent.

The extrusion is a polymer material process by pressing the polymer through a mold cavity arranged magnitude. On the machine, there is a screw extruder that rotates so that the material can lead to the extrusion machine parts ranging from the feed zone and melting zone to the melt-pumping zone. In this work, single screw extruder that only has one screw is placed in the barrel. The material which is shaped like an elongated "noodles", was pulled through the water for cooling and hardening. Inside this machine, the composites were cut into small shape as shown in Figure 3.

Figure 3. Biocomposite granules

3.2 Mechanical properties

Composites granules were then used to produce a helmet using molding machine. Table 2 lists mechanical properties of the helmets. As previously mentioned, testing was conducted according to the SNI. The testing was carried out using three different temperatures. This study found that the impacts of the helmet at a temperature of -20 °C are in the range of 81.3 to 121.7. In addition, their impacts at temperatures of 15 and 50 °C are in the ranges of 84.6 to 152.1 and 113.2 to 137.9, respectively. Moreover, this work confirmed that this helmet was suitable according to the SNI in terms of its impact and head injury criteria.

3.3 Molecular Analysis

FTIR spectra of OPEFB (filler), recycled/virgin ABS (matrix), and biocomposites are depicted in Figures 4 and 5. FTIR spectra of biocomposites were dominated by the peaks of the matrix, which are 2237 cm^{-1} (N≡C) and1600-1900 cm^{-1} (C = O) of the polymer ABS as well as in the range of 670-1000 cm^{-1} ($CH = CH_2$ benzene). In addition, the peak appearing at wave number 2360 cm^{-1} and 1627 cm^{-1} are derived from the filler. Their dominance peaks from recycled ABS are due to the composition of the biocomposite is generally recycled ABS. As previously mentioned that OPEFB are 15% of the total biocomposites.

Furthermore, bond formation between cellulose molecules and ABS can be explained by the theory of orbital hybridization. There are several bonds namely single, double, and triple bonds. This is influenced by the charging of valence electrons in the level of hybridization. For instance, the bond is called hybridization sp if it involves only s and p orbitals. The amount of the bond is determined by the difference between the number of the valence electrons in the antibonding and bonding level. Understanding the formation of bonds in complex molecules is necessary to understand the process of the formation of a hybrid bond in the molecule. For example, based on the calculation of molecular orbital theory, the energy levels of C-O can be seen in Figure 6.

Table 2. Mechanical properties of the helmets.

Code	Temperature (°C)	Clash position	Paron type	The impact in units of gravity	Suitability of SNI	Head injury criteria	Suitability of SNI
SN1	-20	backside	flat paron-1	121.7	suitable	760	suitable
			flat paron -2	103.6	suitable	398	suitable
		on the top	hemispherikal paron	81.3	suitable	335	suitable
	15	backside	flat paron-1	145.0	suitable	1090	suitable
			flat paron-2	152.1	suitable	797	suitable
		on the top	hemispherikal paron	84.6	suitable	373	suitable
	50	backside	flat paron-1	137.9	suitable	799	suitable
			flat paron-2	113.2	suitable	433	suitable
		on the top	hemispherikal paron	127.4	suitable	453	suitable
SN2	-20	backside	flat paron-1	151.7	suitable	1141	suitable
			flat paron-2	133.6	suitable	662	suitable
		on the top	hemispherikal paron	86.1	suitable	334	suitable
	15	backside	flat paron-1	148.8	suitable	1090	suitable
			flat paron-2	153.6	suitable	791	suitable
		on the top	hemispherikal paron	93.7	suitable	415	suitable
	50	backside	flat paron-1	130.3	suitable	980	suitable
			flat paron-2	122.2	suitable	587	suitable
		on the top	hemispherikal paron	101.7	suitable	411	suitable

Bonding C-O involves one σ bond and two π bond (Figure 7). It has a distance of 1.1 Å so consistent with a triple bond. In comparison, a single binding C-O has a distance of 1.43 Å and double bind C-O within Average 1.23 Å. In this study, FTIR spectrum of C-O is around 1000 cm^{-1}. C-O bond was formed on the kind of p orbitals serve as bonding orbitals. From Figure 6, it can be seen that there are 4 valence electrons in the atom C, two in the 3p level and two at level s with a total molecular orbital of 4. In addition, for O, there are 6 valence electrons, 4 in 3p level and two at s lvel with total number of orbital level 4.

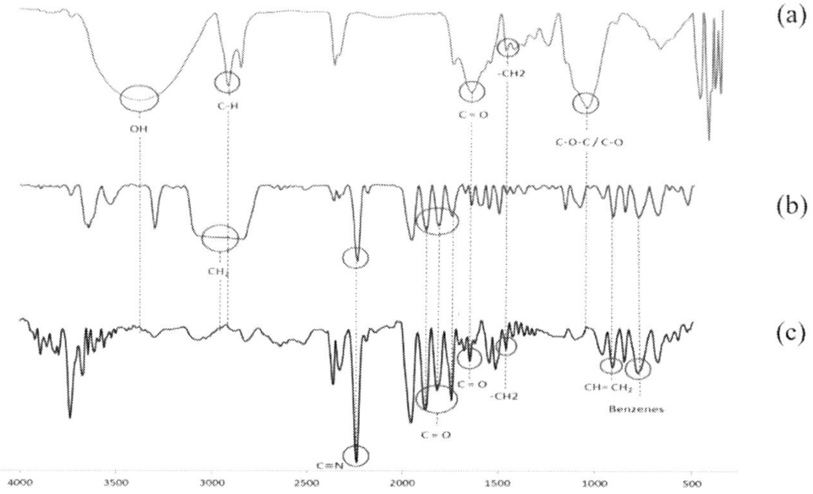

Figure 4. FTIR spectra of (a) OPEFB fiber, (b) recycled ABS, and (c) biocomposite (SN1)

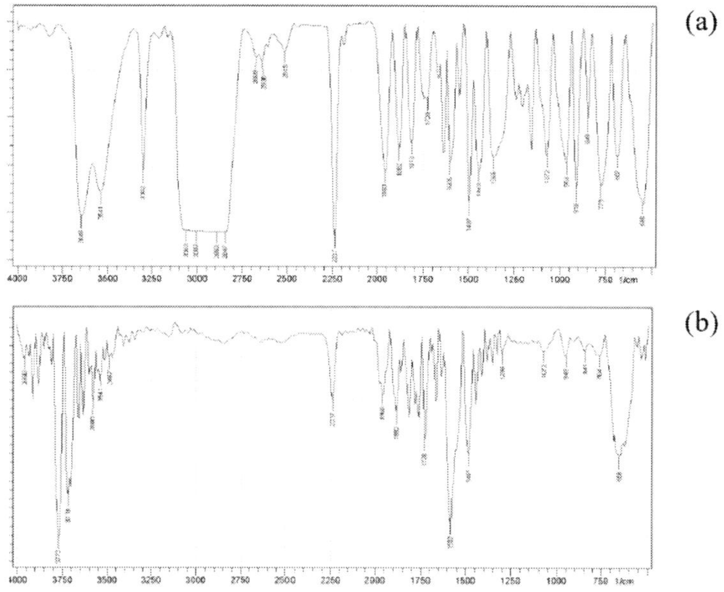

Figure 5. FTIR spectra of (a) virgin ABS and (b) biocomposite (SN2)

After hybridization level, the combined energy is 4 + 4 = 8. Explanation of bond formation for more complex molecules is necessary information about the levels of the complete hybrid. The number of hybrid orbitals level ABS-OPEFB fiber is the summation of the quantities of each molecular orbital ABS and fiber TKKS. The type and amount of the bond that is formed depend on their valence electron.

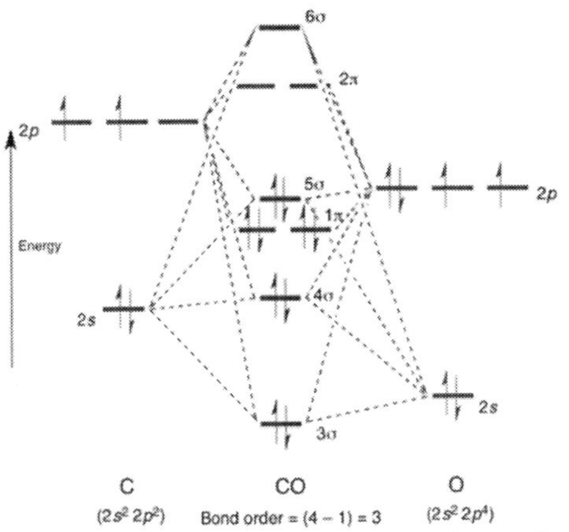

Figure 6. Energy level diagram of C-O [12]

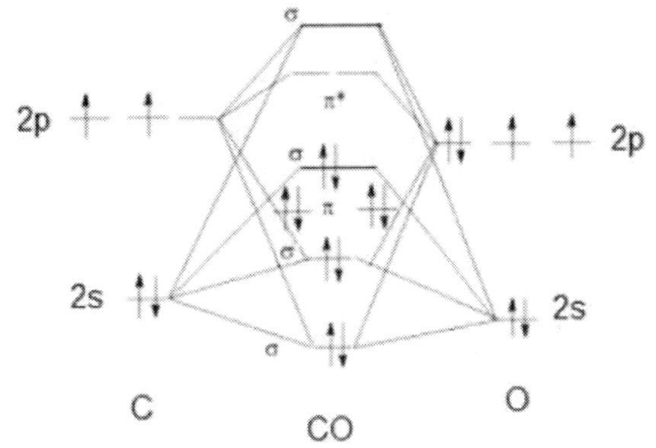

Figure 7. Molecular orbital of diagram of C-O [13]

4. Conclussion

This study aims to investigate mechanical and molecular properties of acrylonitrile butadiene styrene composites filled with OPEFB microfibers. Mechanical properties of helmets are suitable according to the SNI. FTIR spectra confirmed the presence of functional groups from filler and matric composite. Basically, the molecular interaction between the matrix with filler can be explained by the theory of the orbital hybridization. However, explanation bond formation for more complex molecules like this work needs information on levels of the complete hybrid.

Acknowledgement

This study was fully funded by the Indonesia Oil Palm Estate Fund (Badan Pengelola Dana Perkebunan Kelapa Sawit [BPDPKS]), Ministry of Finance, Republic of Indonesia, under contract no. PRJ-59/DPKS/2015.

References

[1] Khalid M, Ratnam C T, Chuah T G, Ali S and Choong T S Y 2008 *Mater. Des.* **29** 173
[2] Nikmatin S, Syafiuddin A and Irwanto D A Y 2016 *BioResources* **12** 1090
[3] Rozman H D, Saad M J and Mohd Ishak Z A 2003 *Polym. Test.* **22** 335
[4] Sreekala M S, George J, Kumaran M G and Thomas S 2002 *Compos. Sci. Technol.* **62** 339
[5] Piarpuzán D, Quintero J A and Cardona C A 2011 *Biomass Bioenerg.* **35** 1130
[6] Alam M Z, Ameem E S, Muyibi S A and Kabbashi N A 2009 *Chem. Eng. J.* **155** 191
[7] Tan I A W, Ahmad A L and Hameed B H 2009 *J. Hazard. Mater.* **164** 473
[8] Thambirajah J J, Zulkali M D and Hashim M A 1995 *Bioresour. Technol.* **52** 133
[9] Alam M Z, Mamun A A, Qudsieh I Y, Muyibi S A, Salleh H M and Omar N M 2009 *Biochem. Eng. J.* **46** 61
[10] Roothaan C C J 1951 *Rev. Mod. Phys.* **23** 69
[11] Ziroff J, Forster F, Schöll A, Puschnig P and Reinert F 2010 *Phys. Rev. Lett.* **104** 233004
[12] Molecular orbitals in carbon monoxide CO. *http://www.chemtube3d.com/orbitalsCO.htm*. University of Liverpool. Retrieved October 10, 2016.
[13] Fourier transform infrared spectroscopy of metal ligand complexes. http://archive.cnx.org/contents. Retrieved October 15, 2016.

Photonic jet μ-etching: from static to dynamic process

A Abdurrochman[4], S Lecler[1], J Zelgowski[1], F Mermet[2], J Fontaine[3] and B Y Tumbelaka[5]

[1] ICube-IPP, Université de Strasbourg, CNRS, Illkirch, Strasbourg, France
[2] IREPA LASER, Illkirch, Strasbourg, France
[3] INSA de Strasbourg, Boulevard de la Victoire, Strasbourg, France
[4] Department of Physics, UniversitasPadjadjaran, Jl. Raya Bandung-Sumedang KM. 21, Jatinangor, Sumedang 45363, Indonesia
[5] Department of Electronics Engineering, UniversitasPadjadjaran, Jl. Raya Bandung-Sumedang KM. 21, Jatinangor, Sumedang 45363, Indonesia

E-mail: a.andri@phys.unpad.ac.id, sylvain.lecler@unistra.fr

Abstract. Photonic jet etching is a direct-laser etching method applying photonic jet phenomenon to concentrate the laser beam onto the proceeded material. We call photonic jet the phenomenon of the localized sub-wavelength propagative beam generated at the shadow-side surfaces of micro-scale dielectric cylinders or spheres, when they are illuminated by an electromagnetic plane-wave or laser beam. This concentration has made possible the laser to yield sub-μ etching marks, despite the laser was a near-infrared with nano-second pulses sources. We will present these achievements from the beginning when some spherical glasses were used for static etching to dynamic etching using an optical fiber with a semi-elliptical tip.

1. Introduction

The photonic jet was coined in 2004 to designate a phenomenon in the vicinity of a dielectric particle when it was illuminated by an electromagnetic plane-waveλ. This phenomenon was observed first in the shadow-side of a micro-cylinder base on the calculation of the resulting electric field obtained through high-resolution of *finite-difference time-domain* (FDTD) computational solution of Maxell's equations [1]. For microspheres, the solution was given by implementing an exact Eigen function series solution using the *Mie theory*[3]. If λ_0 is the wavelength, a photonic jet has five important properties [4]: (1) It propagates as a non-evanescent beam beyond the particle along a length ℓ_{FWHM} that can extend more than $2\lambda_0$ with a subwavelength full-width at half-maximum (FWHM). (2) Its FWHM can be smaller than the diffraction limit, ($<\lambda_0/2$). (3) It is not a resonant phenomenon. (4) It has a high intensity that exceeds significantly the intensity of the illuminating wave. (5) Putting a nano-particle within a photonic jet will generate a perturbation of far-field backscattered power of the illuminated particle that is much higher than its initial Rayleigh scattering, which can be used to reveal the presence of nano-particles. Heifetz et.al has summarized some important applications of photonic jet such as: photonic jet microscopy, small propagation losses waveguide, photonic jet surgery, fluorescence enhancement, nano-particle detector, optical tweezers and optical data storage. This manuscript, especially, will present the application of photonic jet for micro- and submicro-etching. The achievements began with some glass spheres used for static etchings until the success of using an optical fiber with a semi-elliptical tip for dynamic etchings.

2. Parameters of photonic jet to consider for µ-etching

In our researches, we find that four parameters may be considered for photonic jet µ-etching[5]. They are:

- I_{max}: the highest photonic jet intensity generated beyond the sphere.
- f_c: the focal point of photonic jet from the highest intensity localization to the sphere surface.
- Γ_{FWHM}: the photonic jet lateral size as FWHM.
- ℓ_{FWHM}: the photonic jet length where the intensity is at a half maximum ½I_{max}.

These parameters must be adapted for the photonic jet µ-etching application.

3. Photonic jet static-etching

Many articles have suggested using microspheres for material's surface processing and especially laser direct-etching. They have reported lateral feature sizes of etching smaller than a half of the incidence wavelength ($D_{etch} \leq \lambda_0/2$), that was in ultraviolet region. These experiments are done on for different purposes, especially to improve the lateral resolution or to obtain mask-less patterning. Most of them practice a normal directive scheme to realize photonic jet static-etching (figure 1). Our preliminary works also adopted this scheme with some adaptations to meet our experiment-setup [6]. The microspheres were placed at a defocused position of the laser. We used nanosecond pulses of near-IR lasers (28 ns pulses of a 1064 nm Nd:YAG laser and 160 ns pulses of a 1060 nm YLIA laser), four different sizes of glass-microspheres (diameter of 4, 6, 24 and 35 µm) on a silicon wafer plate [6-8].

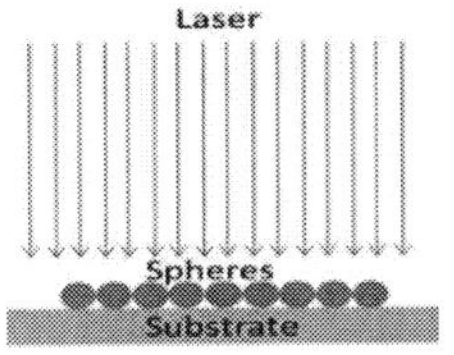

Figure 1. The normal directive laser illumination

Even though we managed to achieve sub-micron etching marks on silicon, our calculations revealed that these etching marks were not yielded by the highest intensity peak at the focal point (Figure 2). They were etched by fluence on the sample surface under the microsphere; with $F_{sur} < F_{PJ}$. Thus, to add a distance and put the material at the focal point of photonic jet will be better to allow the maximum photonic jet fluence etches the material. However, it is not an easy task with microspheres. Therefore, we have proposed applying photonic jet out of curved-tip of a waveguide. In addition to be able to control the sample-sphere distance, it may be possible to move the waveguide above the material to engrave. And so we will perform photonic jet dynamic-etching.

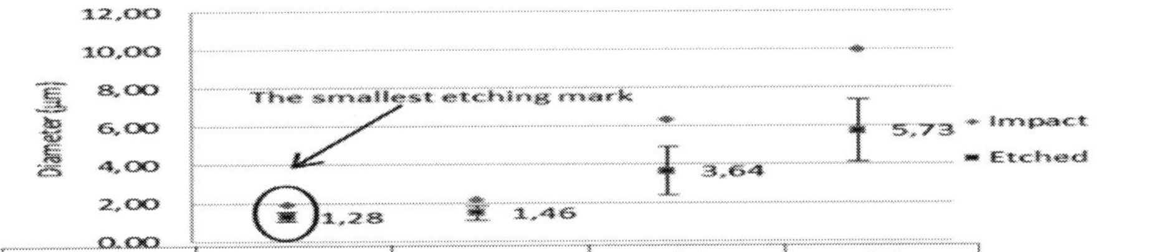

D_s (µm)	4	6	24	35	µ-sphere diameter
Γ_{FWHM} (µm)	0.7	0.6	0.7	0.8	PJ diameter
f_c (µm)	0.15	0.4	3	5	PJ focal point
F_{PJ} (J/cm²)	48	96	511	773	Max. PJ fluence at focal point
F_{sur} (J/cm²)	33	41	123	97	Max. beam fluence on the surface

Figure 2. The photonic jet static-etching on silicon is smaller than computational photonic jet diameter. They are produced by the beam on the surface which has the fluence smaller than photonic jet fluence at the focal point

IOP Conf. Series: Materials Science and Engineering **196** (2017) 012043 doi:10.1088/1757-899X/196/1/012043

4. Photonic jet dynamic-etching

There are two technical issues in applying photonic jet for dynamic etching: (1) how to generate photonic jet out of curved-tip waveguide; (2) how to select the photonic jet parameters for μ-etching. The first technical issue has been solved analytically and observed experimentally in microwave frequency [9, 10]. It is possible to generate photonic jet out of a tipped parallel plate waveguide. We have reported the analytical and experimental works concerning photonic jet out of tipped-end optical fiber [5, 11, 12], as shown in figure 3, which led to consider the second technical issue: how to select the photonic jet parameters for laser μ-etching.

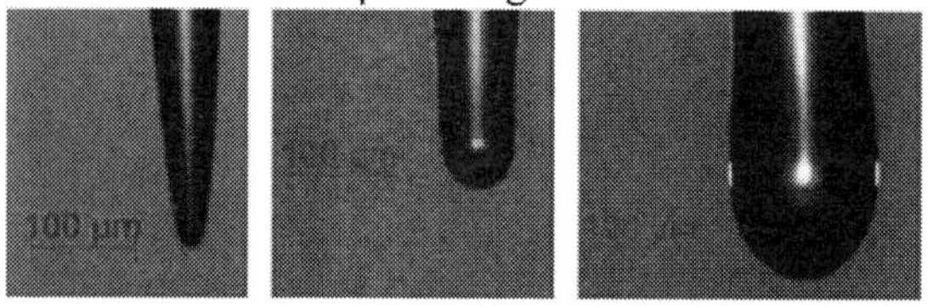

Figure 3. Different shaped tips and core sizes of fibers realized by *Lovalite©*.

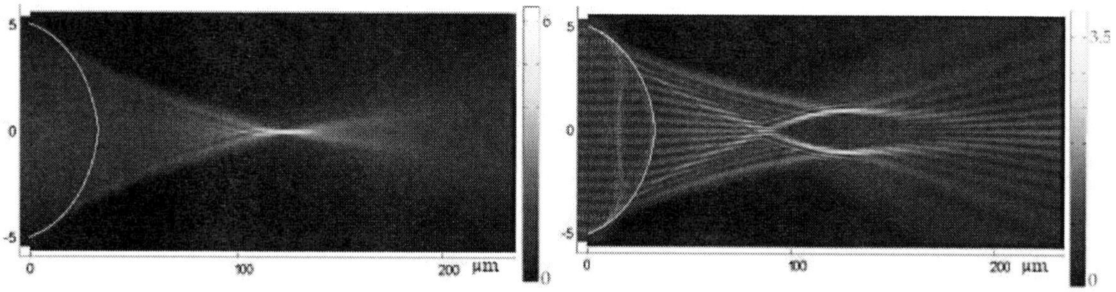

a. b.

Figure 4. Intensity maps computed around the shaped tip of the 100 μm core diameter silica waveguide in 2D with (a) the fundamental mode excitation and (b) the 19th mode excitation.

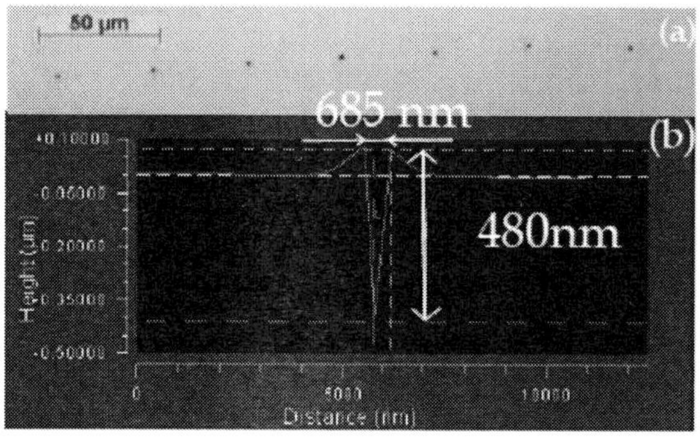

Figure 5. Etching marks on silicon made by photonic jet out of a 100 μm core optical fiber: (a) series of marks observed by an optical microscope and (b) profile measured by an optical profile-meter.

Figure 6. Grating of lines with a period of 5 μm etched on silicon by photonic jet out of a 100 μm core optical fiber.

We identify two main parameters which must be considered in order to have the optimum photonic jet parameters for μ-etching: (1) the optical fiber parameters and (2) the mode of electromagnetic wave (laser beam). We have used a finite element method with, respectively, the silica optical fiber tips, perfectly matched layer absorbing boundary condition around free space and scattering boundaries in the optical fiber cladding. Two optical fiber parameters must be selected to achieve optimum photonic jet parameters for μ-etching: the diameter of the fiber core and the shape of the tip [5]. Large core diameter has more guided electromagnetic modes; the photonic jet is concentrated further from the tip with larger beam spot (Γ_{FWHM}). Elliptical elongated tips generate higher photonic jet concentration closer to the tip than circular-shape or obtuse tip. Figure4 shows optical fibers which have been tested, with diameter respectively of 20, 50 and 100 μm, shaped tip and different curvatures. These optical fibers were placed in our experimental scheme in [5]. And we learn that the energy distribution on electromagnetic modes must be taken into account in optical fiber [5, 12]. Illustrated in figure 4, a fundamental mode is focused at the focal point, while higher mode (the 19th mode) is spread around the focal point.

Our first etching marks on silicon, with too many energy, were not as small as the photonic jet diameter in simulation (figure 5a); especially for the larger fiber core, since only the fundamental mode was taken into account in our simulations. In multimode fiber, the whole propagated energy is not used to ablate the material but is spread around the focal point as in figure 5b.

Naturally, a 100 μm core fiber is multimode at 1064 nm (around 70 groups of modes can propagate inside the fiber). Nevertheless, we made static etchings with a diameter of 685 ± 5 nm (figure6) using 35 pulses of laser VGEN ISP 1-40-30 (1064 nm, 100 ns) and 64 μJ injected in to the 100/140 all silica fiber[11]. The same fiber has been used to perform dynamic etchings, showing the interest to use optical fiber for such a photonic jet etching. A grating of lines with a period 5 μm is presented figure 6. In dynamic, due to the overlapping the line width measures around 2 μm.

5. Conclusion

The development of photonic jet static-etching processes using microspheres, and photonic jet dynamic-etching processes using shaped tip optical fiber have been reviewed and discuss, including the influence of the different parameters. We have shown that sub-wavelengthμ-etching can achieve with a nanosecond near infrared laser. It is possible to engrave lines using the overlapping of the etching-marks in line, since we used an optical fiber with three-axis motorized micro-stage.

Acknowledgments: SATT-Conectus Alsace projects of France and the scholarships of Directorate General of High Educations of Indonesia in 2011-2015.

References

[1] Chen Z, Taflove A and V. Backman 2004 *Op. Exp.* **12** 1214
[2] Li X, Chen Z, Taflove A and V. Backman 2005 *Op. Exp.* **13** 526
[3] Lecler S, Takakura Y and Meyrueis P2005 *Op. Lett.* **30** 2641
[4] Heifetz A, Kong S C, Sahakian A V, Taflove A and Backman V 2009 *J. Comp. Theor. Nanosci.* **6** 1979
[5] Abdurrochman A, Zelgowski J, Lecler S, Mermet F, Tumbelaka B Y and Fontaine J 2016 *AIP Conf. Proc.* **1712** 030001
[6] Abdurrochman A, Lecler S, Mermet F, Tumbelaka B Y, Serio B and Fontaine J 2014 *App. Op.* **53** 7202
[7] Abdurrochman A, Lecler S, Fontaine J, Mermet F, Meyrueis P, Tumbelaka B Y and Montgomery P 2014 *Proc. SPIE Phot. Europe* Ed. Mackenzie J I *etal* (Brussels) **9135** 913523
[8] Abdurrochman A, Lecler S, Mermet F, Fontaine J and Tumbelaka B Y 2014 *Int. Con. Phys.* (Yogyakarta,Indonesia) doi:10.2991/icp-14.2014.2
[9] Halaq H, Lecler S, Takakura Y, Robert S and Sauviac B 2010 *Workshop EM IEEE* (Saint-Malo, France)
[10] Ammar N, Aguili T, Baudrand H, Sauviac B and Ounnas B. 2015 *IEEE Transactions on Antennas and Propagation* **63** 4857
[11] Zelgowski J, Abdurrochman A, Mermet F, Pfeiffer P, Fontaine J and Lecler S 2016 *Op. Let.* **41** 2073
[12] Zelgowski J, Lecler S, Abdurrochman A, Pfeiffer P, Mermet F, Fontaine J and Tumbelaka B Y 2016 *SPIE-Photonic Europe* (Brussels)

Synthesis of Mesoporous $CaCO_3$ Powder using Natural Limestone as Ca Ion Source with Solution Mixing Method

Z Arifin, Q Lailiyah, M Zainuri, Darminto

Advanced Materials Research Group, Department of Physics, Faculty of Mathematics and Natural Sciences, Institut teknologi Sepuluh Nopember (ITS), Kampus ITS Sukolilo, Surabaya 60111, Indonesia.

E-mail: zaenalw18@gmail.com

ABSTRACT. In this study, the $CaCO_3$ powder has successfully been synthesized by mixing $CaCl_2$ derived from natural limestone and Na_2CO_3 in the same molar ratio. The mixing process of two solutions was performed by employing the molar contents of 0.25, 0.5, 0.75 and 1 M at varying temperatures of 35, 50 and 80°C. The produced $CaCO_3$ microparticles were characterized by scanning electron microscopy (SEM), X-ray diffraction (XRD) and Brunauer-Emmett-Teller (BET). The highest content of vaterite phase of the samples is around 96 wt%, resulted from the process using solution of 0.25 M at 35°C. All the samples prepared show mesoporous solids with pore sizes of 3-4 nm and specific surface area of 81- 90 m^2/g.

1. Introduction

The controlled synthesis of inorganic materials of specific size and morphology is a key aspect in fields as diverse modern materials, catalysis, medicine, electronics, ceramics, pigments and cosmetics [1–4]. The strategy of using organic templates or modifiers with complex functionalization patterns to control the nucleation, growth, and alignment of inorganic crystals has widely been adopted for the biomimetic synthesis of inorganic materials with complex form [4–8]. Several porous minerals have been used including synthetic zeolite [9], silica xerogel materials [10-13], porous hollow silica nanoparticle [14], porous hydroxyapatite [15], porous silica–calcium phosphate composite [16], porous calcium carbonate microparticle as a template for the assembly of polyelectrolyte layer by layer [17-18] and other porous ceramics.

In this work, we have successfully prepared the vaterite $CaCO_3$ microspheres through the precipitation reaction of $CaCl_2$ with Na_2CO_3 at molar contents of 0.25, 0.5, 0.75 and 1 M at temperatures of 30, 50 and 80 ^oC. The effect of the solution concentration and temperature on the formation of vaterite phase, morphology, porous size and specific suface area will be investigated and discussed

2. Materials and Methods

2.1. Materials

The limestone from a quarry in Tuban (East Java, Indonesia), as the starting material, was calcined for 5 hours at 900 °C. The calcined limestone was then dissolved in a solution of 10 M HCl and stirred at a constant rate of 500 rpm for 30 minutes. Furthermore, the resulted $CaCl_2$ solution was diluted to be of 0.25, 0.5, 0.75 and 1 M. Moreover, the as-received Na_2CO_3 was dissolved in demineralized water to prepare Na_2CO_3 solutions with the same molar content as $CaCl_2$.

2.2. Preparation of porous $CaCO_3$ powder

The $CaCO_3$ particles were prepared by rapid mixing of equal volume of $CaCl_2$ and Na_2CO_3 aqueous solutions with the same volume. Typically, 0.25 M $CaCl_2$ was rapidly poured into 0.25 M Na_2CO_3 solution and then agitatated by stirring with a magnetic stirrer for 2 hours at 35 ^oC. The precipitate was filtered off, thoroughly washed with aquaest, and dried in air. The course of the reaction was observed with a light microscope. Other variations of the solution is proceeded in the same way with concentrations of 0.5, 0.75 and 1 M in temperatures of 35 ^{O}C, 50 ^{O}C dan 80 ^{O}C. This simple procedure results in highly homogeneous spherical $CaCO_3$ powders.

2.3. Characterization

The powder diffraction patterns were recorded using XRD (Philips X'Pert MPD system, monochromated Cu-K_{α} radiation), employing record in the diffraction angle (2θ) range of 20 – 70°.
The morphology of powder was observed using SEM (FEI Inspect S50, at the acceleration voltage of 15 kV), by dropping the particle suspension on a glass slide, drying overnight, then sputtering with gold-palladium in a vacuum evaporator (Emitech SC7620 *sputter cooter*). The specific surface area, pore volume and size distribution of $CaCO_3$ powders were determined following the Brunauer–Emmett–Teller (BET) method of nitrogen adsorption/desorption at −196 ^{O}C with an ASAP-2010 surface area analyzer.

3. Results and discussion

The X-ray diffraction pattern of $CaCO_3$ precipitate with 0.25 M concentration at different temperature, as shown in Fig 1a. This figure shows calcite and vaterite phases at 35 - 80°C. For a comparison, we also included the powder processed at 90 ^{O}C showing an aragonite phase. The formation of vaterite phase was the highest of ~ 96 wt% and calcite phase of ~4 wt% which was achieved by the process at 35°C. Because the vaterite phase is only dominantly formed at 35 - 80 ^{O}C, then the reaction of vaterite $CaCO_3$ is only performed at this temperature with a solution concentration variation 0.25 M, 0.5 M, 0.75 M and 1 M.

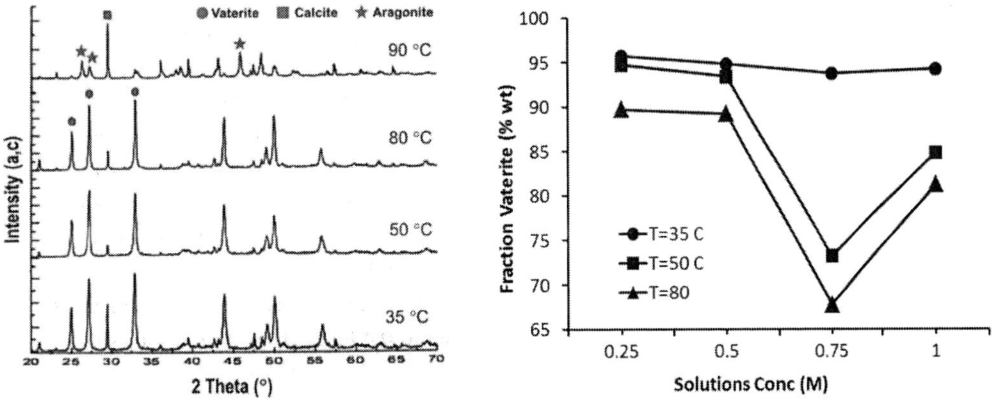

(a) (b)

Figure 1. (a) The XRD spectra of the precipitated $CaCO_3$ prepared using solution of 0.25 M at different temperatures as specified. (b) The resulted vaterite phase fraction depending on solution content of all samples prepared.

The synthesis of $CaCO_3$ microparticles at low concentration, 0.25 M and low temperature, T = 35 °C shows ~ 97 wt% of vaterite phase. The fraction of vaterite phase was decreased with the increased of concentration and temperature, as exhibited in Fig. 1b. Calsite is solid particle with higher density than that of vaterite phase. Vaterite was composed by the aggregation of nano spheroid with large porous spheroid. The saturated formation of vaterite will be reached faster in the increasing temperature. Therefore, at higher concentration and temperature, vaterite can probably be re-transformed into calcite [19]. The formation rate of vaterite from amorphous particle is higher than the transformation rate of vaterite into calcite. The formation of vaterite will occur when the concentration of bicarbonate ion (HCO_3^{2-}) in the solution is dominant. This condition was achieved by 1M solution at processing temperature of 35, 50 dan 80 °C. The Vaterite has a large surface area (porous), soluble particles and unstable thermodynamic properties. Hence, it is easy to transform into calcite with the increasing of solution concentration and temperature. At low concentration of 0.25 M, the weight fraction of vaterite phase was ~96 wt% and was decreased to 94.2% for 1 M solution. The increasing weight fraction of calcite phase was not significant and not observed in Fig. 1b. This indicates that the transformation of vaterite phase to calcite phase at 35 ^{O}C was very small.

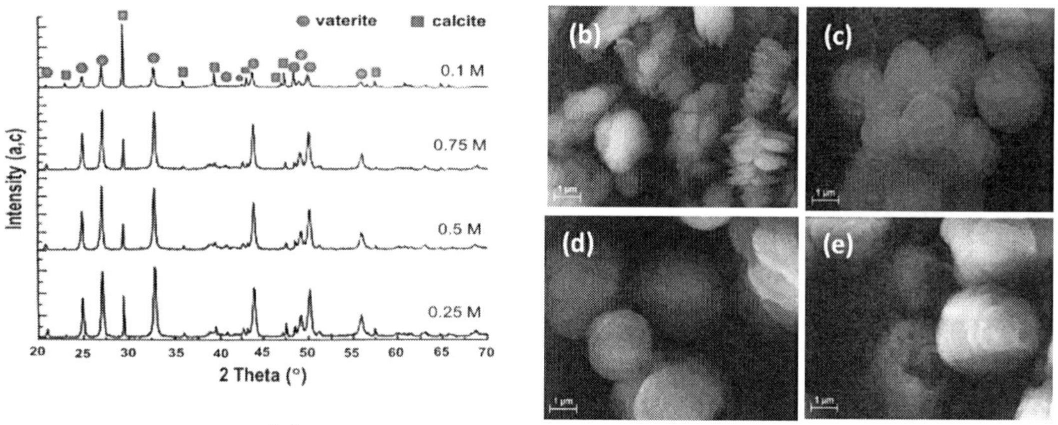

(a)

Figure 2. (a) The XRD spectra of $CaCO_3$ precipitates prepared at 35°C with solution molar content as specified. The SEM images of corresponding sampels with molar contents of : (b) 0.25 M, (c) 0.5 M (d) 0.75 M (e) 1 M.

Fig. 2a shows the XRD spectra, exhibiting the effect of concentration at 35°C on the formation of vaterite phase, which is not significant and due to the formation of $CaCO_3$ from $CaCl_2$ and Na_2CO_3 with the same molar content, 1:1. It was also confirmed by Kedra, et al. [20] and Montez, et al. [21], who reported that the size, phase and morphology of $CaCO_3$ were affected by the amount Ca^{2+}dan CO_3^{2-}ions. Moreover, the morphology of the corresponding samples is given in Fig. 2b-2e. One may observe in those figures that the particle shape for all samples is generally spherical with aggregation. The pile of spheres may construct the bigger aggregates creating pores inside.

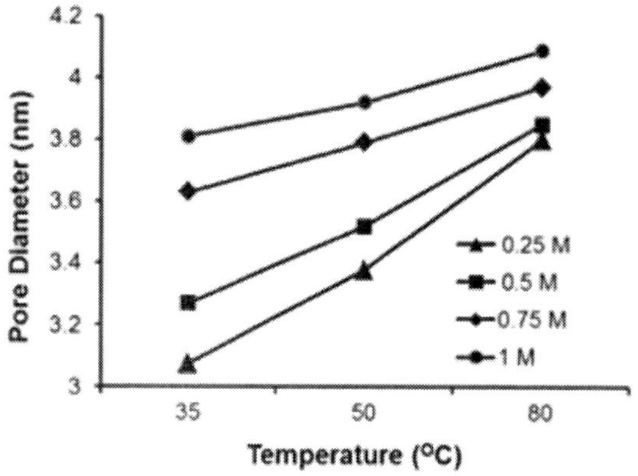

Figure 3. The pore size of $CaCO_3$ powders prepared with varying processing parameters as specified.

The resulted hollow $CaCO_3$ particles can be divided into three types, namely microporous, mesoporous, and macroporous with pore size are < 2 nm, 2 – 50 nm and > 50 nm, respectively. The BET measurement results show the pore size of $CaCO_3$ particle using solution of 0.25 – 1 M at different temperature, as shown in Fig. 3. The pore size of $CaCO_3$microparticle was obtained to be 3 - 4 nm with the specific surface area of 81– 90 m^2/g. So, the prepared $CaCO_3$ is a mesoporous powder. The pore size was increased with the increasing solution concentration and the mixing temperature.

4. Conclusion

The $CaCO_3$ powder has succesfully been synthesized utilizing natural limestone as Ca ion source employing a solution mixing method with Na_2CO_3 counterpart solution. The maximum weight fraction of the synthesized vaterite $CaCO_3$ micro-sized powders is around 96 wt%, obtained from the mixing process at 35 °C using solution with molar content of 0.25 M. The resulted $CaCO_3$ precipitate is mesoporous powder with the pore size of 3 – 4 nm and the specific surface area of 81 – 90 m^2/g.

Acknowledgements
This research was partially supported by "PUPT" Research Grant, DRPM, Kemenristekdikti, 2015 - 2016.

References
[1] Wei H, Shen Q, Zhao Y, Zhou Y, Wang D, Xu D, 2004. *J. Cryst. Growth.* **264** 424.
[2] Yu J, Yu J.C, Zhang L, Wanga X, Wu L, 2004. *Chem. Commun.* **24** 14.
[3] Yu S.H, Cölfen H, Xu A.W, Dong W, 2004. *Cryst. Growth.* **4** 33.
[4] Hosoda N, Sugawara A, Kato T, 2003. *Macromolecules.* **36** 6449.
[5] Zhang S, Gonsalves K.E, 1998. *Langmuir.* **14** 676.
[6] Shen Q, Chen Y, Wei H, Zhao, Wang D, Xu D, 2005. *Cryst. Growth.* **5** 1387.
[7] Guo Y, Yang L, Yang X, Zhang X, Zhu S, Jiang K, 2003. *Macromol. Biosci.* **3** 163.
[8] Sugawara A, Ishii T, Kato T, Angew J, 2003. *Chem. Int. Ed.* **42** 5299.
[9] Fisher K.A., Huddersman K.D, Taylor M.J, 2003. *Chem. Eur. J.* **9** 5873.
[10] Suzuki K, Yumura T, Tanaka Y, Akashi M, 2001. *J. Control.* **75** 183.
[11] Kortesuo P, Ahola, M, Kangas, M, Leino, T, Laakso, S, Vuorilehto, L, Yli-Urpo, A, Kiesvaara, J, Marvola, M, 2001. *J. Control.* **76** 227.
[12] Otsuka M, Tokumitsu K, Matsuda Y, 2000. *J. Control.* **67** 369.
[13] Charnay C, Begu S, Tourne-Peteilh C, Nicole L, Lerner D.A, Devoisselle J.M, 2004. *Eur. J. Pharm. Biopharm.* **57** 533.
[14] Li Z.Z, Wen, L.X, Shao, L, Chen, J.F, 2004. *J. Control.* **98** 245.
[15] Kim H.W, Knowles J.C, Kim H.E, 2004. *Biomaterials.* **25** 1279.
[16] El-Ghannam A, Ahmed K, Omran M, 2005. *J. Biomed. Mater. Res. B: Appl. Biomater.* **73** 277.
[17] Volodkin D.V, Larionova N.I, Sukhorukov G.B, 2004. *Biomacromolecules.* **5** 1962.
[18] Volodkin D.V, Petrov A.I, Prevt M, Sukhorukov G.B, 2004. *Langmuir.* **20** 3398.
[19] Han Y.S, Hadiko G, Minoru Takahashi, 2006. *Journal of Crystal Growth.* **289** 269
[20] Kedra-Krolik K, Gierycz P, 2009. *of Cryst. Growth.* **311** 3674.
[21] Montes-Hernandez, Renard F, Geoffroy N, Charlet, Pironon J, 2007. *J.Cryst.Growth.* **308** 228.

Effect of Mould Coating on Skin Formation and Nodule Characteristics of Thin Wall Ductile Iron Casting

D Dhaneswara[1], B. Suharno[1], A Aprilio[1], R D S Ariobimo[2], N Sofyan[1]

[1]Department of Metallurgical and Materials Engineering, Faculty of Engineering, Universitas Indonesia, Depok, 16424 Indonesia
[2]Mechanical Engineering Department, Faculty of Industrial Technology, Trisakti University, Jakarta, Indonesia

E-mail: donanta.dhaneswara@ui.ac.id

Abstract. Thin wall ductile iron (TWDI) has the potential alternative for lightweight aluminium use in automotive parts. The main problem in TWDI, however, is the formation of skin during the casting, which may reduce its mechanical properties. This casting skin is formed by the decomposition of nodular graphite at the mould interface during the casting process. One of the ways to work around this problem is by using mould coating to control the cooling process. In this work, three variables of mould coatings were used, i.e. graphite, MgO, and MgO/graphite double layers. The results showed that the average casting skin thickness in double layer coating was the lowest (30.41 μm), 57% lower than that of in MgO (71.46 μm) and 60% lower than that of graphite (74.44 μm). The reduction of casting skin thickness increased the mechanical properties of TWDI (346 MPa), 69% higher than that of MgO (223 MPa) and 26% higher than that of graphite (297 MPa). The same is true for ductility (2.7%), which was higher than that of MgO (1.43%) and that of graphite (1.43%).

1. Introduction

Thin wall ductile iron (TWDI) is produced due to the need for a light weight material but with low cost consideration. In this case, the thin wall casting method makes the possibility for ductile iron to compete with the light weight of aluminium [1]. Compared with aluminium, the ductile iron has several advantages of high yield strength, low production energy and thus low production cost [2].

The appropriate technology in making automotive parts using TWDI is by using thin wall casting method. The TWDI casting only needs simple production process and tools, which are affordable for low budget manufacturers, however, the casting thickness standard for this thin wall casting has not been established. The main problem frequently found in TWDI is the formation of skin during the casting process, which significantly reduces the mechanical properties of TWDI [3].

Casting skin is a graphitic layer found at the TWDI surface. This phenomenon is caused by several factors, i.e. inadequate processing temperature, inappropriate carbon equivalent value, and inefficient liquid treatment during the casting process [4]. This skin formation is formed in regular cast iron and is normally eliminated by machining process, however, this machining process would not be possible when the casting product is very thin of only 1 to 10 mm thickness.

In this work, the effect of mould coating on the formation of casting skin and nodule characteristics of thin wall ductile iron by using three variables of mould coatings, i.e. graphite, MgO, and MgO/graphite double layers is reported and discussed.

IOP Conf. Series: Materials Science and Engineering **196** (2017) 012045 doi:10.1088/1757-899X/196/1/012045

2. Experimental

The casting method used in this work was based on the previous work [5]. The plate design was 150 x 75 mm with 1 mm thickness, as can be seen schematically in Figure 1. The mould was made of furan sand and the metal cast was 1400 kg of ductile iron grade FCD 45 scrap.

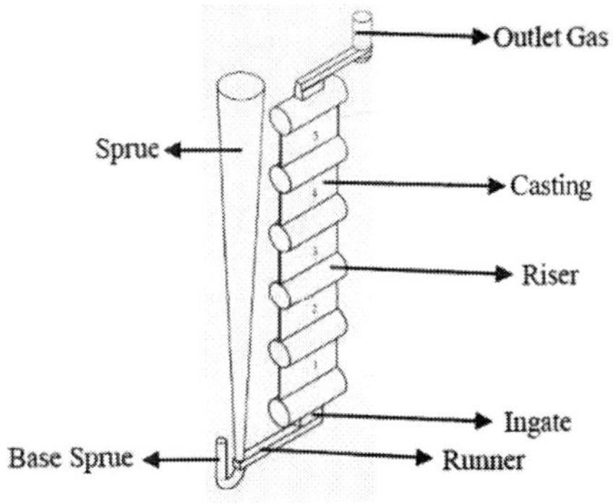

Figure 1. Schematic of the mould template

Mould coatings preparation began with the mixing of coating materials with ethanol, specific gravity measurement, manual brush of the mould, and solvent evaporation. The coating substance was poured into a bucket contained ethanol and mixed to achieve a value of 55 in Baume scale. The mix was applied on the mould surface using a manual brush after the furan sand cured perfectly. The applied coating was then heated to vaporize the solvent from the mixture and the coating would attach perfectly on the mould surface. These steps were applied on all samples. Slightly different with sample C, which has double-layer coating, MgO coating was firstly applied on the mould surface. After the MgO was set in, graphite coating was then applied on top of the MgO layer using the same method. Figure 2 shows the grey coloured of the graphite coating layer was firmly attached on top of creamy coloured of MgO coating layer.

Chemical composition check was conducted before and after liquid treatment. Nodulizing agent was 7.7 kg of FeMgSi-6% with some amount of S70-type inoculant in the liquid treatment process. Different mould coatings were used, i.e. graphite (Sample A), MgO (sample B), and double-layer of graphite/MgO (sample C).

Figure 2. Visualization of the mould after the application of double-layer coating

3. Results and Discussion

Chemical composition of the molten metal examined by using spectrometry is given in Table 1. Based on this composition, the carbon equivalent before and after liquid treatment was calculated to be 4.1625 and 4.4683, respectively, whereas holding and pouring time of the molten metal into the mould is given in Table 2. The pouring process was at 1406 °C performed in sequence started from sample C, sample B, and sample A. The FCD 45 metal was poured and was used as the same source as of liquid treatment batch to eliminate the difference in chemical composition and carbon equivalent (CE). There was an occurrence of the pouring time in sample B in which after the pouring of molten metal had finished and the molten metal was seen filled the mould perfectly, while the mould A was still in pouring condition the molten metal level in mould B dropped down. The molten metal in mould B was then refilled after the pouring process of mould A had finished. The holding and pouring time for all the samples are given in Table 2.

Table 1. Chemical composition of the molten metal before and after liquid treatment

Element	Before	After	Standard
C	3.6933	3.6964	3.50 – 3.90
Si	1.6294	2.6003	2.40 – 2.80
Mg	0.0000	0.0841	0.03 min
Mn	0.4866	0.4214	0.30 – 0.50
P	0.0085	0.0098	0.03 max
S	0.0232	0.0200	0.02 max
Cu	0.3732	0.1413	0.15 max
Cr	0.0042	0.0142	0.15 max
Ni	0.0022	0.0032	0.15 max
Al	0.0033	0.0106	0.03 max

Table 2. Holding and pouring time

Sample	Holding Time	Pouring Time
A	163 seconds	11 seconds
B (unfinished)	142 seconds	12 seconds
B (refilled)	185 seconds	7 seconds
C	115 seconds	15 seconds

Figure 3 shows layer of the mould coating schematically. In this figure, the double-layer coating is used in sample C. As can be seen in the figure, the compound in the layer will react with the sulphur existed in the sand at the mould surface, and hence reducing the sulphur content on the mould interface. Meanwhile, the graphite coating layer, which is in direct contact with the molten metal, would function as local carburizer without any interruption from the sulphur at the molten metal interface and could prevent graphite depletion phenomenon, and thus reducing the casting skin thickness.

Figure 4 shows that the double layer coating method in sample C has proven to reduce casting skin thickness of up to 60% when compared to that of sample A, which used standard graphite coating only. At the same time, it is seen that sample B, which used single layer MgO coating only, seems to have the most skin formation. The anomaly of casting thickness in sample B is expected to be due to the pouring method described previously. The delayed solidification in sample B could make the inoculant in the molten metal faded away and thus causing severe skin formation.

Figure 5 shows the nodule count formed in each sample. The highest nodule count was found in sample C followed by sample B and sample A with the least nodule count. There is a trend that this nodule count is pertinent to the thickness of casting skin formation. As can be seen in the figure, the thicker the casting skin, the less the nodule count. Sample C has an average 2014 nodule count per mm^2, followed by B with 1900 and A with 1671 nodule count. The 60% difference of casting skin thickness of sample C from sample A shows that reduced casting skin of sample C is related to the increase of nodule count in sample C 20% more than nodule count of sample A. In this instance, the use of mould coating affects the nodule count as the mould coating helps to reduce the casting skin formation, which correlated to the number of nodules formed in the metal.

IOP Conf. Series: Materials Science and Engineering **196** (2017) 012045 doi:10.1088/1757-899X/196/1/012045

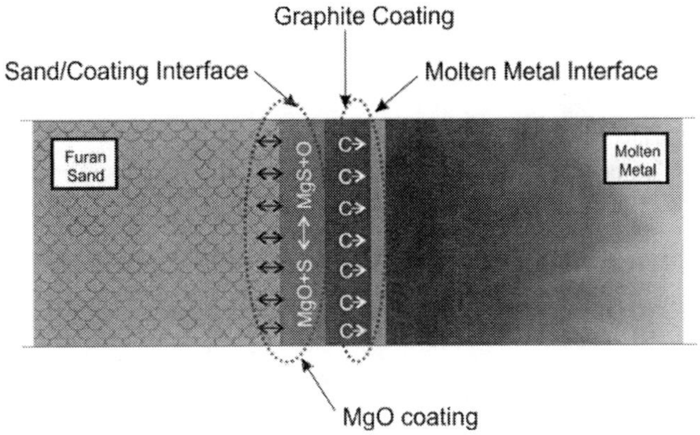

Figure 3. Schematic of double layer mould coating mechanism

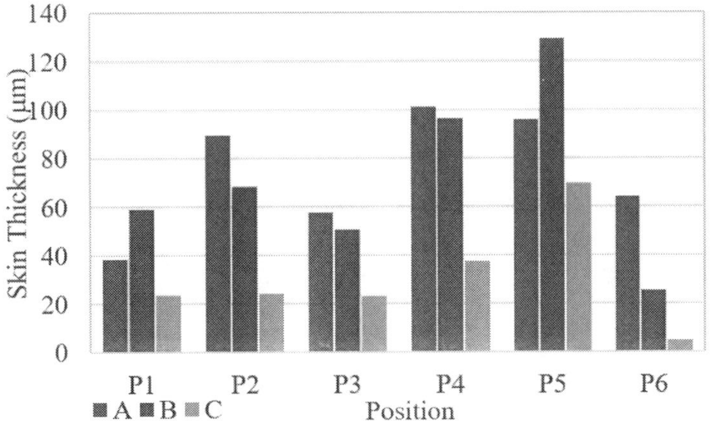

Figure 4. Skin thickness formation at different casting positions

Figure 6 shows the nodularity of each specimen. Sample B has the highest nodularity, followed by sample C and sample A with the lowest nodularity. In average, nodularity in sample B has the highest value because the single layer mould coating of MgO directly functions as a late nodulizer [3]. The MgO coating at the molten metal interface will react with the molten metal as a nodulizer, which take place inside the mould. This will then make the nodule to form by increasing the residual amount of Mg content in molten metal during solidification process. Sample C also has the MgO coating, however, the MgO in sample C cannot react directly with the molten metal, instead it will function as sulphur reducer at sand/coating interface, making the nodularity in sample C is slightly lower than that of sample B. Table 3 shows the tensile test results of the TWDI samples of three variables. The ultimate tensile strength of sample C has the highest value, 26% higher than that of sample A and 69% higher than that of sample B. The same is true for the elongation, in which sample C is 93% higher than that of sample A and sample B.

These results were supported by the secondary electron images of the fracture area in every sample as can be seen in Figure 7. All samples show dimple fractures on the surface as an indication of ductile characteristic. The characteristic differences among these samples are in nodule count, nodule size and shape, and distribution of the nodules.

Surface fracture in sample A shows the characteristics of a dimple indicating a ductile fracture. The nodules distribute evenly, but some nodules formed in sample A looks larger than the other two samples. The size of the nodules is about 50 micrometres, and is of a primary graphite form. The number of nodules formed in the sample A are fewer compared to that of in the sample B and sample C.

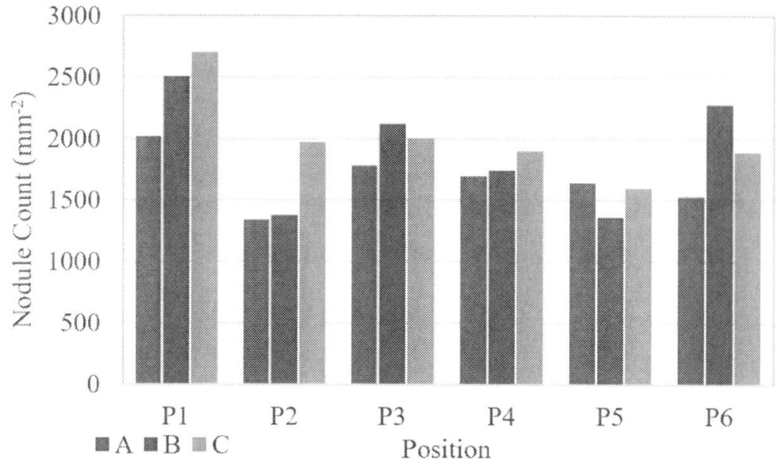

Figure 5. Nodule count at different casting positions

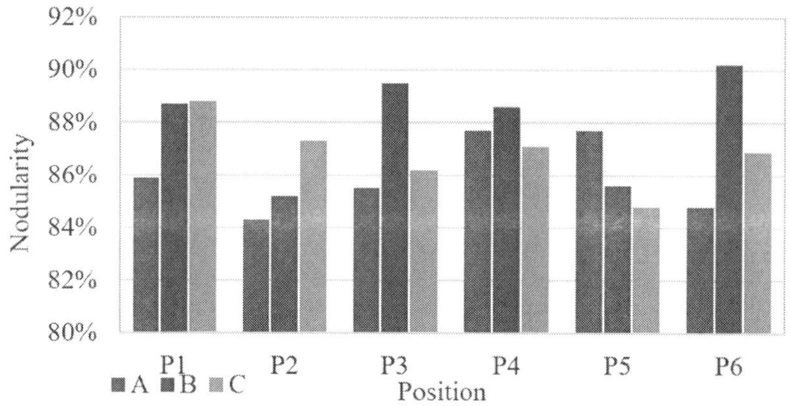

Figure 6. Nodularity at different casting positions

Table 3. Tensile test results

Sample	Strength (MPa)	Elongation (%)
A	297	1.43
B	223	1.43
C	376	2.76

More nodules are found in B compared to that of in sample A. Nodules overall diameter in sample B is larger than that of sample A. The nodules in sample B has much irregularities in shapes, which could be caused by the less amount of residual inoculant that affected the nodule characteristics. This is consistent with quantitative observations nodule diameter discussed earlier.

In sample C, surface fracture has a dimple characteristic and the nodules spread evenly. There are more number of nodules found in C compared to that of in the sample A and sample B but smaller in size. These characteristics are the optimum preferences for thin wall ductile iron. This finding is also in agreement with the qualitative and quantitative data discussed earlier. This fact is also supported by the data from ultimate tensile strength (UTS). Referring to JIS standard of elongation of ferritic ductile iron, the above results are certainly still very far from the standard elongation of 10%. This is due to the persistence of the skin layer contained in the casting, the anomalies graphite such as graphite primer and exploded graphite, carbide formation.

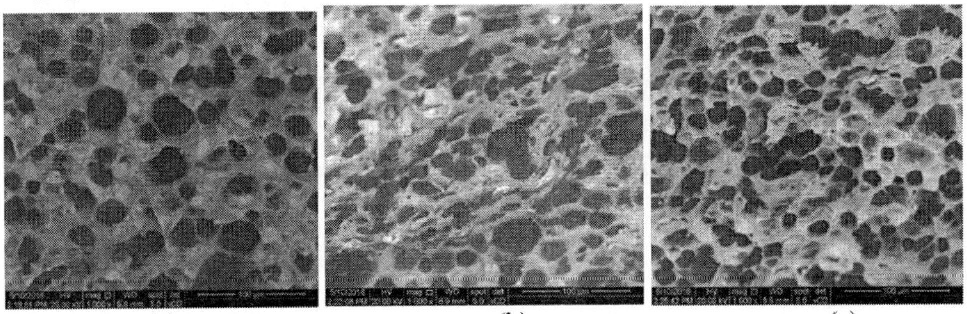

(a) (b) (c)

Figure 7. Secondary electron image from tensile test fracture surface (a) sample A, (b) sample B, and (c) sample C. Bar scale is 100 microns

4. Conclusion

In this work, the casting skin thickness in double-layer coating of MgO and graphite is of 30.41 μm, 57% thinner than that of using single-layer of graphite only (71.46 μm) and 60% thinner than that of single-layer of MgO only (74.44 μm). Double-layer mould coating efficiently reduces casting skin thickness by reducing sulphur content at the mould surface and stabilized the carbon content of the molten metal inside the mould. The nodule count was high when the casting thickness was formed at the lower value. Nodularity of the sample with single-layer of MgO only was the highest (87.97%), compared with the sample with single-layer of graphite only (85.98%) and double-layer of MgO and graphite (86.85%), indicating that the MgO directly functions as late nodulizer. Tensile strength of the sample from mould coating with double-layer was the highest (376 Mpa), 69% higher than that of sample with single-layer MgO only (223 MPa) and 26% higher than that of the sample with single-layer of graphite only (297 MPa). Elongation of the sample with double-layer (2.76%) was higher than that of the sample with single layer of graphite only (1.43%) and the sample with single layer of MgO only (1.43%).

Acknowledgement

The authors would like to thank DRPM Universitas Indonesia for the research grants through Hibah PITTA No: 2080/UN2.R12/HKP.05.00/2016.

Reference

[1] Javaid A 1999 *AFS Transactions* **107** 441
[2] Stefanescu D M, Dix L P, Ruxanda R E, Crobitt-Coburn C and Piwonka T S 2002 *AFS Transactions* **110** 1149.
[3] Auferheide R C, Showman R. E, Hysell M A 2005 *AFS Transactions* **113** 567.
[4] Dix L P, Ruxanda R, Torrance J, Fukumoto M and Stefanescu D M 2003 *AFS Transaction* **107** 1149.
[5] Stefanescu D, Wills S, Massone J, Duncan F 2009 *AFS Transactions* **117** 687.

IOP Conf. Series: Materials Science and Engineering **196** (2017) 012046 doi:10.1088/1757-899X/196/1/012046

Time of Growth Dependent of ZnO-Nanorods by Self-Assembly Methods and its Structural Properties

A Aprilia, A Bahtiar, L Safriani, Ayunita C C, Afifah N, N Syakir, Risdiana, T Saragi, S Hidayat, Fitrilawati and R E Siregar

Department of Physics, Faculty of Mathematics and Natural Sciences, Universitas Padjadjaran, Jl. Raya Bandung-Sumedang KM 21. Jatinangor, 45363, West Java, INDONESIA

E-mail: a.aprilia@phys.unpad.ac.id

Abstract. ZnO-nanorods (ZnO-Nrs) have been successfully prepared on glass substrate using self-assembly method by varying deposition time. Zn acetate dehydrate and 2-methoxyethanol was used as raw material and solvent respectively (for ZnO seed layer preparation), meanwhile Zn nitrate hexahydrate and hexametylenetetramine (HMTA) dissolved in deionized water used as solution growth of ZnO-Nanorods (ZnO-Nrs). In this work, deposition times of ZnO-Nrs were varied by 120 min, 150 min, 180 min and 210 minutes at 100°C of annealing temperature. In order to investigate the physical properties of resulting ZnO, several measurements such as x-ray diffraction (XRD), ultra-violet visible spectroscopy and scanning electron microscopy (SEM) were carried out. Based on ZnO nanorods SEM image with varying time depositions, seems that increasing deposition time the nanorod size decrease but when the time reach 210 min, the average size of nanorods turned back increase. From XRD measurement, the average grain size and lattice constant (*c*) which is assemble the nanorod structure and lattice constant (*c*) was determined by Debye-Scherrer formula and Bragg law's respectively. The growth process of ZnO nanorod by 180 min time deposition was known as an appropriate time to produce ZnO nanorods with high crystal quality due to sharp peak intensity of XRD spectrum.

1. Introduction

Now days, metal oxide materials have been attractive many researchers according to their semiconductor properties and board application in optoelectronic devices. Titanium dioxide (TiO_2), Vanadium oxide (V_2O_3), Molybdenum trioxide (MoO_3) and Zinc Oxide (ZnO) were commonly used in many kinds of electrical devices such as thin film transistor, gas-chemical-biomelecular sensors, light emitting diode, ultraviolet (UV) detector, and solar cells [1]. Besides offer the uniquely properties, the structure of these metal oxides could be easy controlled as needed by using various preparation methods such as vapor based deposition (metal oxide chemical vapour deposition-MOCVD), pulsed laser deposition (PLD), epitaxy and solution-chemical based deposition. Among these methods, the simplest one is using solution-chemical based deposition such as sol-gel method, spray-pyrolysis, hydrothermal, and also self-assembly.

ZnO, one of the semiconducting material, has attracted considerable in optoelectronics field because of its wide bandgap (~ 3.4 eV), wide resistivity range (10^{-4} – 10^{12} Ω.cm) high exciton binding energy (~ 60 meV), and already claimed show a long-term stability [2, 3]. Many variations of ZnO micro/nano structured have been developed and already reported as photoelectrode in light based devices, particularly in solar cells technology [4]. ZnO nanoparticles as electron transport material in organic or hybrid solar cell has been reported in past ten years [5,6]. Besides that, this metal oxide also generally used in emerging photovoltaic devices, such as perovskite solar cell and dye sensitized solar cells [7,8]. Silicon solar cell technology also use this kind metal oxide as anti-reflectance coatings in order to enhance solar irradiation adsorption by using special nano-structure moth eye facet of ZnO which is shows more than 40% increments on short circuit current and conversion efficiency [9]. The nanostructured of ZnO in photoelectrode or photoanode system for solar cell technology, played an important role in charge carrier collection and drive the recombination rate. The appropriate nanostructures such as nanorod, nanowire, nanocone, nanobelt or even nanoforest of ZnO have been claimed could be increase the solar cell performance according to high efficiency in charge transport and reduce the recombination rate [10].

In general, ZnO-nanorods growth using solution based preparation and thermal treatment has been well reported [11]. Kyung et.al. prepared the ZnO-nanorod structure using chemical solution deposition and shows that the annealing temperature influences the alignment and length of the resulting nanorod. It was also reported that ZnO seed layer properties such as crystallinity, surface morphology, thickness and preparation condition provide high influence on ZnO nanorod growth [3]. In contrary, D. Polsongkram et.al claimed that their procedure by using hydrothermal method allows the growth of ZnO nanorods without any seed and/or surfactant but they also imply that the precursor concentration and deposition time have a big influence on the morphology and ordering ZnO nanorods [12]. In this work, the ZnO with nanorods structure were prepared using an easy and low-cost self-assembly growth technique at certain annealing temperature and variation of deposition time.

2. Experimental Methods

All chemicals used in this study were analytical grade without further purifications. Zinc acetate dehydrate ($Zn(CH_3COO)_2.2H_2O$) and zinc nitrate hexahydrate($Zn(NO_3)_2.6H_2O$) purchased from Merck used as raw material for ZnO seed layer preparation and ZnO nanorods growth solution respectively. Two layers of ZnO seed layer were deposited on glass substrate using spin coating technique at 1500 rpm. Before used glass substrate was cleaned in ethanol and acetone by ultrasonic cleaner equipment. ZnO seed layer precursor made from 0.5 mM zinc acetate dehydrate in 2-methoxyethanol anhydrous (Sigma-Aldrich) and diethanolamine (Sigma Aldrich) as a solvent and stabilizer respectively. A transparent precursor solution was achieved and continued with deposits it onto cleaned glass substrate. Two step annealing treatment at 250°C and 500°C was followed to forming the ZnO particles as seed layer. These all methods following the procedure from our previous research [5].

ZnO nanorods (ZnO-Nrs) were grown by self-assembly method in isolated weighting bottle, with glass substrate/ZnO seed layer side was facing down at 45°. The growth solution was prepared by dissolving equimolar zinc nitrate hexahydrate and hexametylenetetramine in deionized water. Growth temperature was fixed at 100°C for 120 min, 150 min, 180 min and 210 minutes inside the regularly laboratory oven. After that, ZnO nanorods was washed in ethanol and deionized water several times and then annealed at 500°C for 30 min on isolated hotplate.

The scanning electron microscopy (JEOL JSM-6510A) was used to identify the surface morphology of different time of growth ZnO-Nanorods. In order to study the absorption spectrum during deposition time, ultra-violet visible spectroscopy (T70+ PG Instrument) was carried out for each sample. To confirm the nanorod structure, the equipment of X-pert Pro, PW3050/60 x-ray diffractometer was used and continued with data analysis to determine crystals lattice of ZnO-Nrs by Debye-Scherer formula.

3. Results and Discussion

Figure 1 shows the morphology of ZnO-Nanorods (ZnO-Nrs) resulted that prepared with self-assembly method at 100°C annealing temperature by 120 min, 150 min, 180 min, and 210 minutes' time deposition. Up to 180 min time deposition the average diameter seems decrease but back to increase at 210 min. It is clear that ZnO forms hexagonal wurtzite but unfortunately the diameter size shows a broad range. For 120 min deposition time the average of diameter size is about 100 nm-200 nm, 150 min is 100 nm-160 nm, 180 min is 80-140 nm and for 210 min is 100 nm-200 nm (figure 3(b)). Although the range of diameter size not significantly differ between each samples this images show that increasing time deposition the surface of ZnO-Nrs growth denser. In figure 1 the lower time deposition (120 min) shows a large empty space compare with 180 min (figure 1 (c)) and 210 min (figure 1(d)) however the hexagonal structure for sample with 210 min time deposition appears to be defected.

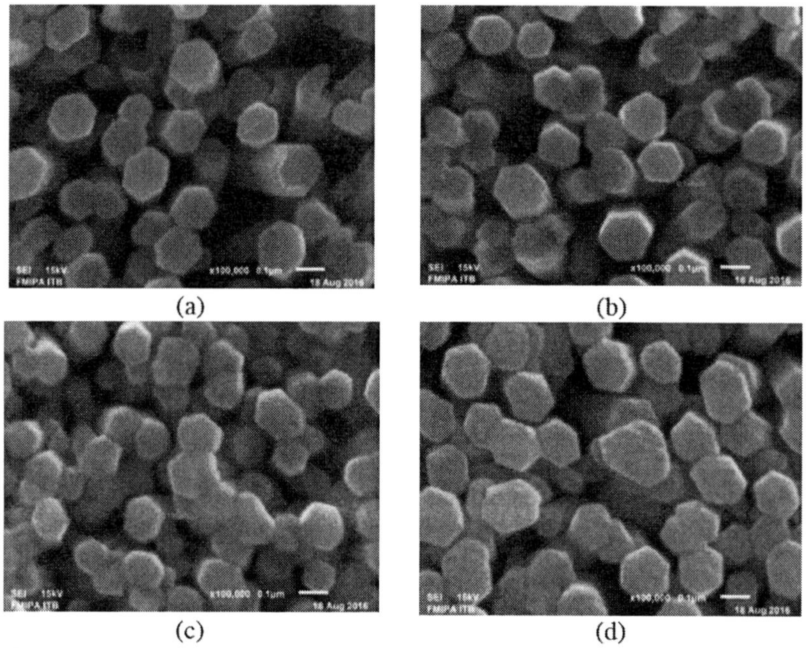

Figure 1. Scanning electron microscope (SEM) images of ZnO-Nanorods grown on glass substrate by variation growth time during self-assembly deposition (a). 120 minutes (b) 150 minutes (c) 180 minutes and (d) 210 minutes.

To verify the nanorod structure, x-ray diffractometer was used to investigate the structure and the results are shown in figure 2. All diffraction peaks can be indexed to hexagonal wurtzite structure in accordance with JCPDS 36-1451 (Joint Committee on Powder Diffraction Standards) of ZnO ($a = 3.249$ Å, $c = 5.206$ Å). All samples show highly possessed c-axis orientation according to narrower and highly dominant peak at (002). Other peak that attributed to others crystal orientation of ZnO is very small so that the rod structure is dominant. The sharp and narrow diffraction peak indicate that this metal oxide has good crystallinity as shown in figure 2. The peak intensity on XRD spectrum for deposition time up to 180 min was increase in contrast with 210 min, indicating that for the sample with 210 min time deposition the quality crystal was reduce. This situation can be related to the concentration of [Zn^{2+}] and [OH^-] during increments of time which is facilitating the supersaturated condition of precursor aqueous solution [12]. When the time increase more than 180 minutes, the condition is over supersaturated thus the existence of separated clusters $Zn(OH)_2$ were run out and the longer time in deposition process will deteriorates crystal structure growth.

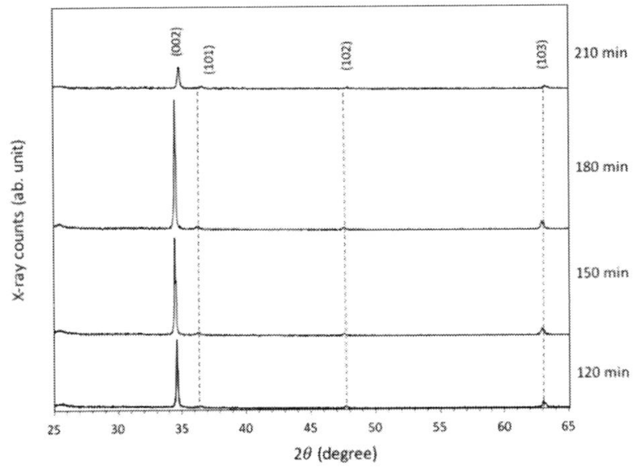

Figure 2. XRD spectra of zinc oxide nanorods layer on glass substrate by variations deposition time. All peaks were agreeing with JCPDS data for hexagonal wurtzite zinc oxide structure. A dominant (002) preferred orientation was appeared as time of growth increase. Particularly for samples up to 180 min growing time. Indicates that there was an optimum time for deposit nanorod structure at certain precursor concentration.

The defect on crystal growth at 210 minutes of deposition time, clearly seen at figure 2, all peaks position for this sample were shifted approximately 0.1°-0.4° compared with JCPDS ZnO data. Table 1 shows the XRD data from the JCPDS card compared with the XRD ZnO results. The average grain size can be simply determined using Scherrer's formula and already reported elsewhere [13]:

$$D = \frac{k\lambda}{B\cos\theta} \tag{1}$$

where k is the Scherrer constant, which is dependent on the crystallite shape and can be considered as 0.9 [14]; the x-ray wavelength of the incident Cu Kα radiation (λ) is 0.154 nm; B is the FWHM (full width at half maximum) of the respective peak; and θ is diffraction peak angle at certain plane. The calculated crystallite (grain) size is presented in table 1. The result showed that the ZnO-Nrs grain size tend to be decrease as time deposition increase. The ZnO particle nucleation process was beginning when the solution is supersaturated and continued to merge each other to form ZnO stuctures [12]. An appropriate time deposition will facilitate the supersaturation solution and produce a higher crystals quality (shows with sharp peak intensity on XRD spectrum) and smaller grain size (decreased FWHM) [3]. Smaller grain size was also caused by the ZnO molecule at the surface are energetically less stable, thus a longer deposition time was estimated to produces ZnO-Nanorods with high aspect ratio (ratio between length and diameter of ZnO) [12]. The lattice constant a and c of the ZnO wurtzite structure can be measured using Bragg's law [13]:

$$a = \frac{1}{\sqrt{3}}\frac{\lambda}{\sin\theta} \tag{2}$$

$$c = \frac{\lambda}{\sin\theta} \tag{3}$$

Unfortunately, the lattice constant a cannot be directly calculated using equation (2) because the peak of (100) plane does not appeared for all samples in XRD spectra. This situation may be related to the growth ZnO-Nrs behavior in this case.

Table 1. XRD parameter of ZnO-Nrs

Crystal plane		**(002)**		**(101)**	**(102)**	**(103)**
Deposition time	D (nm)	c (Å)		2θ (°)		
120	24.8	5.18	34.61	36.37	47.77	63.05
150	25.6	5.20	34.49	36.27	47.61	62.89
180	22.1	5.19	34.52	36.41	47.63	62.97
210	16.0	5.14	34.85	36.69	47.97	63.25
JCPDS	-	5.20	34.42	36.25	47.53	62.86

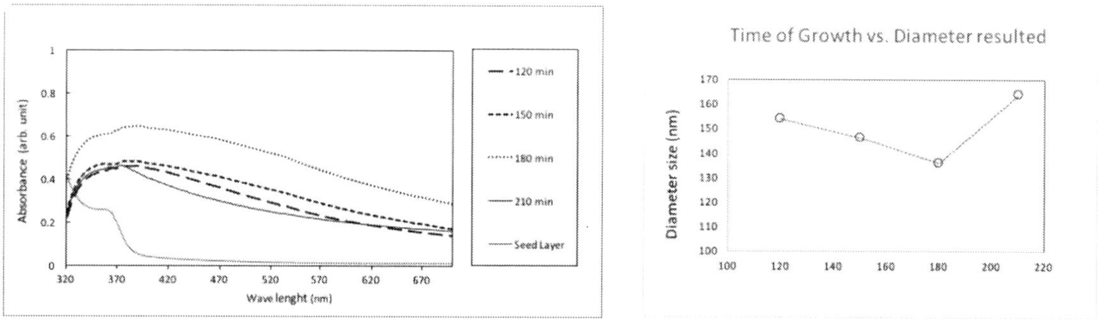

(a) (b)

Figure 3. (a) Ultraviolet (UV) visible spectra of ZnO-Nanorods layers with variation of deposition time. (b) Graph of deposition time versus average diameter ZnO nanorod.

Figure 3(a) represent the result of ultra violet (UV) visible spectroscopy characterization in the wavelength range of 300 – 750 nm for each sample, including ZnO seed layer. The result shows that during time deposition increases the absorbance increased, except for the sample with 210 minutes time deposition. This situation can be explained by increments time of growth at certain concentration and temperature while the colloidal suspension of the precursor has been over supersaturated, the crystal growth was stopped/freezers and discontinue the altitude growth and the quantity of ZnO crystallite/grain. The absorbance spectrum of ZnO seed layer exhibit strong exitonic absorption peak at 360 nm and high transparency (>95%) at visible range. This seed layer is critical point for growing ZnO nanorod structure as mentioned by H.J. Lee et.al whose reported that the lack quality of seed layer can results the randomly oriented of ZnO nanorod [15].

4. Conclusions

In this study, the effect of time deposition for growing ZnO nanorod structure was investigated. From the experimental results, it can conclude that deposition of ZnO-Nrs on ZnO seed layer/glass substrate using self-assembly method at 100 °C deposition temperature, has limited time on deposition process (approximate at 180 minutes'). From SEM images, increasing time deposition also reducing the average diameter size of nanorods as shown in Fig.3(b), but when the time is over than 180 min in this case at 210 minutes; the rod diameter seems back increase. XRD measurement reveal that the grain size of ZnO nanorods tend to reduce as time deposition increase. All samples exhibit preferred orientation along c-axis (002) plane with sharp and narrow peak while other peaks have low intensity, indicate that crystal structure at (002) plane has good crystal quality in the other words it has been present a rod structure. The transparency of ZnO nanorods is decrease for all samples up to 180 minutes' time deposition related to quantity of ZnO-Nrs. Unfortunately, the aspect ratio (L/r; length/radius) of nanorod structure cannot be estimated because the lack of data nanorod length (cross-sectional SEM images).

5. Acknowledgment

This work was financially supported by Universitas Padjadjaran under Academic Leadership Program (ALG 1-1-6) year of 2016.

References

[1] Schmidt M L and MacManus D J L 2007 *Materialstoday* **10** No. 5 p40-48
[2] Gyu C Y, Chunrui W and Won Il P 2005 *Semicond. Sci. Technol.* **20** S22–S34
[3] Kyung H K, Kazuomi U, Yoshio A and Midoeri K 2014 *Materials* **7** 2522-33
[4] Michael G 2006 Thin film solar cells: Fabrication, characterization and applications *Nanocrystalline Injection Solar Cells* ed Jef P and Vladimir A (Jhon Wiley & Sons Publisher) chapter 9 pp 363-385
[5] Annisa A, Priastuti W, Veinardi S, Herman, Rahmat H, Akihiko F and Masanori O 2013 *Solar Energy Materials & Solar Cells* **111** 181-188
[6] Waldo J B, Wienk M M and Jansen R A J 2006 *Advanced Functional Materials* **16** 1112-16
[7] DaeYon S, Jeong-Hyeok I, Hui-Seon K and Nam-Gyu P 2014 *J. Phys.Chem. C.* **118** (2014) 16567-73
[8] Hafiz M A J, Wenxiu Q, Xingtian Y, Yonglei Xing, Junyou S and Ling Bing K 2016 *Journal of Alloys and Compounds* **685** 610-18
[9] Rahul D, Stefan F, V Benno MR, Yasemin O, Saeed H and Dietmar K 2012 *Bioinsprir. Biomim.* **7** (8pp)
[10] Jalal R, Mohammad H M, C H Raymond O, Shahrom M and Mohamad R M 2015 *PLOS ONE* **April 13**, 1-14
[11] Pichanan T, S Kumar and Joydeep D 2012 *Int. J. Electrochem. Sci.* **7** 4988-99
[12] Polsongkram D, P Chamninok, S Purkird, L Chow, O Lupan, G Chai, H Khallaf, S Park and A Schulte 2008 *Physica B* **403** 3713-3717
[13] Kai long F, Uda H, Kashif M and Chun Hong V 2014, *Nanoscale Research Letters* **9**:429
[14] Oh H, Krantz J, Litzov I, Stubhan T, Pinna L and Brabec C J 2011 Solar Energy Materials and Solar Cells **95** 2194-99
[15] Hyukjae L, Jun-Ho S, Jisu C, Jung Bin K, Tae-Heui K and Kyung B P 2013 *Electronic Materials Letters* **9** 357-362

μSR Study of Organic Superconductor λ-$(BETS)_2GaCl_4$

D P Sari[1,2], R Asih[1,2], S S Mohm-Tajudin[6], N Adam[6], K Hiraki[3], Y Ishii[4], T Takahashi[3], T Nakano[2], Y Nozue[2], S Sulaiman[3], M I Mohamed-Ibrahim[3], and I Watanabe[1,2,5]

[1]RIKEN Nishina Center, Hirosawa, Wako, Saitama 351-0198, Japan,
[2]Department of Physics, Osaka University, Toyonaka, Osaka 560-0043, Japan,
[3]Department of Physics, Gakushuin University, Toshima, Tokyo 171-8588, Japan,
[4]College of Engineering, Shibaura Institute of Technology, Saitama 337-8570, Japan,
[5]Department of Physics, Hokkaido University, Sapporo 060-0810, Japan
[6]Universiti Sains Malaysia, Pulau Pinang, 11800, Malaysia

E-mail: dita.puspitasari@riken.jp

Abstract. Muon-spin-relaxation (μSR) measurements in the transverse-field (TF) of 30 G were carried out from 7 K down to 1.8 K on the non-magnetic anion-based organic superconductor λ-$(BETS)_2GaCl_4$. The TF-μSR time spectrum showed a significant increase with the Gaussian-type damping behavior below the superconducting transition temperature T_C = 5 K confirming the bulk SC state with the full volume fraction. The zero-field (ZF) μSR time spectrum did not show any change against the temperature down to 1.7 K, suggesting that the time reversal symmetry of the Cooper pair might not be broken.

1. Introduction

The quasi two-dimensional organic superconductor, λ-$(BETS)_2GaCl_4$ (where BETS stands for $C_{10}S_4Se_4H_8$, bisethylenedithiotetraselenafulvalene) is metallic and undergoes the superconducting (SC) state below T_C = 5 K. Fig. 1 shows the crystal structure and a photograph of λ-$(BETS)_2GaCl_4$ (BGC). This system attracts on-going interest as this system contains non-magnetic anions. If we change the closed shell of the Ga ion to be the Fe one ion, it shows predictably similar SC properties in the field of 33 T even though the crystal sturucture does not change [1]. The first report of the specific heat measurement on BGC suggested a conventional full-gap superconductor [2]. But more recent high-resolution thermodynamic measurement by using a utilized ruthenium oxide temperature sensor, down to 0.6 K, suggesting the *d*-wave SC state [3]. Other experiments such as the flux flow resistance [4] and the scan tunneling microscopy [5] supported conventional and unconventional possibilities. One of the good methods of investigating the SC state of organic superconductors is muon spin relaxation (μSR). TF-μSR can be used to determine the magnetic penetration depth λ which related to superfluid density. The temperature dependence of $\lambda(T)$ provides an important test of the gap symmetry. This technique has been succesfully applied to studies of the previous generation of organic superconductors $(TMTSF)_2ClO_4$ and κ-$(BEDT\text{-}TTF)_2Cu(NCS)_2$ [6]. Additional advantage of using μSR is that the measurement in the zero field. An injucted muon is a sensitive probe of local internal magnetic fields of ≈ 0.1 G which corresponds to that caused by nuclear dipole moments.

Thus, it is an ideal probe to search spontaneous internal fields yielding as a result of the breaking of the time-reversal symmetry of the Cooper pair in the unconventional SC [7]. In this paper, we report a preminarily μSR study on BGC.

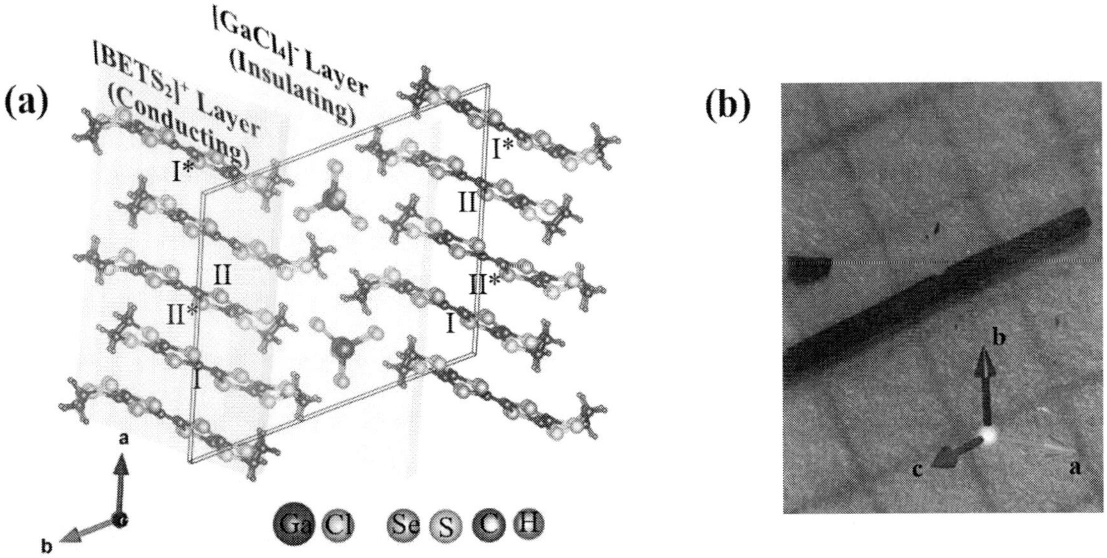

Figure 1: (a) Crystal structure of λ-$(BETS)_2GaCl_4$. There are four BETS molecules (BETS-I, I*, II and II*) in a unit cell which are containing of two types of those, Those four BETS molecules are crystallographic ally equivalent and are in the inversion symmetry with each other and in the same direction along the fourfold quasi-stacking structure. This alignment makes the crystal morphology needle-like prism. (b) Photograph of black λ-$(BETS)_2GaCl_4$ crystal. The average crystal dimensions are 3 x 0.26 x 0.13 mm^3.

2. Experimental Details

Single crystal samples of BGC were syntheiszed by the electrochemical oxidation method [8]. The zero-resisivity below 5.2 K and the diamangetization with the full volume fraction were confirmed from the resitivity and magnetic susceptibility measurements. Single crystals are crushed for μSR studies making the sample to be in the coarsed state. μSR measurements were carried out at the RIKEN-RAL Muon Facility in the UK.

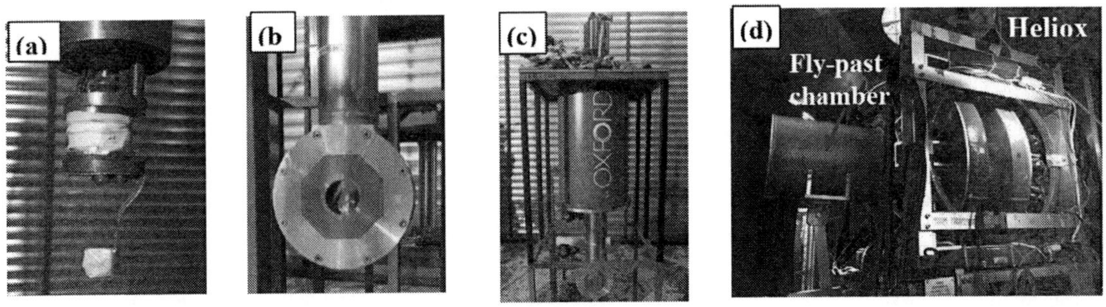

Figure 2. (a) Sample mounting. Coarsed crystals with the 120 mg mass are packed by the using 25 μm thickness and high purity (99.999 %) silver foil. The sample holder is designed for the fly-past setup for the ^{3}He cryostat. (b) The fly-past set up for the ^{3}He cryostat named HELIOX. The sample is located at the center of circular part of the tail. (c) Whole view of HELIOX with the fly-past setup. (d) The vaccume chammber extension for the fly-past setup installed on the μSR spectrometer in Port-2 of RIKEN-RAL Muon Facility, named ARGUS.

A ^{3}He cryostat was used for TF-μSR to cool the sample down to 1.8 K. Figure 2 show pictures of the experimental setup. We applied TF in perpendicular to the initial muon-spin polarization in the normal state at 7 K. And then, we did the field-cooling down to 1.8 K. A ^{4}He flow-type cryostat was used for ZF-μSR to cool the sample down to 1.7 K. The zero-field calibration was done to maintain the zero-field condition to be less than 10 mG.

3. Results and Discussions

Figure 3 shows the TF-μSR time spectrum at 7 K (in the normal state) and at 1.8 K (in the SC state). The damping behavior of the muon-spin precession due to inhomogenious internal fields at the muon site was observed. The damping behavior at 7 K is due to the distribution of local fields at the muon which is broadenend mainly by randomly oriented nuclear moments surronding the muon stopping site [9].

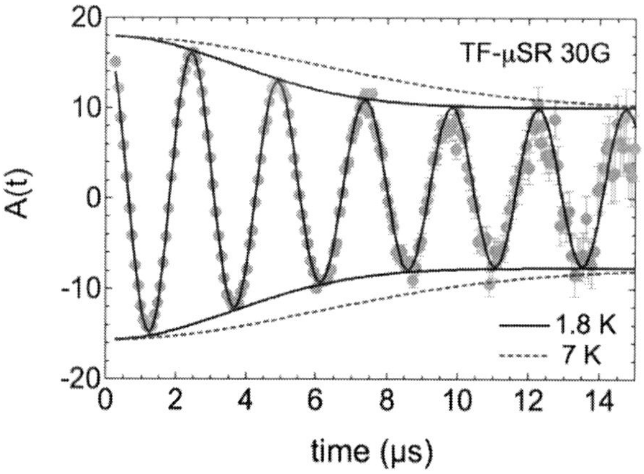

Figure 3. TF-μSR time spectrum of λ-(BETS)$_2$GaCl$_4$ under TF= 30 G at 1.8 K (close blue circle) and the best-fit result of the data by using Eq. (2). The black solid line is the damping part of the fitting result, whose rate value is σ, at 1.8 K and the broken red line is that of the one at 7 K.

We analyse the time spectra by using the following function;

$$P^{TF}(t) = A_1{}^{(-\sigma t)^2}\cos(\gamma_\mu H_{ext_1} t + \phi) + A_2\cos(\gamma_\mu H_{ext_2} t + \phi). \quad (1)$$

Here, σ is the Gaussian damping rate, $H_{ext_1}$is the avaraged field at the muon side in the SC state. The A_1 and A_2 are asymmetry parameters of the Gayssian-type damping and the background components, respectively. The A_2 was fixed to be that achieved at the base temperature. H_{ext_2} is TF and ϕ is the phase of the muon-spin precession. In the normal state, σ was estimated to be 0.1172 + 0.0023 μs^{-1}. The Gaussian damping rate in the SC state, σ_{SC}, can be calculated from the relation $\sigma^2 = \sigma_{SC}^2 + \sigma_{NM}^2$ where σ_{NM} is the Gaussian damping rate of the background signal, and estimated to be 0.1708 + 0.0022 μs^{-1}. This decrease in the σ_{SC} compared in the SC state is caused by the appearance of the flux state which produces the distribution of the penetrated magnetic fields in the sample. The detail analysis on temperature dependence of σ_{SC} is on-going at this moment and will be publised in a separated paper.

Figure 4(a) shows the ZF-μSR time spectra masured above/at/below T_C. No change oscillation in the time spectra was confirmed within the experimental tollerance. Time spctrra were analyzed by the streached exponential function;

$$P^{ZF}(t) = A_1\left(\exp(-\sigma t)^{\beta}\right) + A_2. \quad (2)$$

The β was estimated to be 0.88 which was close to that of the Lorentzian function. This Lorentzian-like depolarization behavior tends to be well observed in other organic superconductors [9]. No change in the time sprctrume below T_C in ZF means that no spontaneous magnetic field appeared in the SC state and the time reversal symmetry of the Cooper pair is not broken in BGC. In the case of that the Cooper pair has the *p*-wave pairing symmetry, sponteneous internal field appears in the SC state. This sponteneous internal field is small in the order of a couple of Gauss but can be picked up by the muon. Therefore, our current results can conclude that the *p*-wave pairing symmetry in inlikely to explain the SC state of BGC. More detailed μSR investigations are now going on by us at the RIKEN-RAL Muon Faclity and the results will be published very soon. [11]

(a) **(b)**

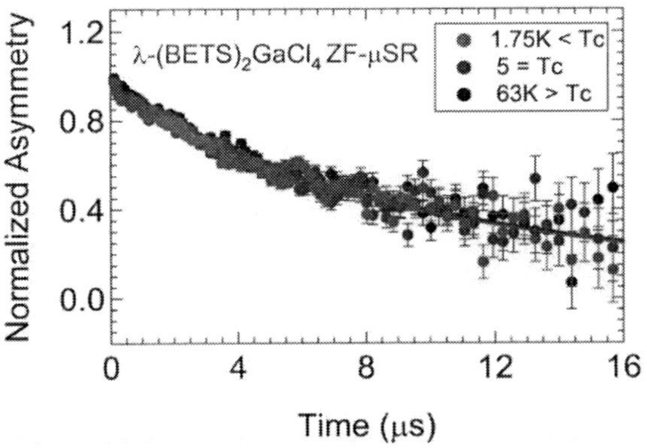

Figure 4. a) ZF-μSR time spectra at 63 K (above Tc), 5 K (at Tc) and 1.75 K (below Tc). b) Temperature independence of the relaxation rate obtained from the best fit results of time spectra by using Eq.(2).

Conclusions

ZF- and TF-μSR invesigations werep performed on the organic superconductor, λ-$(BETS)_2GaCl_4$ in order to study the superconducting state. TF-μSR under 30 G down to 1.8 K confirmed the appearance of full superconducting volume fraction with Tc = 5 K. Furthermore, from ZF- μSR, the time reversal symmetry of the Cooper pair was confirmed not to be broken reducing a possibility of the *p*-wave pairing symmetry.

References

[1] S. Uji *et al,* Global Phase Diagram of the Magnetic Field-Induced Organic Superconductors λ-$(BETS)_2Fe_xGa_{1-x}Cl_4$, J. Phys. Soc. Jpn., 72 no. 2 (2003) 369-373
[2] Y. Ishizaki *et al.,* Specific Heat of Organic Superconductor λ-$(BEDT\text{-}TSS)_2GaCl_4$, Synth. Metals, 133-134 (2003) 219-220
[3] S. Imajo *et al.,* Thermodynamic Evidence of *d*-wave Superconductivity of the Organic Superconductor λ-$(BETS)_2GaCl_4$, J. Phys. Soc. Jpn., 85 (2016) 043705
[4] S. Yasuzuka *et al,* J. Phys. Soc. Jpn., In-Plane Anisotropy of Flux-Flow resistivity in layered Organic Superconductor λ-$(BETS)_2GaCl_4$, 83 (2014) 013705
[5] K. Clark *et al,* Superconductivity in just four pairs of $(BETS)_2GaCl_4$ molecules, Nanotechnol, 5 (2010) 5
[6] F. Pratt, J. Phys. Soc. Jpn., Superconductivity and Magnetism in Organic Material Studied with μSR, 85 (2016) 091008
[7] G. M. Luke *et al.,* Time-reversal Symmetry-Breaking Superconductivity in Sr_2RuO_4, Nature, 394 (1998) 558
[8] H. Kobayashi, A New Organic Superconductor, λ-$(BEDT\text{-}TSF)_2GaCl_4$, Chem. Lett. (1993) 1559-1562
[9] F. L. Pratt *et al.,* Low-Field Superconducting Phase of $TMTSF_2ClO_4$, Phys. Rev. Lett., 110 (2013) 107005
[10] F. L. Pratt and S. J. Blundell, Universal Scaling Relation of Molecular Superconductor Phys. Rev. Lett., 94 (2005) 097006
[11] D. P. Sari *et al.* (to be submitted).

IOP Conf. Series: Materials Science and Engineering **196** (2017) 012048 doi:10.1088/1757-899X/196/1/012048

Improvement of Catalytic Performance over $LaMnO_3$/Nanographene Platelets Composites

Nur Afifah[1,2] and Rosari Saleh[1,2,a]
[1]Departemen Fisika, Fakultas MIPA, Universitas Indonesia, Kampus UI Depok, 16424 Depok, Indonesia
[2]Integrated Laboratory of Energy and Environment, Fakultas MIPA, Universitas Indonesia, Kampus UI Depok, 16424 Depok, Indonesia
[a]Corresponding's author: rosari.saleh@gmail.com, rosari.saleh@ui.ac.id

Abstract. In this work $LaMnO_3$ – NGP composites with various NGP wt. % were prepared using co-precipitation method. The prepared $LaMnO_3$ – NGP composites were characterized by X-Ray Diffraction (XRD), Brunauer–Emmett–Teller (BET) surface area analysis, and Fourier Transform Infrared (FT-IR) spectroscopy. The prepared samples were then used to remove methylene blue (MB) as a model of an organic pollutant under visible light and ultrasonic irradiation separately and simultaneously. The obtained results revealed that the catalytic performance of $LaMnO_3$ – NGP composites show better performance than $LaMnO_3$ nanoparticles. The photosonocatalytic activity also shows the highest degradation efficiency. Furthermore, the main influence factors on the catalytic activity were studied and discussed.

1. Introduction

Advanced oxidation processes (AOPs) including photocatalytic and sonocatalytic are known as an alternative method to conventional method for effective complete oxidation of various organic pollutants [1-2]. The generation of reactive radicals such as superoxide ($\bullet O_2^-$) and hydroxyl ($\bullet OH$) radicals in the each catalytic process could attack a number of organic pollutants in wastewater [3]. Moreover, the combination of photocatalytic and sonocatalytic, i.e. the so-called photosonocatalytic seem to enhance the degradation efficiency because of the increase in the generation of reactive radicals [4]. Although the photocatalytic and the sonocatalytic have been intensively explored individually for the degradation application, very few reports are available on their combination (i.e. photosonocatalytic) [5-6], especially using perovskite materials as a catalyst.

Perovskite especially $LaMnO_3$, has also been used in the catalytic application. As a catalyst, $LaMnO_3$ has been chosen due to its high catalytic activity, low cost, and environmental friendliness [7-8]. We had already reported the catalytic performance of $LaMnO_3$ nanoparticles previously [9]. However, $LaMnO_3$ nanoparticles could not completely degrade the methylene blue organic pollutant because of its high recombination of the photogenerated electron-hole pairs. Therefore, to suppress the recombination of the photogenerated electron-hole pairs in $LaMnO_3$, it can be loaded with nano graphene platelets (NGP) since the NGP has superior electrical and mechanical properties, excellent electron mobility, good thermal stability and large specific surface area [10-12]. The aim of this work was to study the effect of NGP incorporation on the catalytic performance of $LaMnO_3$/NGP composites using methylene blue (MB) as a model of organic pollutant. Furthermore, the main factors on the catalytic activity were studied and discussed.

2. Experimental details

2.1. Materials

Lanthanum (III) chloride heptahydrate ($LaCl_3 \cdot 7H_2O$), Manganese (II) chloride tetrahydrate ($MnCl_2 \cdot 4H_2O$), and sodium hydroxide (NaOH) purchased from Merck (Kenilworth, NJ, USA) were used to synthesize $LaFeO_3$ nanoparticles. Nanographene platelets (NGP), purchased from Angstron Materials (Dayton, OH, USA), were then used to synthesize $LaMnO_3$/NGP composites. Distilled water was also used for all of the solutions.

2.2. Synthesis of the $LaMnO_3$/NGP composites

$LaMnO_3$ nanoparticle was synthesized using the co-precipitation method based on our previous work [9]. The obtained $LaMnO_3$ nanoparticle was then used to synthesize $LaMnO_3$/NGP composites. First, NGP was dissolved into distilled water and ethanol through ultrasonic treatment for 2 h. Then, the prepared $LaMnO_3$ nanoparticles were added to the NGP solution, followed by stirring for 2 h. Finally, the suspension was then centrifuged and dried at 70 °C for 12 h. A series of $LaMnO_3$/NGP composites were prepared by varying the weight ratio of NGP to $LaMNO_3$ at 3, 5, and 10 wt%.

2.3. Characterization

The samples were characterized by X-Ray Diffraction (XRD), Brunauer–Emmett–Teller (BET) surface area analysis, and Fourier Transform Infrared (FT-IR) spectroscopy

2.4. Catalytic Activity

The photo-, sono, and photosono- catalytic activity of $LaMnO_3$/NGP composites were evaluated by using methylene blue (MB) as the model of orgnaic pullutant. The detailed procedures in the catalytic experiments have been described in our previous work [9]. The degradation of MB was measured using a UV-visible spectrophotometer. The results was presented as C_t/C_0, where C_t is the MB concentration in each irradiated time interval (15 min) and C_0 is the MB initial concentration after dark condition (30 min). In order to obtain the main active species in the catalytic process, diammonium oxalate, sodium sulfate, and tertbutyl alcohol were used as scavengers for a hole, electron, and hydroxyl radical species, respectively.

3. Results and Discussion

The FT-IR measurements were used to confirm the existence of NGP in the $LaMnO_3$/NGP composite. Figure 1a shows the FT-IR spectra of the $LaMnO_3$/NGP composites with various wt% NGP. As a comparison, the FT-IR spectrum of NGP is also shown. The FT-IR spectra of NGP and its composites show a strong absorption at wavenumber 3440 cm^{-1}, which corresponds to the O–H stretching vibration [13] and an absorption at 1460–1750 cm^{-1}, which is attributed to C=O stretching vibration [13-15]. The incorporation of $LaMnO_3$ in the composites was also confirmed by the presence of metal-oxygen stretching vibration on the absorption peak at wavenumber 539 cm^{-1} [16].

The structural properties of the samples were characterized using XRD measurements. Figure 1b shows the XRD pattern of NGP, $LaMnO_3$ nanoparticle, and $LaMnO_3$/NGP composites with different wt% NGP. The XRD pattern of the composites exhibit characteristic peak assigned to (101), (020), (111), (200), (002), (210) (220), (022), (202), (040), (222), (321), (123), and (004) crystallography planes of the orthorhombic structure of $LaMnO_3$ [17-18] and (200) planes of the graphite structure of NGP [19]. The diffraction peak of NGP in the composites increased with increasing wt% NGP. No other phases or impurities were detected in the XRD pattern of all samples. The crystal size and lattice parameters of all samples are tabulated in Table 1. The lattice parameter obtained by Rietveld refinement method using MAUD program and crystal size <D> of the samples calculated using the Scherrer's formula [20]. The BET method was also examined to obtain the specific surface area of all samples. The obtained specific surface area values are then summarized in Table 1. The results showed the specific surface area of the composite increase with the incorporation of NGP content.

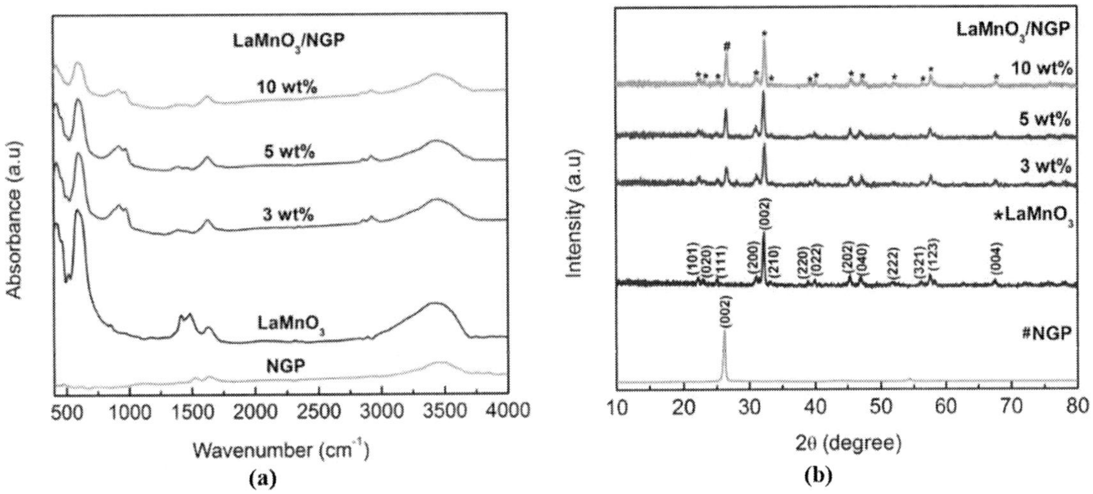

Figure 1. (a) FTIR spectra; and (b) XRD pattern of $LaMnO_3$/NGP composites with different wt% NGP.

Table 1. The lattice parameter, grain size <D>, and surface area of $LaMnO_3$/NGP composites.

Sample	Lattice Parameter			<D> (nm)	Surface area (m^2/g)
	a (Å)	b (Å)	c (Å)		
$LaMnO_3$	5.7419	7.7215	5.5403	44	17
$LaMnO_3$/3wt% NGP	5.7335	7.6998	5.5272	16	23
$LaMnO_3$/5wt% NGP	5.7586	7.7090	5.5464	23	34
$LaMnO_3$/10wt% NGP	5.7448	7.7056	5.5191	21	42

The photo-, sono-, and photosono- catalytic activity of $LaMnO_3$/NGP composites for degrading MB are illustrated in the Figure 2. For comparison, the photo-, sono-, and photosono- catalytic activity of $LaMnO_3$ nanoparticles are also shown in the figure. It can be seen that MB degradation in the presence of $LaMnO_3$ nanoparticles increases with increasing irradiation time. Moreover, the incorporation of NGP in the composites could further increase the degradation efficiency. The degradation rate of MB was also calculated using the pseudo-first-order kinetic rate equation [21]. As clearly seen in the inset of Figure 2, the obtained result revealed that 5 wt% NGP shows the maximum degradation rate in all of the catalytic processes. In addition, the MB degradation under photosonocatalytic activity exhibits faster degradation rate compared to photocatalytic or sonocatalytic. To detect the main active species responsible in the process of MB degradation, the experiment was repeated using $LaMnO_3$/NGP composite (5 wt%) with the addition of active species like an electron, hole, and hydroxyl radicals. $Na_2S_2O_8$, diammonium oxalate, and tertbutyl alcohol were used as an electron, hole, and hydroxyl radical scavengers. The addition of scavenger (presented in Figure 3) could decrease the degradation rate of MB: the largest decrease is obtained with the present of hole scavenger. It indicates that the hole plays the most important role in the degradation process of MB.

The higher degradation rate with the NGP incorporation could depend on several factor: the first is electron and hole separation efficiency and the second is the surface area of the catalyst. NGP, with its electron transport layer capacity could inhibit the photo-generated electron and hole of $LaMnO_3$ to recombine by transferring photo-generated electrons in the conduction band of $LaMnO_3$ into NGP layers [7-8]. Therefore, the photo-generated electron and hole could react with oxygen molecules (O_2) and hydroxyl groups (OH^-) before recombining, to generate superoxide radicals ($\bullet O_2^-$) and hydroxyl

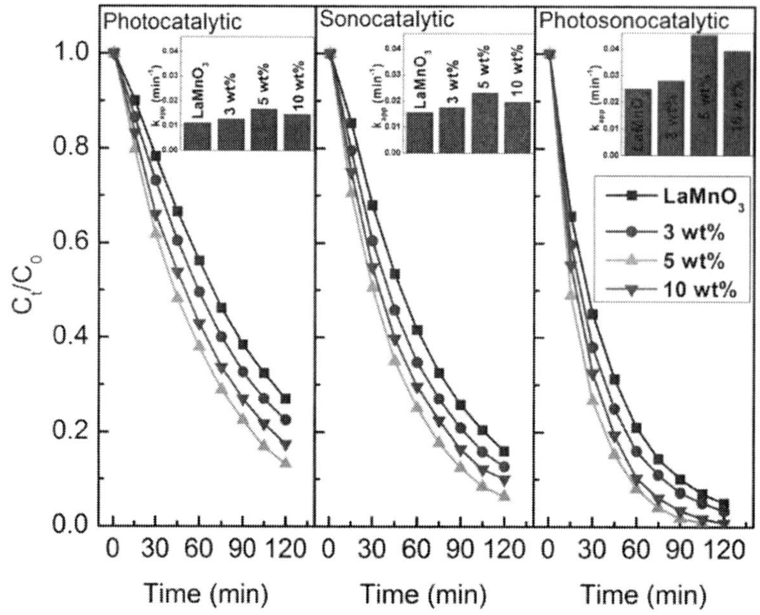

Figure 2. The photo-, sono- and photosono- catalytic activity of $LaMnO_3$/NGP composites with diffrent wt % NGP.

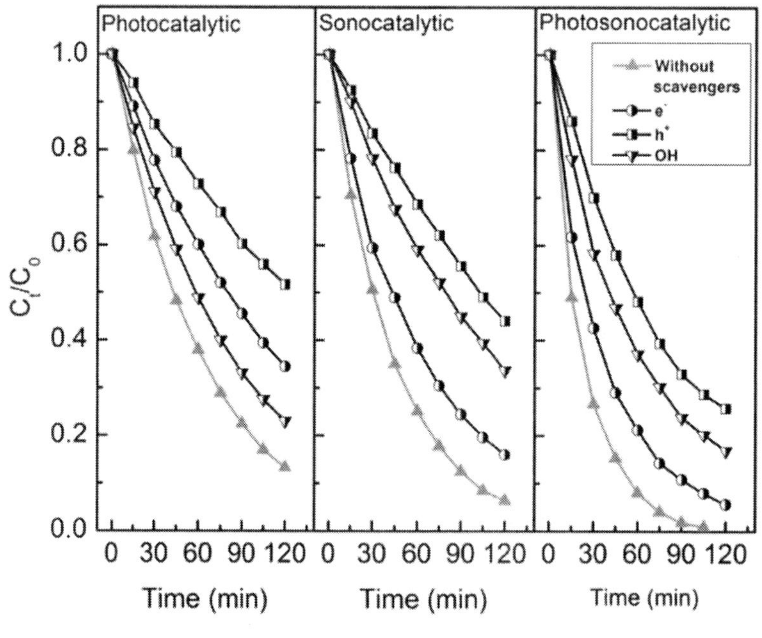

Figure 3. The effect of scavangers on the photo-, sono- and photosono- catalytic degradation of MB using $LaMnO_3$/5wt% NGP composite.

radicals (•OH) radicals, respectively. These reactive radicals are responsible for degradation of the complex chemical bonds of organic pollutants such as MB. NGP with its large surface area could increase the contact probability between the catalyst and the light, thus increasing the rate of oxidation process [7]. Furthermore, the different mechanism on the photo-, sono-, and photosono- catalytic activity have been explained in our previous work [9]. It was reported that the higher concentrations of reactive radicals (especially •OH radicals) were produced during the photosonocatalytic process [3-4], so that it may explain the fastest degradation rate of MB.

4. Conclusion

The influence of NGP incorporation on the photo-, sono-, and photosono- catalytic activity of $LaMnO_3$/NGP composites have been investigated under light or/and ultrasonic irradiation separately and simultaneously. The results showed that NGP could inhibit the electrons and holes recombination and enhance the catalytic performance. The addition of 5 wt% NGP gives maximum degradation rate of MB. In addition, the photosonocataytic showed faster degradation compared to photocatalytic and sonocatalytic alone.

References

[1] Jagannathan M, Grieser F, and Ashokkumar M 2013 *Sep. Purif. Technol.* **103** 114
[2] Madhavan J, Kumar P S S, Anandan S, Grieser F, dan Ashokkumar M 2010 *Sep. Purif. Technol.* **74** 336
[3] Markovic D, Saponjic Z, Radoicic M, Redetic T, Vodnik V, Potkonjak B, and Redetic M 2015 *Ultrason. Sonochem.* **24** 221
[4] Vinoth R, Karthik P, Devan K, Neppolian B, and Ashokkumar M 2016 *Ultrason. Sonochem.* In Press
[5] Ertugay N and Acar F N 2014 *Appl. Surf. Sci.***318** 121
[6] Zhou M, Yang H, Xiang T, Li R S, Zhang H M, and Wang X X 2015 *J. Hazard. Mater.* **289** 149
[7] Hu J, Ma J, Wang L and Huang H 2014 *J. Alloy. Compd.* **583** 539
[8] Hu J, Men J, Ma J and Huang H 2014 *J. Rare Earth.* **32** 1126
[9] Afifah N and Saleh R 2016 *Mater. Sci. Forum* **864** 99
[10] Aleksandrzak M, Adamski P, Kulkulka W, Zielinska B and Mijowska E 2015 *Appl. Surf. Sci.* **331** 193
[11] Morales-Torres S, Pastrana-Martínez L M, Figueiredo J L, Faria J L, and Silva A M 2013 *Appl. Surf. Sci.* **275** 361
[12] Machado B F and Serp P 2012 *Catal. Sci. Techn.* **2** 54
[13] Tang L, Nguyen V H, Lee Y R, Kim J and Shim J J 2015 *Synthetic Met.* **201** 54
[14] Taufik A and Saleh R 2016 *AIP Conf. Proc.* **1725** 020089
[15] Fujii T, Matsusue I, Nakatsuka D, Nakanishi M and Takada M 2015 *Mater. Chem. Phys.* **129** 805
[16] Wenwei W, Jinchao C, Xuehang W, Sen L, Kaituo W, and Lin T 2013 *Adv. Powder Technol.* **24** 154
[17] Galal A, Atta N F, and Ali S M 2011 *Electrochim. Acta* **56** 5722
[18] Zhang C, Guo Y, Lu G, Boreave A, Retailleau L, Baylet A, and Giroir-Fendler A 2014 *Appl. Catal. B: Environ.* **148–149** 490
[19] Lin W-D, Liao C-T, Chang T-C, Chen S-H, and Wu R-J 2015 *Sensor. Actuator. B* **209** 555
[20] Köferstein R, Jäger L and Ebbinghaus S G 2013 *Solid State Ionics* **249–250** 1
[21] Saleh R and Djaja N F 2014 *Superlattice. Microst.* **74** 217

Characteristics of Thermally Reduced Graphene Oxide Thin Film as DSSC Counter Electrode

F Yuliasari, A Aprilia, N Syakir, L Safriani, T Saragi, Risdiana, S Hidayat, A Bahtiar, R Siregar, and Fitrilawati

Departement of Physics Padjadjaran University, Jalan Raya Jatinangor Km 21 Sumedang 45363 West Java Indonesia

Email: fitrilawati@phys.unpad.ac.id

Abstract. We report characteristics of reduced graphene oxide (RGO) as a counter electrode for dye-sensitized solar cell (DSSC). The RGO thin films were prepared on FTO (Fluorine-doped Tin Oxide) substrates and followed by a reduction process. The RGO film was used as a counter electrode in a DSSC device, with a structure of FTO/TiO_2/ruthenium dye/mosalyte/RGO/FTO. UV-Vis measurements show an increasing absorption spectrum of RGO film after thermal reduction process and the FT-IR spectrum confirms a removal of the oxygen containing groups after thermal reduction process. The efficiency (η) of the DSSC that applied RGO film as a counter electrode is 0.96%.

1. Introduction

Graphene is attracting a lot of attention due to its promising properties for electronic devices [1-4]. The high optical transmittance (~97.7%) and good conductivity of graphene made it as an excellent candidate for f transparent conductive electrodes [5,6]. A low cost method involving thermal reduction of graphene oxide (GO) has attracted considerable attention. However, there are many intercalated impurities, such as carboxyl and hydroxyl groups, embedded in the reduced graphene oxide that must be removed through reduction processes prior to use.

Dye-sensitized solar cell (DSSC) that is a promising alternative to conventional silicon or compound semiconductor solar cells [7]. The basic DSSC structure consists of dye that adhered to nanocrystalline TiO_2 photoanode, an electrolyte solution, and Pt counter electrode [8]. The role of counter electrode is catalyzing the reduction of the redox species in the electrolyte that used as a mediator in regeneration of dyes after electron injection [7]. Pt is commonly used in as counter electrode. Recently, carbon materials have been considered as candidates of counter electrodes for DSSC [9]. In this work we study a reduction method using thermal heating of graphene oxide (GO) obtain graphene-like that known as reduced graphene oxide (RGO) and then used it as a DSSC counter electrode.

2. Experiments

Thin films of GO were prepared from 4 mg/ml GO dispersed in water (Graphenea). The GO dispersion was sonicated (Branson 1800 Series) for 1 hour at room temperature for exfoliation prior to use. The GO dispersion was deposited on FTO substrate by using spincoating technique (Chemat Technology Spin-coater KW-4A) at 1000 rpm for 30 second and dried at 30 °C to remove residual

IOP Conf. Series: Materials Science and Engineering **196** (2017) 012049 doi:10.1088/1757-899X/196/1/012049

solvent . In order to obtain reduced graphene oxide (RGO) films, the GO films were thermally reduced by heating at 200°C for 1 hour.

The RGO film was then used as counter electrode in DSSC, with a structure of FTO/TiO_2/ruthenium dye/mosalyte/RGO/FTO. Besides a DSSC with RGO counter electrode ($DSSC_{RGO}$), we also prepared a DSSC with GO as counter electrode ($DSSC_{GO}$), and a DSSC with Pt counter electrode ($DSSC_{Pt}$) as a control. For DSSC assembly, the TiO_2 mesoporous was created by coating titania paste Ti-Nanoxide T/SP (Solaronix SA) on FTO substrate using screen printing technique and continued by sintering at 500°C for 30 minutes to convert anatase to rutile phase. In order to trap incident light, the Ti-Nanoxide MC/SP (Solaronix SA) was coated on FTO/TiO_2-mesoporous with the same deposition and sintering procedure. The FTO/TiO_2 was then immersed overnight in N719 dye solution. The dye solution was prepared from N719 and chenodeoxycolic acid with ratio of 1:10 that dissolved in ethanol. After dye immersion, the FTO/TiO_2/N719 was washed using acetonitrile to remove residual dye particles. *Surylin* (solaronix SA) 25 μm thicknesses was used as spacer between photoanode and counter electrode. Mosalyte as ionic liquid (solaronix SA) was injected through the hole at the counter electrode and then sealed with a transparent tape to avoid leaking.

RGO film was characterized by Ultra-violet Visible spectroscopy (T70+ UV/Vis Spectrometer PG Instrument, Ltd) and Fourier Transform Infrared spectroscopy (FT-IR Shimidzu). The DSSC performances were characterized by using DC voltage (Yokogawa GS200) integrated with digital multimeter (Yokogawa 7555) under 36.5 mW/cm^2 light irradiation. Device efficiency was calculated from ratio of maximum power of the device (P_{max}) and light power irradiation (P_{light}), while P_{max} is obtained from an open circuit voltage (V_{oc}), a short circuit current density (J_{sc}), and a fill factor (FF).

3. Results and Discussion

The prepared GO films have yellow-brown color and when they have been converted to RGO, the films become grey-black color. The UV-Vis absorbance spectrum of GO film was changed after thermal reduction process as shown in Figure 1. Since the FTO substrate absorbs light in the wavelength region of 200-300 nm, the absorption peak of GO and RGO films could not be identified in the spectra. An absorption difference between FTO and the GO film indicate a presence of GO layer on the FTO substrate. The absorbance spectrum of GO film was increased after thermal reduction process indicates that the RGO film more conductive compare to the GO film. It is related to a recovery of conjugated bonds due to a thermal reduction process that removed oxygen functional group on the GO film.

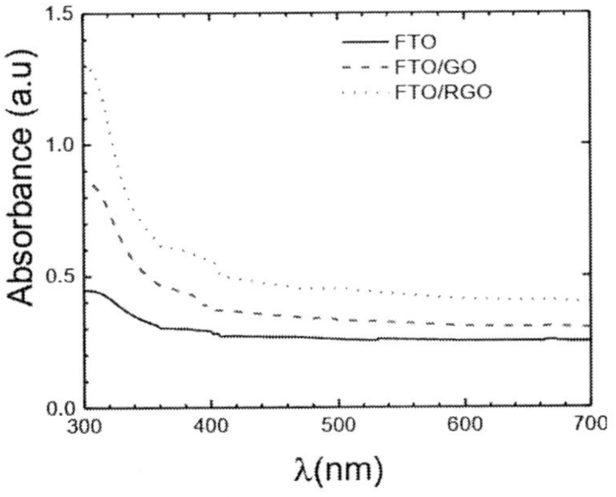

Figure 1. Thermal reduction process has increased UV-Vis absorption spectra of GO film, bared FTO substrate (solid line), FTO/GO film (dashed line), and FTO/RGO film (dotted line).

Figure 2 shows the FT-IR spectra of the GO film (solid line) and the RGO film (dashed line). From the figure it is obvious that thermal reduction process has removed some oxygen containing groups. The intensity of hydroxyl O-H group at 3400 cm^{-1} are considerably reduced and C=C vibration at 1600 cm^{-1} are significantly increased. However, several carboxyl and carbonyl C=O groups at 1735 cm^{-1}, carbonyl 1430 cm^{-1}, epoxy C-O at 1270 cm^{-1}, alkoxy C-O at 1042 cm^{-1} are only reduced and still remained on the RGO film, The increasing of C=C vibration intensity indicates a recovery of conjugated bonds in the RGO film. This result is in-line with the increasing of UV-Vis absorption of RGO film. Some remaining oxygen functional groups still appeared on the RGO film suggest that an improvement of reduction process is still needed.

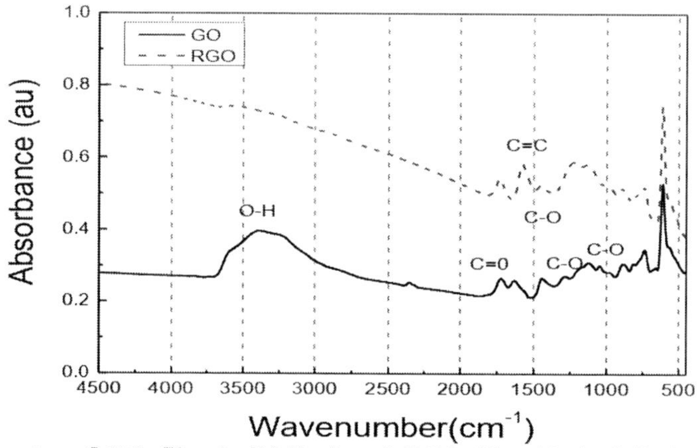

Figure 2. FT-IR Spectra of GO film (solid line) and RGO film (dashed line) shows a removal of several oxygen functional groups after at 200°C for 1 hour.

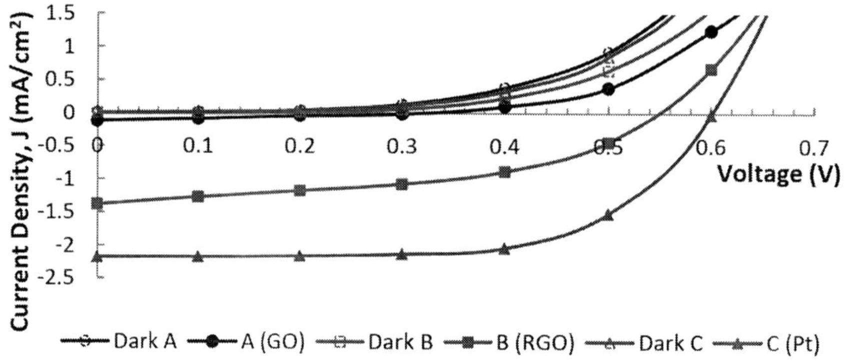

Figure 3. Photocurrent density-voltage (J-V) curves of DSSC with varied counter electrode, GO (circle), RGO (square), and Pt (triangle).

Figure 3 shows J-V curves of DSSC with varied counter electrode (CE), namely $DSSC_{pt}$, $DSSC_{RGO}$ and $DSSC_{GO}$. The photovoltaic parameters of J_{sc}, V_{oc}, FF of the DSSC were extracted from J-V curves and η was calculated, and all results are shown in Table 1. We found that the DSSC with RGO film as counter electrode has a better performance compared to the GO film. This result implies that a thermal reduction process was reduced the oxygen functional groups and it contributes to the changes of insulating state of GO film become semiconducting state of RGO film. The photocurrent density was significantly increased from 0.12 mA/cm^2 in the device with GO film as counter electrode to 1.38

mA/cm^2 in the device with RGO film. The change indicates that there is an increment of charge transfer inside the cell. Besides, the value of V_{oc} and FF were also increased from 0.3 volt to 0.58 volt and from 29% to 44%, respectively. Increasing of V_{oc} and FF actually attributed to a better interfacing between all layers inside the cell. Performance differences between DSSC that employed GO and RGO film as counter electrode, are mainly influenced by amount of oxygen impurities. Our experimental results show that $DSSC_{Pt}$ that employ Pt as counter electrode still has better performance compare to the $DSSC_{RGO}$ and $DSSC_{GO}$. Since the FTIR spectra of the RGO film confirm an existence of remained oxygen functional group, it is expected that an improvement of thermal reduction process can further reduce the oxygen content of RGO film. With an improvement of RGO film quality, the performance of DSSC that employed RGO film as a counter electrode can be further enhanced.

Table 1. Photovoltaic parameters of DSSC with structure of FTO/TiO2/N719dye/CE/FTO using various counter electrode.

Counter Electrode (CE)	V_{oc} (V)	J_{sc} (mA/cm^2)	FF (%)	η (%)
GO	0.30	0.12	29	0.03
RGO	0.58	1.38	44	0.97
Pt	0.60	2.18	64	2.30

4. Conclusions

Thermal reduction process of GO film at 200°C for 1 hour can convert the material into RGO that has less oxygen content. The RGO film can be used as a DSSC counter electrode although its performance still below the one that employed Pt as a counter electrode. The efficiency of DSSC with RGO film as counter electrode that has V_{oc} = 0.58 volt , J_{sc} = 1.38 mA / cm^2, and FF = 44% is η=0.96%. Since the oxygen functional group contained in the RGO still can be reduced by improving reduction process, the efficiency of DSSC with RGO counter electrode can be improved.

Acknowledgment

This work was funded by Academic Leadership Grant (ALG) Universitas Padjadjaran contract number 480/UN6.RKT/Kep/HK/2016

References

[1] Geim A K, Novoselov K S (2007). *Nat. Mater.* 6(3), 183.
[2] Balandin A A, Ghosh S, Bao W Z, Calizo I, Teweldebrhan D, Miao F, Lau C N (2008), *Nano Lett.* 8(3), 902
[3] Bolotin K I, Sikes K J, Jiang Z, Klima M, Fudenberg G, Hone J, Kim P, Stormer H L (2008), *Solid State Commun.* 146(9-10), 351.
[4] Morozov S V, Novoselov K S, Katsnelson M I, Schedin F, Elias D C, Jaszczak J A, Geim A K (2008), *Phys. Rev. Lett.* 100(1), 016602.
[5] Cai W W, Zhu Y W, Li X S, Piner R D, Ruoff R S (2009), *Appl. Phys. Lett.* 95(12), 123115
[6] Li X S, Zhu Y W, Cai W W, Borysiak M, Han B Y, Chen D, Piner R D, Colombo L, Ruoff R S (2009), *Nano Lett.* 9(12), 4359
[7] Jang HS , Yun J M, Kim D Y, Park D W, Na S I, and Kim S S (2012), *Electrochim. Acta*, 81, 301
[8] O'Regan B & Gratzel M (1991), *Lett. To Nat.*, 353, 737
[9] Kay A, and Grätzel M, *Sol. Energy Mater. Sol. Cells* (1996), 44 (1), 99

Enhancement of Sonocatalytic Activity and Stability of Magnetic $LaFeO_3/Fe_3O_4$ Nanocomposites

Nur Afifah[1,2] and Rosari Saleh[1,2]

[1]Departemen Fisika, Fakultas MIPA, Universitas Indonesia, Kampus UI Depok, 16424 Depok, Indonesia
[2]Integrated Laboratory of Energy and Environment, Fakultas MIPA, Universitas Indonesia, Kampus UI Depok, 16424 Depok, Indonesia

E-mail: rosari.saleh@ui.ac.id

Abstract. Recyclable catalyst using magnetic $LaFeO_3/Fe_3O_4$ nanocomposite has been successfully synthesized using co-precipitation method. The structural and magnetic properties of the samples were investigated using X-Ray Diffraction (XRD, Brunauer–Emmett–Teller (BET) surface area analysis, and Vibrating Sample Magnetomfeter (VSM) Spectroscopy. The XRD results revealed that the $LaFeO_3/Fe_3O_4$ nanocomposite shows orthorhombic and cubic spinel structure from $LaFeO_3$ and Fe_3O_4, respectively. The saturation magnetization of $LaFeO_3$ increases with the increasing of Fe_3O_4 content in the nanocomposite. The sonocatalytic performance and recyclability of the sample were evaluated by degrading methylene blue (MB) as a model of organic pollutant. Compared to $LaFeO_3$ nanoparticles, the sonocatalytic activity of $LaFeO_3/Fe_3O_4$ nanocomposite showed better performance. The stability of the catalyst was also checked by removing the catalyst from the MB solution simply using an external magnetic field. The result shows that the sample has good stability.

1. Introduction

Since 1990, the use of ultrasonic irradiation to destroy organic contaminant in wastewater has drawn special attention which offers several advantages such as safety, cleanness, high penetrability in an aqueous medium, high degradation efficiency and energy conversion without generating any secondary pollutant [1-4]. However, the use of ultrasonic irradiation alone may not be suitable for all kinds of pollutant and various variables such as the power of ultrasonic, the chemical structures, and the size of pollutant molecules. In addition, the ultrasonic process consumes a high amount of energy and time that makes it expensive [5]. Applying the catalyst in the sonocatalytic process is one of the promising ways to improve the performance of the ultrasonic irradiation in degrading the pollutants. The sonocatalytic process can be explained by the mechanism of sonoluminescence and hot spot. When an ultrasonic irradiates a catalyst, electron in the valence band of catalyst are excited to the conduction band by the sonoluminescence and hot spot, leaving the holes in the valence band. The sono-excited electrons and holes are participated in a redox reaction to degrade organic pollutant, which makes catalyst behave like photocatalyst. Therefore, in sonocatalytic activity, ultrasonic irradiation not only induces sonolysis but also generates electron and hole pairs in the catalyst which could enhance the catalytic performance by providing additional nucleation sites for development of cavity microbubble and subsequently increase the pyrolytic degradation and reactive radical generation [6].

$LaFeO_3$ with unique optoelectronic and a narrow optical gap has received intensive attention because of its strong catalytic activity [7]. The sonocatalytic activities of $LaFeO_3$ nanoparticles have been reported in our previous work [8]. The results showed that $LaFeO_3$ nanoparticles had good sonocatalytic performance and stability. However, the stability experiment could not be carried out easily by an applied external magnetic field. Therefore, in this work magnetite (Fe_3O_4) that has strong magnetic property is introduced to $LaFeO_3$ to enhance the sonocatalytic performance and the stability of the catalyst. The samples were synthesized using the co-precipitation method and were characterized by several measurements. The sonocatalytic performance and stability were also checked using methylene blue (MB) as a model of organic pollutant.

2. Experimental details

2.1. Synthesis of the Nanoparticles

Lanthanum (III) chloride heptahydrate ($LaCl_3{\cdot}7H_2O$), Iron (II) chloride tetrahydrate ($FeCl_2{\cdot}4H_2O$), and sodium hydroxide (NaOH) were used to synthesize $LaFeO_3$ nanoparticle using the co-precipitation method based on our previous work [8]. Iron (II) sulfate heptahydrate ($FeSO_4.7H_2O$), sodium hydroxide (NaOH), ethanol, ethylene glycol, and acetic acid (20%) were used to prepare Fe_3O_4 nanoparticles using the sol-gel method [9].

2.2. Synthesis of the Nanocomposites

The prepared $LaFeO_3$ and Fe_3O_4 nanoparticles were then used to synthesize the $LaFeO_3/Fe_3O_4$ nanocomposite with the variation of the molar ratio of 1:0.1, 1:0.3, 1:0.5, and 1:1, respectively. The synthesized nanocomposite was obtained through the following steps: first, the $LaFeO_3$ nanoparticle was dissolved in distilled water, while Fe_3O_4 was dissolved in ethanol. Then, these two solutions were mixed and ultrasonicated for two hours. The obtained mixtures were then centrifuged to separate the precipitations and solutions. Thereafter, the precipitations obtained were allowed to stand at room temperature and then heated 100°C for one hour under vacuum conditions.

2.3. Characterization

The samples were characterized by X-Ray Diffraction (XRD), Brunauer–Emmett–Teller (BET) surface area analysis, and Vibrating Sample Magnetometer (VSM).

2.4. Sonocatalytic Activity

The detailed sonocatalytic procedure have been reported in our previous work [8]. Methylene blue (MB) was selected as a model of organic pollutant and the sonocatalytic activity were carried out under ultrasonic irradiation. The degradation of MB was monitored using a Hitachi (UH5300) UV-visible spectrophotometer. The percentage degradation of MB was recorded as C_t/C_0, where C_t is the concentration of MB at each measurement interval during irradiation and C_0 is the initial concentration of MB after reaching adsorption–desorption equilibrium. To further study the recyclability of the catalyst, $LaFeO_3/Fe_3O_4$ nanocomposite was used repeatedly for four-times. In the each run, the catalyst could be easily separated by an applied magnetic field.

3. Results and Discussion

The magnetic separation process in the sonocatalytic is a very important issue for its practical application. The catalyst is expected to be easy recovering from the aqueous organic dyes and recycling for several times. Therefore, to study the magnetic properties of the samples, Figure 1a presents the magnetic hysteresis curves of $LaFeO_3/Fe_3O_4$ nanocomposite with different molar ratio of Fe_3O_4. The magnetic hysteresis curves of $LaFeO_3$ and Fe_3O_4 nanoparticles are also shown as a comparison. The results show that all samples exhibit ferromagnetic behavior at room temperature. The corresponding saturation magnetization values are also summarized in Table 1. The saturation magnetic values of $LaFeO_3/Fe_3O_4$ nanocomposite increase with the increasing of Fe_3O_4 contents. It indicates that $LaFeO_3/Fe_3O_4$ nanocomposite could be further applied for the separation process. It can

IOP Conf. Series: Materials Science and Engineering **196** (2017) 012050 doi:10.1088/1757-899X/196/1/012050

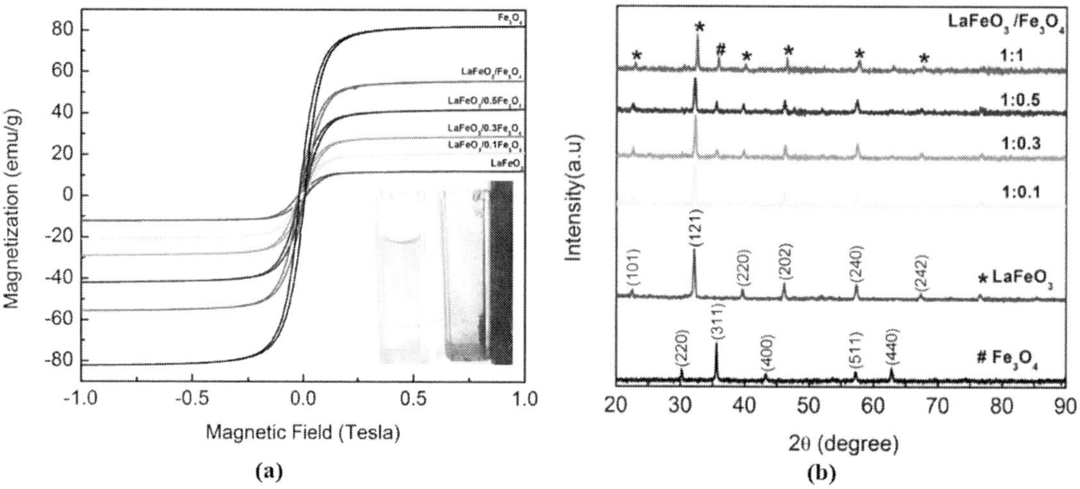

Figure 1. (a) VSM spectra; (b) XRD pattern of $LaFeO_3/Fe_3O_4$ nanocomposites with different molar ratio of Fe_3O_4.

Table 1. The lattice parameter, grain size <D>, magnetic saturation (M-S) and surface area of $LaFeO_3/Fe_3O_4$ nanocomposite.

Sample	Lattice Parameter $LaFeO_3$ a (Å)	b (Å)	c (Å)	Lattice Parameter Fe_3O_4 a=b=c (Å)	<D> $LaFeO_3$ (nm)	<D> Fe_3O_4 (nm)	M-S (emu/g)	Surface Area (m^2/g)
$LaFeO_3$	5.5436	7.8456	5.5528	-	58	-	12.08	31
Fe_3O_4				8.3638	-	45	82.13	3
$LaFeO_3/0.1Fe_3O_4$	5.5692	7.8564	5.5623	8.3258	24	27	20.87	24
$LaFeO_3/0.3Fe_3O_4$	5.5324	7.8497	5.5660	8.3288	30	24	28.93	19
$LaFeO_3/0.5Fe_3O_4$	5.4580	7.8126	5.5550	3.3299	34	21	41.96	13
$LaFeO_3/Fe_3O_4$	5.4312	7.8096	5.5509	8.3307	36	20	55.58	9

be seen in the inset of Figure 1a, the catalyst could be easily separated by an applied external magnetic field.

Figure 1b displays the XRD pattern of $LaFeO_3/Fe_3O_4$ nanocomposite with different molar ratio of Fe_3O_4. As a comparison, the XRD pattern of $LaFeO_3$ and Fe_3O_4 nanoparticles are also shown in the figure. For $LaFeO_3/Fe_3O_4$ nanocomposites, the diffraction peaks at $2\theta = 22.6°$, 32.2°, 39.8°, 46.2°, 57.5°, 67.4°, and 76.7° which index to the (101), (121), (220), (202), (240), (242), and (204), planes, respectively, are corresponded to the orthorhombic structure of $LaFeO_3$ [10]. While the existence of diffraction peaks at $2\theta = 35.5°$ which represent to the (311) planes correspond to the cubic spinel structure of Fe_3O_4 [9]. No other segregation of phase was found, indicating that the desired synthesized sample was obtained. The lattice parameter and crystal size <D> of the samples, obtained by Rietveld refinement method using MAUD program and calculated using the Scherrer's formula [11], are summarized in Table 1. The specific surface area measured using BET method is also tabulated in Tabel 1.

The sonocatalytic degradation of methylene blue (MB) using $LaFeO_3$ nanoparticles as a catalyst is shown in Figure 2a. The effect of Fe_3O_4 incorporation on the sonocatalytic performance of $LaFeO_3/Fe_3O_4$ nanocomposites is also illustrated in the figure. It can be seen that MB degradation in the presence of $LaFeO_3/Fe_3O_4$ nanocomposites increases with the increasing nanocomposites molar ratio up to 1:0.5 and decreases when the nanocomposites molar ratio further increase. However, all nanocomposites samples still show better performance than $LaFeO_3$ nanoparticles. The

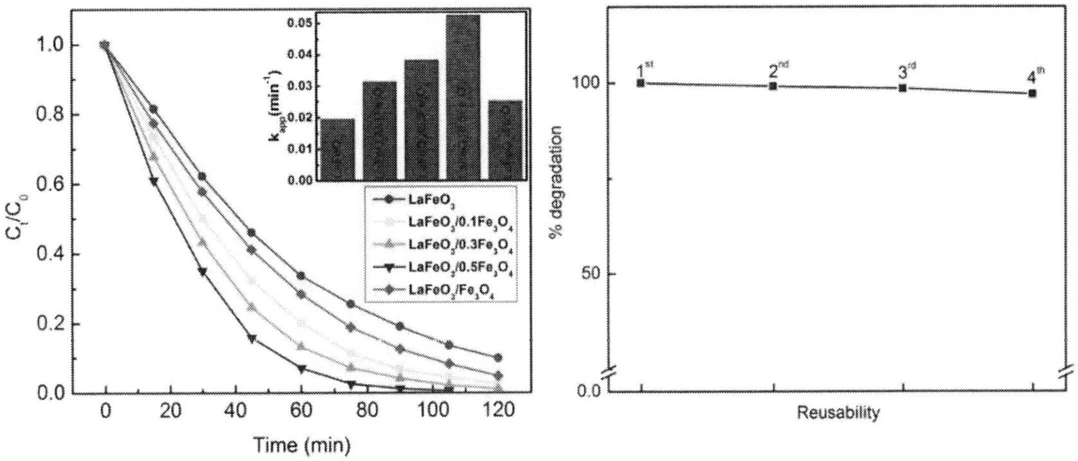

(a) (b)

Figure 2. (a) The sonocatalytic activity of $LaFeO_3/Fe_3O_4$ nanocomposites with different molar ratio of Fe_3O_4; (b) The stability of catalyst in the sonocatalytic process using $LaFeO_3/0.5\ Fe_3O_4$ nanocomposites.

sonodegradation of MB using $LaFeO_3/Fe_3O_4$ nanocomposite obeys pseudo-first order kinetics with respect to the concentration of MB: $\ln(C_t/C_0)= k_{app}\, t$, where k_{app} is the apparent reaction constant for catalysis used as the basic parameter for different catalysts [12]. The apparent reaction constant of $LaFeO_3/Fe_3O_4$ nanocomposites was calculated and the results are presented in the inset of Figure 2a. It can be seen clearly in the figure, the $LaFeO_3/Fe_3O_4$ nanocomposite with a molar ratio of 1:0.5 exhibits maximum sonocatalytic performance for degrading MB.

In the sonocatalytic process, it is known that ultrasonic irradiation not only induces sonolysis but also generate electron and hole pairs in the catalyst [6]. The present of Fe_3O_4 in the nanocomposite could improve the sonocatalytic performance by trapping the photo-generated electrons from the conduction band of $LaFeO_3$ [13]. The photo-generated electrons could then react with Fe^{3+} ions to form Fe^{2+} ions. After that, Fe^{2+} ions would react with molecular oxygen (O_2) to yield Fe^{3+} ions and superoxide radical ($\bullet O_2^-$) as following this equation: $Fe^{2+} + O_2 \rightarrow Fe^{3+} + \bullet O_2^-$ [14]. The produced superoxide radicals are known as one of the strong reduction and oxidation agents that could destroy the complex chemical bonds of organic pollutants such as MB [15].

The stability of the catalyst was also checked using $LaFeO_3/Fe_3O_4$ nanocomposites with maximum degradation rate. The removed catalyst as depict in the inset of Figure 2b, was used to degrade MB solutions four-times under the same treatment conditions. After four repetitions (as showed in the Figure 4b), insignificant losses of activity were observed, indicating that the catalyst has excellent stability and recyclability in the sonocatalytic process.

4. Conclusion

The orthorhombic and cubic spinel structure of $LaFeO_3/Fe_3O_4$ nanocomposites show better sonocatalytic performance than $LaFeO_3$ nanoparticles. The 1:0.5 of $LaFeO_3/Fe_3O_4$ nanocomposite exhibits maximum sonocatalytic performance for degrading MB. The samples also show ferromagnetic behavior at room temperature which makes the separation process in the stability experiment become easier. The results show that the catalyst has excellent stability and recyclability in the sonocatalytic process.

References

[1] Pang Y L, Abdullah A Z, and Bhatia S 2011 *Desalination* **277** 2011
[2] Petrier C, Jiang Y, and Lamy M F 1998 *Environ. Sci. Technol.* **32** 1316
[3] Suslick K S, Doktycz S J, and Flint E B 1990 *Ultrasonic* **28** 280

[4] Wang J, Sun W, Zhang Z, Zhang X, Li R, Ma T, Zhang P, and Li Y 2007 *J. Mol. Catal. A Chem.* **272** 84
[5] Ertugay N and Acar F N 2014 *Appl. Surf. Sci.***318** 121
[6] Zhou M, Yang H, Xiang T, Li R S, Zhang H M, and Wang X X 2015 *J. Hazard. Mater.* **289** 149
[7] Tang P, Tong Y, Chen H, Cao F and Pan G 2013 *Curr. Appl. Phys.* **13** 340
[8] Afifah N and Saleh R 2016 *J. Phys. Conf. Ser.* **710** 012030
[9] Arifin S A, Jalaludin S and Saleh R 2015 *Mater. Sci. Forum* **827** 49
[10] Abazari R and Sanati S 2013 *Superlattice. Microst.* **64** 148
[11] Köferstein R, Jäger L and Ebbinghaus S G 2013 *Solid State Ionics* **249–250** 1
[12] Saleh R and Djaja N F 2014 *Superlattice. Microst.* **74** 217
[13] Wang J, Yang J, Li X, Wang D, Wei B, Song H, Li X, and Fu S 2016 *Physica E* **75** 66
[14] Ambrus Z, Balazs N, Alapi T, Wittmann G, Sipos P, Dombi A, and Mogyorosi 2008 *Appl. Catal. B.* **81** 27
[15] Yu J G, Xiang Q J, and Zhou M H 2009 *Appl. Catal. B.***90** 595

IOP Conf. Series: Materials Science and Engineering **196** (2017) 012051 doi:10.1088/1757-899X/196/1/012051

Temperature Dependent Raman Studies of $Pr_2Zr_2O_7$ Single Crystal

K D Jatmoko[1], A Prasetyo[1], V Suendo[1], S Nakatsuji[2], A A Nugroho[1, 2]

[1] Faculty of Mathematics and Natural Sciences, Institut Teknologi Bandung, Jl, Ganesha 10, Bandung, Indonesia
[2] Institute for Solid State Physics, The University of Tokyo, Kashiwa 277-85881, Japan

E-mail: nugroho@fi.itb.ac.id

Abstract. The temperature-dependent Raman studies of $Pr_2Zr_2O_7$ has been performed on single crystals samples in the temperature range from 80 K to 820 K. By using polarization measurement, the vibration modes of A_{1g}, E_g, and $4T_{2g}$ have been assigned on 295.9 cm^{-1}, 497.3 cm^{-1}, and 102.4 cm^{-1}, 309.2 cm^{-1}, 670.6 cm^{-1}, 878 cm^{-1}, respectively. The ambiguity vibration mode around 300 cm^{-1} in pyrochlore compounds can be resolved. The 380 cm^{-1} mode showed sharpening upon heating up from 370 K which support the present of Oxygen distortion towards vacant site.

1. Introduction

The studies of cubic pyrochlore oxides, $A_2B_2O_7$ with A= rare earth, and B= transition metal, have been intensively carried out for over thirty years because of the novel phenomena lies on its geometrical-frustrated systems. Raman spectroscopy has been utilized in pyrochlore compounds to investigate the oxygen sublattice's disorder [1, 2]. The Raman vibration modes in pyrochole are well establish as A_{1g} + E_g + $4T_{2g}$, however the assignment for the intense peak around 300 cm^{-1} remains unclear whether it belongs to modes A_{1g} or T_{2g}. Furthermore, the temperature dependent of the pyrochlore spectra above the room temperature is still missing, in particular for $Pr_2Zr_2O_7$.

Here, we report the temperature dependence of polarized Raman spectra of $Pr_2Zr_2O_7$ ranging from 80K to 820 K. Based on the symmetry analysis and polarized spectra of for the intense peak can be resolved. Our results are consistent with the present understanding of pyrochlore spectra which strongly influenced by the local change of oxygen state. We observed an appearance of Raman mode around 380 cm^{-1}. The polarized measurements have been performed on single crystal of $Pr_2Zr_2O_7$. For the first time, the assignment of the vibration modes is completely consistent respect to the selection rules.

2. Experimental

Single crystal of $Pr_2Zr_2O_7$ was synthesized by the floating zone method with a high-temperature optical image furnace equipped with four xenon arc lamps (Crystal Systems inc.) under oxygen atmosphere reported by Kimura, et.al. [3]. The as-grown crystal have a dark brown color. Polarized Raman Spectra were recorded using Bruker Raman Senterra at 850 K to 80 K temperature range with a cooling/ heating stage (Linkam TM600E Tadworth UK). A diode laser (785 nm) was used and focused with 50x microscope objective. The spectral resolution of the measurement was around 3-5 cm^{-1}. The measurement were carried out after temperature stabilization with laser power of 50 mW and the scan rate of 0.1 s^{-1}. The spectra were subtracted by taking into account the Bose-Einstein occupation factor and fitted with Lorentzian model. The polarization was adjusted on both the incident and scattered beam. The measurement were performed parallel ($x(yy)\underline{x}$ or $x(zz)\underline{x}$) and perpendicular ($x(zy)\underline{x}$ or $x(yz)\underline{x}$), along axes of [011] plane.

3. Result and Discussion

According to group theory, the Raman actives vibration for the pyrochlore with the *Fd3m* space group are $A_{1g} + E_g + 4T_{2g}$ modes. Figure 1 shows the Raman spectra of $Pr_2Zr_2O_7$ for the parallel and cross polarization configurations. The peaks at 295.9 cm^{-1} and 497.3 cm^{-1} observed using the parallel polarization are assigned for the A_{1g} and E_g vibration modes, while the peaks at 102.4 cm^{-1}, 309.2 cm^{-1}, 670.6 cm^{-1}, and 878 cm^{-1} observed using the cross polarization are all assigned for T_{2g} vibration modes. At around 300 cm^{-1}, the peak for A_{1g} and T_{2g} are clearly distinguishable using different polarization configurations. According to the symmetry analysis, the A_{1g} and E_g modes are related to the oxygen with the site symmetry of *48f*, while the T_{2g} modes are related to the oxygen with site symmetry of *48f* and *8a*. For the parallel polarization, a broad peak near A_{1g} mode and a shoulder at E_g mode are also observed. This may due to a non uniform local distortions between Zr-O(*48f*)-Zr bonds which might due to the lack of occupation sites of oxygen or zirconium.

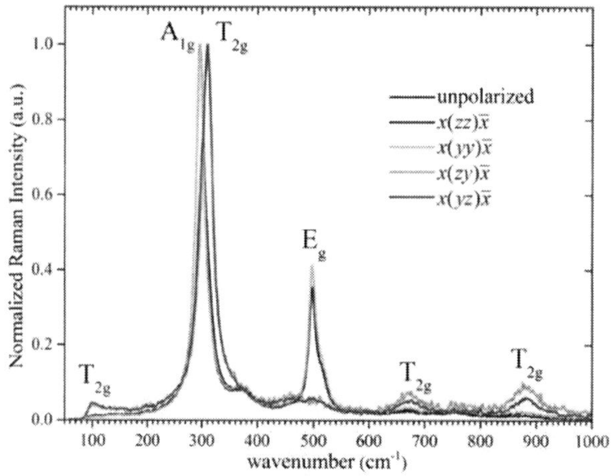

Figure 1. Room Temperature Raman spectra of $Pr_2Zr_2O_7$ single crystal in all polarization configuration.

The temperature dependent Raman Spectra of $Pr_2Zr_2O_7$ from 140 K to 820 K using parallel and cross polarization are displayed in Fig.2. It is noted that upon increasing temperature, the broad peak near the A_{1g} (at 380 cm^{-1}) become sharper for the parallel polarization while only the T_{2g} modes around 350 cm^{-1} persist for the cross polarization. The broadening of peaks become more pronounce above 370 K. In contrast to the temperature dependent of Bose statistic, the vibration modes of A_{1g} and T_{2g}, around 300 cm^{-1}, show a hardening upon decreasing temperature. This hardening is a peculiar feature which could be associated with the particular distortion of the local structure. The softening of the E_g and rest of T_{2g} modes upon decreasing temperature are consistent with the observation in other pyrochlore measurements [4,5,6]. The disappearance of the T_{2g} modes upon increasing temperature indicates that the vibrations involving Pr-O(*8a*)-Pr bond and Pr-O(*48f*)-Zr bond become weaker. This is in contrast to the emerging peak around 380 cm^{-1}. Both the T_{2g} and A_{1g} can be related to the local structure around the Pr site. The oxygens around the Pr site are shown in Fig. 3.

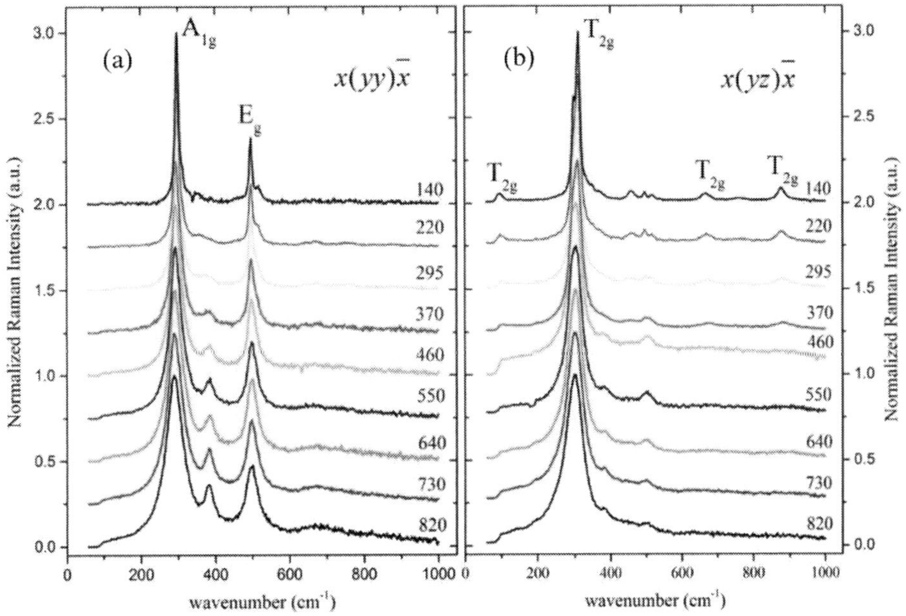

Figure 2. Polarized Raman Spectra of $Pr_2Zr_2O_7$ single crystal as a function of temperature for (a) parallel and (b) cross polarization.

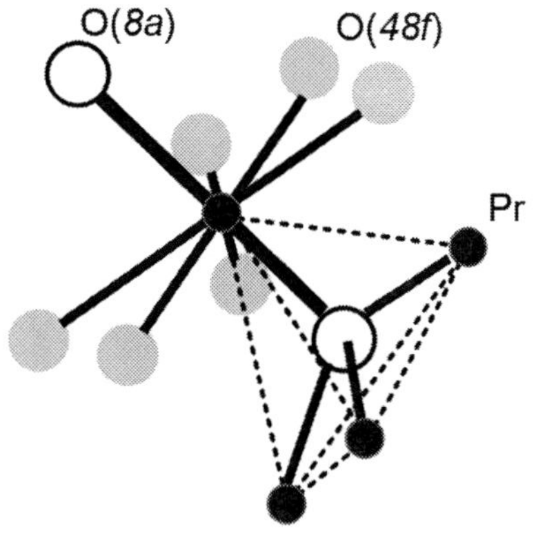

Figure 3. Schematic picture of oxygens surrounding Pr site. The O(*8a*) bridge the bond between Pr-O-Pr while the O(*48f*) bridge the bond between Pr-O(*48f*)-Zr [7].

4. Conclusion

The polarized Raman measurements of $Pr_2Zr_2O_7$ have shown that all expected modes from group analysis can be observed. The vibration modes of A_{1g}, E_g, and $4T_{2g}$ have been assigned on 295.9 cm^{-1}, 497.3 cm^{-1}, 102.4 cm^{-1}, 309.2 cm^{-1}, 670.6 cm^{-1}, and 878 cm^{-1} respectively. For the vibration mode around 300 cm^{-1}, the A_{1g} mode is evidenced to have lower energy compared to the T_{2g} mode which should be general for the pyrochlore systems. At high temperature, The 380 cm^{-1} mode showed hardening upon heating up from 370 K. This emerging peak Raman mode is attributed to local distortions around the Pr site.

5. Acknowledgement

We acknowledge Institut Teknologi Bandung through Dana Riset KK 2016.

References

[1] Glerup M, Nielsen O F and Poulsen F W 2001 *J. Solid State Chem.* **160** 25
[2] Hess N J, Begg B D, Conradson S D, McCready D E, Gassman P L and Weber W J 2002 *J. Phys. Chem. B* **106** 4663
[3] Kimura K, Nakatsuji S and Nugroho A A 2013 *J. Kor. Phys. Soc.* **63** 719
[4] Mączka M, Hanuza J, Hermanowicz K, Fuentes A F, Matsuhira K and Hiroi Z 2008 *J. Raman Spectrosc.* **39** 537
[5] Saha S, Muthu D V S, Singh S, Dkhil B, Suryanarayanan R, Dhalenne G, Poswal H K, Karmakar S, Sharma S M, Revcolevschi A and Sood A K 2009 *Phys. Rev. B* **79** 134112
[6] Saha S, Prusty S, Singh S, Suryanarayanan R, Revcolevschi A and Sood A K 2011 *J. Solid State Chem.* **184** 2204
[7] Vandenborre N T, Husson E and BrussetH 1981 Spectrochimica Acta **37A** 118